Lecture Notes in Artificial Intelligence   7961

Subseries of Lecture Notes in Computer Science

T0223892

Jacques Carette   David Aspinall
Christoph Lange   Petr Sojka
Wolfgang Windsteiger (Eds.)

Intelligent
Computer Mathematics

MKM, Calculemus, DML, and Systems and Projects 2013
Held as Part of CICM 2013
Bath, UK, July 8-12, 2013
Proceedings

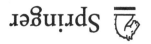 Springer

Volume Editors

Jacques Carette
McMaster University, Hamilton, ON L8S 4K1, Canada
E-mail: carette@mcmaster.ca

David Aspinall
University of Edinburgh, Edinburgh EH8 9AB, UK
E-mail: david.aspinall@ed.ac.uk

Christoph Lange
University of Birmingham, Edgbaston, Birmingham B15 2TT, UK
E-mail: math.semantic.web@gmail.com

Petr Sojka
Masaryk University, 60200 Brno, Czech Republic
E-mail: sojka@fi.muni.cz

Wolfgang Windsteiger
Johannes Kepler University, 4040 Linz, Austria
E-mail: wolfgang.windsteiger@risc.jku.at

ISSN 0302-9743   e-ISSN 1611-3349
ISBN 978-3-642-39319-8   e-ISBN 978-3-642-39320-4
DOI 10.1007/978-3-642-39320-4
Springer Heidelberg Dordrecht London New York

Library of Congress Control Number: 2013941551

CR Subject Classification (1998): I.1, F.4.1, I.2.2-3, I.2.6, I.2, I.7, F.3.1, D.2.4, F.3, H.3.7, G.4, H.2.8

LNCS Sublibrary: SL 7 – Artificial Intelligence

Typesetting: Camera-ready by author, data conversion by Scientific Publishing Services, Chennai, India

Printed on acid-free paper

Springer is part of Springer Science+Business Media (www.springer.com)

Preface

Continued advances in computer science and in information technology are contributing to the forward march of the use of computers in mathematics. While they used to play a tremendous supporting role through computation, proof assistants, and as mathematical publishing tools, this has now spread to more social aspects of the mathematics process. The series of Conferences on Intelligent Computer Mathematics (CICM) host collections of co-located meetings, allowing researchers and practitioners active in related areas to share recent results and identify the next challenges.

The sixth in this series of Conferences on Intelligent Computer Mathematics is held in Bath, UK, in 2013. Previous conferences, all also published in Springer's *Lecture Notes in Artificial Intelligence* series, have been held in the UK (Birmingham, 2008: LNAI 5144), Canada (Grand Bend, Ontario, 2009: LNAI 5625), France (Paris, 2010: LNAI 6167), Italy (Bertinoro, 2011: LNAI 6824), and Germany (Bremen, 2012: LNAI 7362). CICM 2013 included three long-standing international meetings:

- 12th International Conference on Mathematical Knowledge Management (MKM 2013)
- 20th Symposium on the Integration of Symbolic Computation and Mechanized Reasoning (Calculemus 2013)
- 6th workshop/conference on Digital Mathematics Libraries (DML 2013)

Since 2011, CICM has also been offering a track for brief descriptions of systems and projects that span the MKM, Calculemus, and DML topics, the "Systems & Projects" track. The proceedings of the meetings and the Systems & Projects track are collected in this volume.

CICM 2013 also contained the following activities:

- Demonstrations of the systems presented in the Systems & Projects track
- Less formal "work in progress" sessions

We used the "multi-track" features of the EasyChair system, and our thanks are due to Andrei Voronkov and his team for this and many other features. The multi-track feature also allowed transparent handling of conflicts of interest between the Track Chairs and submissions. There were 73 submissions, 17 of which were withdrawn. Each of the remaining 56 submissions was reviewed by at least three, and most by four, Program Committee members. The committee decided to accept 30 papers. However, this is a conflation of tracks with different acceptance characteristics. The track-based acceptance rates were:

MKM: 7 acceptances out of 18 submissions

Calculemus: 5 acceptances out of 12 submissions

DML: 6 acceptances out of 8 submissions

S & P: 12 acceptances out of 16 submissions

Invited talks, this year accompanied by full papers included in these proceedings, were given by:

Ursula Martin with the full paper on "Mathematical Practice, Crowdsourcing, and Social Machines" coauthored with Alison Pease:

For centuries, the highest level of mathematics has been seen as an isolated creative activity, to produce a proof for review and acceptance by research peers. Mathematics is now at a remarkable inflexion point, with new technology radically extending the power and limits of individuals. Crowdsourcing pulls together diverse experts to solve problems; symbolic computation tackles huge routine calculations; and computers, using programs designed to verify hardware, check proofs that are just too long and complicated for any human to comprehend.

Mathematical practice is an emerging interdisciplinary field which draws on philosophy, social science and ethnography, and the input of mathematicians themselves, to understand how mathematics is produced. Online mathematical activity provides a rich source of data for empirical investigation of mathematical practice — for example, the community question-answering system mathoverflow contains around 40,000 mathematical conversations, and polymath collaborations provide transcripts of the process of discovering proofs. Such investigations show the importance of "soft" aspects such as analogy and creativity, alongside formal deduction, in the production of mathematics, and give us new ways to think about the possible complementary roles of people and machines in creating new mathematical knowledge

Social machines are a new paradigm, identified by Berners-Lee, for viewing a combination of people and computers as a single problem-solving entity, and the subject of major international research endeavors. We outline a research agenda for mathematics social machines, a combination of people, computers, and mathematical archives to create and apply mathematics, with the potential to change the way people do mathematics, and to transform the reach, pace, and impact of mathematics research.

Assia Mahboubi presented "The Rooster and the Butterflies":

This paper describes a machine-checked proof of the Jordan-Hölder theorem for finite groups. The purpose of this description is to discuss the representation of the elementary concepts of finite group theory inside type theory. The design choices underlying these representations were crucial to the successful formalization of a complete proof of the Odd Order Theorem with the Coq system.

Moreover, Patrick D. F. Ion spoke on "Mathematics and the World Wide Web":

Mathematics is an ancient and honorable study. It has been called The Queen and The Language of Science. The World Wide Web is something brand-new that started only about a quarter of a century ago. But the World Wide Web is having a considerable effect on the practice of mathematics, is modifying its image and role in society, and can be said to have changed some of its content. There are forces at work in the Web that may be changing our world not necessarily for the better. I will be exploring some of the issues this raises.

May 2013

Jacques Carette
David Aspinall
Christoph Lange
Petr Sojka
Wolfgang Windsteiger

Organization

CICM 2013 was organized by the Conference on Intelligent Computer Mathematics steering committee, which was formed at CICM 2010 as a parent body to the long-standing Calculemus and Mathematical Knowledge Management special interest groups. The conferences organized by these interest groups continue as special tracks in the CICM conference. After the 2012 meeting, the DML workshop became a conference, and joined the other two as another track. These tracks and the Systems & Projects track had independent Track Chairs. This year, rather than having independent Programme Committees for each track, an experiment was tried in having a unified PC for all of CICM. Local arrangements, the life-blood of any conference, were handled by the Department of Computer Science of the University of Bath, Bath, UK, in the capable hands of James Davenport.

CICM Steering Committee

Secretary

Michael Kohlhase   Jacobs University Bremen, Germany

Calculemus Delegate

Renaud Rioboo   ENSIIE, France

Treasurer

William Farmer   McMaster University, Canada

DML Delegate

Thierry Bouche   Université Joseph Fourier, Grenoble, France

MKM Delegate

Florian Rabe   Jacobs University Bremen, Germany

CICM PC Chair 2012

Johan Jeuring   Utrecht University and Open University,
The Netherlands

CICM PC Chair 2013

Jacques Carette   McMaster University, Ontario, Canada

CICM 2013 Officers

General Program Chair

Jacques Carette   McMaster University, Canada

Local Arrangements

James Davenport   University of Bath, UK

MKM Track Chair

David Aspinall   University of Edinburgh, UK

Calculemus Track Chair

Wolfgang Windsteiger   RISC Linz, Johannes Kepler University, Austria

DML Track Chair

Petr Sojka   Masaryk University Brno, Czech Republic

S & P Track Chair

Christoph Lange   University of Birmingham, UK

Program Committee

Mark Adams	Proof Technologies Ltd., UK
Akiko Aizawa	National Institute of Informatics, The University of Tokyo, Japan
Jesse Alama	CENTRIA, FCT, Universidade Nova de Lisboa, Portugal
Rob Arthan	Queen Mary University of London, UK
Andrea Asperti	University of Bologna, Italy
David Aspinall	University of Edinburgh, UK

Jeremy Avigad	Carnegie Mellon University, USA
Thierry Bouche	Université Joseph Fourier (Grenoble), France
Jacques Carette	McMaster University, Canada
John Charnley	Imperial College London, UK
Janka Chlebikova	University of Portsmouth, UK
Simon Cotton	Imperial College London, UK
Leo Freitas	Newcastle University, UK
Deyan Ginev	Jacobs University Bremen, Germany
Gudmund Grov	University of Edinburgh, UK
Tom Hales	University of Pittsburgh, USA
Yannis Haralambous	Telecom Bretagne
Jonathan Heras	University of Dundee, UK
Hoon Hong	North Carolina State University, USA
Predrag Janicic	University of Belgrade, Serbia
Cezary Kaliszyk	University of Innsbruck, Austria
Manfred Kerber	University of Birmingham, UK
Adam Kilgarriff	Lexical Computing Ltd.
Andrea Kohlhase	Jacobs University Bremen, Germany
Michael Kohlhase	Jacobs University Bremen, Germany
Temur Kutsia	Johannes Kepler University of Linz, Austria
Christoph Lange	University of Birmingham, UK
Paul Libbrecht	Karlsruhe University of Education, Germany
Christoph Lüth	DFKI, University of Bremen, Germany
Till Mossakowski	DFKI, University of Bremen, Germany
Magnus O. Myreen	University of Cambridge, UK
Florian Rabe	Jacobs University Bremen, Germany
David Ruddy	Cornell University, USA
Jiří Rákosník	Institute of Mathematics, Academy of Sciences CR, Czech Republic
Carsten Schürmann	IT University of Copenhagen, Denmark
Petr Sojka	Masaryk University, Czech Republic
Volker Sorge	University of Birmingham, UK
Hendrik Tews	TU Dresden, Germany
Frank Tompa	University of Waterloo, Canada
Josef Urban	Radboud University, The Netherlands
Stephen Watt	University of Western Ontario, Canada
Makarius Wenzel	Université Paris-Sud 11, France
Wolfgang Windsteiger	RISC Institute, JKU Linz, Austria
Richard Zanibbi	Rochester Institute of Technology, USA

Additional Reviewers

Couto Vale, Daniel
Delahaye, David
Felgenhauer, Bertram
Iancu, Mihnea
Jung, Jean
Komendantskaya, Ekaterina
Lee, Jae Hee
Llano Rodríguez, María Teresa
Liška, Martin

Obua, Steven
Pease, Alison
Ramezani, Ramin
Růžička, Michal
Sabeghi Saroui, Behrang
Solovyev, Alexey
Tankink, Carst
Thiemann, René
van der Walt, Paul

Table of Contents

Calculemus

MKM

J. Carette et al. (Eds.): CICM 2013, LNAI 7961, pp. 1–18, 2013.

The Rooster and the Butterflies

Assia Mahboubi

Microsoft Research - Inria Joint Centre

Abstract. This paper describes a machine-checked proof of the Jordan-Hölder theorem for finite groups. This purpose of this description is to discuss the representation of the elementary concepts of finite group theory inside type theory. The design choices underlying these representations were crucial to the successful formalization of a complete proof of the Odd Order Theorem with the Coq system.

1 Introduction

The Odd Order Theorem due to Feit and Thompson [7] is a major result of finite group theory which is a cornerstone of the classification of finite simple groups. Originally published in 1963, this was considered at its time as a demonstration of an uncommon length and intricacy, whose 255 pages filled an entire volume of the Pacific Journal of Mathematics. Later simplified and improved by a collective revision effort [3, 20], it remains a long and difficult proof, combining a broad panel of algebraic theories. In September 2012, the Mathematical Components team, lead by Georges Gonthier, completed [10] a formalization of this result using the Coq system [1, 4].

This achievement is evidence of the maturity of *proof assistants*, a family of software systems pioneered by N. G. de Bruijn's AUTOMATH system [6], that aims at "doing mathematics with a computer". The ambition of proof assistant is to realize an old dream: automate the verification of mathematical proofs. Check a theorem with a proof assistant consists in providing a description of the statement and of its candidate proof in formal logic and then having a generic and relatively small program checking the well-formedness of this proof with respect to the elementary rules of logic. A proof assistant provides the support necessary to obtain both a high confidence in proof checking and the mandatory set of tools required to ease the process of describing statements and proofs.

For the last decade, proof assistants have been successfully employed in a variety of contexts, from hardware and software verification to combinatorics or number theory. However the distinguishing feature of the complete formal proof of the Odd Order Theorem is that the corresponding libraries of formalized mathematics cover a range of algebraic theories that is both wide and deep. The proof of the Odd Order Theorem actually relies on a number advanced results that necessitate non-trivial combinations of arguments arising from several areas of mathematics.

When assisting the user in his verification task, the proof assistant is not expected to invent new results or new justifications. Yet a substantial part of the effort required by such a large scale endeavor consists in reworking the mathematics described in the standard literature so that it can be organized in a satisfactory and modular manner. The software engineering effort leading to (re)usable and composable libraries of formalized mathematics hence also involves re-thinking the mathematical definitions and proof methods. The formalization of the basics has to accommodate the variety of its usage in more advanced parts of the theory.

In this paper, we outline how elementary concepts of finite group theory have been revisited in the low-level libraries of this formal proof of the Odd Order Theorem. We do not claim novelty here: part of the material exposed here has been already described at various levels of detail in other venues that we list in the preamble of each section. Previous publications were mostly written for readers familiar with the Coq system, and most of them deal with programming issues. By contrast, we have tried here to provide a mathematical documentation of a few Coq libraries[1], as distant as possible from Coq syntax, since this intuition might be difficult to grasp from the documentation headers of the corresponding files. These documentation headers can be browsed on-line at: http://ssr.msr-inria.fr/~jenkins/current/index.html

In the next sections, by *formalized*, we mean implemented in the Coq system. The words *formal* and *formally* refer to Coq syntax. The words *informal* and *informally* refer to the corresponding mathematical notations we use throughout the paper to improve the rendering. We however maintain a precise correspondence between formal syntax and informal notations. We also use the collective "we" pronoun to refer to the team of authors of these libraries [10].

The rest of the paper is organized as follows. Section 2 explains the representation adopted for finite types and finite sets. Section 3 is devoted to the definition of finite groups and their morphisms. Section 4 describes the formalization of the quotient operation of group theory. Finally, section 5 illustrates how this material is used in the proof of a standard result of finite group theory, the Jordan-Hölder theorem.

2   Preliminaries

In this section, we recall the formal definitions of the preliminary notions we rely on. This material has already been presented in earlier publications [12, 8, 10, 11], with emphasis on the techniques used for their definition in Coq. These lower layers of formalized mathematics are quite constrained by the features of the logic underlying the Coq system and, in particular, by its constructiveness. A significant effort is put in the formalization of the theory of objects that behave mostly the same in either a classical or a constructive setting. The purpose, and the challenge, of the corresponding libraries is to provide enough infrastructure for the user to safely ignore the choices adopted for the implementation of these lower-level definitions. When these patterns of reasoning are effective, using an

---

[1] https://gforge.inria.fr/frs/?group_id=401

excluded-middle argument or performing a choice operation should be as convenient on top of these axiom-free CoQ libraries as in the setting of a classical logic, like the one assumed by most of the mathematical literature.

2.1 Types with Decidable Equality

The type theory implemented by the CoQ proof assistant is a constructive framework: the excluded middle principle is not allowed for an arbitrary statement without postulating a global axiom. Reasoning by case analysis on the validity of a predicate is valid constructively when it is possible to implement a (total) boolean function which decides this validity: the type of boolean values reflects the class of statements on which the excluded middle principle holds constructively. For instance, we do not need any axiom in order to reason by case analysis on the equality of two arbitrary natural numbers because equality on natural numbers is decidable. A decidable predicate is a predicate that can be (and in our libraries, that is) formalized as a function with boolean values.

The prelude libraries of the system, automatically imported when a CoQ session is started, define an equality predicate parametrized by an arbitrary type T, which is the smallest binary reflexive relation on T. This equality is often referred to as *Leibniz equality*. However, not all types are a priori equipped with an associated total and boolean comparison function, testing the validity of a Leibniz equality. However, the vast majority of the data we manipulate can be modeled with types equipped with such an operator. This operator legitimates constructive reasoning and programming by case analysis on the equality of two objects of such a type and witnesses the decidability of the associated Leibniz equality predicate. The library hence defines a structure for types with a decidable equality, formally called eqType. This structure packages a type with a binary boolean function on this type, plus a proof that the boolean test faithfully models Leibniz equality. For instance, finite types, natural numbers, rational numbers, and real or complex algebraic numbers are instances of this structure. Moreover, pairs, sequences or subtypes of instances of this structure are also canonically types with a decidable equality.

Another important feature of types with a decidable equality is the fact that they enjoy the property of *uniqueness of identity proofs* [14]. This plays a crucial role in our formalization but is out of the scope of the present paper. The interested reader can refer to previous publications [12, 11, 8, 10] for more information on the formalization of this structure.

In all that follows, and unless explicitly stated, by *type* we always implicitly mean *type with a decidable equality*. Hence the reader can safely forget about the constructiveness issues mentioned in this section: case analysis on the equality of two objects is allowed as well as on the membership of an object to a sequence, etc.

Libraries. The corresponding file to this subsection is eqtype.v.

2.2 Finite Types, Finite Sets

The library also defines an interface for *finite types*, which are types with a finite number of elements. Formally this structure is called `finType`. It packages a sequence enumerating exhaustively the elements of the type[2] and a proof that this sequence is duplicate-free. Finite types are a instance of a more general interface for types equipped with a choice operator: for an arbitrary non-empty decidable predicate, the choice operator outputs a canonical witness. In the case of a finite type, the choice operator just inspects the enumeration and picks the first witness encountered.

This representation of finite types is especially convenient to define functions with a finite domain. Our motivation here is to craft a datastructure for functions so that they provably verify the so-called extensionality principle:

$$\forall x, f x = g x \quad \Leftrightarrow \quad f = g$$

which states that the point-wise equality of two functions f and g on their whole domain type is equivalent to the Leibniz equality of these functions. Again, in Coq's type theory, this principle is not valid in general: two programs that output the same values on the same inputs are not necessarily identified by the Leibniz equality predicate. By contrast, we would like for instance to work with a definition of sets which allows us to equate sets that have point-wise characteristic functions.

Let us consider a finite type F, with e the enumeration of its elements. Let $|e|$ be the length of e. We represent a total function f with arguments in F and values in a type T by a finite sequence Im_f of length $|e|$, of elements in type T. Hence the value of f at e_i, the element at position $i < |e|$ in the enumeration e, is the i-th element of the sequence Im_f. We call such a function a *finite function*. This representation validates the extensionality principle: the right-to-left implication is trivial and the left-to-right implication holds because according to our definition of a finite function the Leibniz equality of two finite functions really *is* the Leibniz equality of their respective finite graphs Im_f and Im_g. This equality is granted by the hypothesis of point-wise equality. Note that we do not need to assume any finiteness property on the codomain type. If (aT : finType) is a finite type and (rT : Type) an arbitrary type, the type of finite functions with (finite) domain aT and codomain rT is formally denoted by {ffun aT -> rT}. Most of the theory of finite functions however assumes that the codomain type is an instance of type with decidable equality.

Finite functions with boolean values represent characteristic functions of sets of elements of their domain type. In other words, a finite set over a finite type F is coded by a sequence of boolean values which is a mask on the enumeration of F: *true* values select the elements that belong to the set. Now two finite sets with point-wise equal characteristic functions are (Leibniz) equal by the previous extensionality principle.

---

[2] These points are objects of a previously known type with decidable equality.

For any finite type F, {set F} formally denotes the type of finite sets of elements of type F. Remark that type {set F} has itself a finite number of elements and is hence an instance of finite type. We can therefore form the type {set {set F}}, which is the powerset associated to the finite type F. The library on finite sets provides definitions and notations for the standard concepts related to sets. For instance x \in A denotes the (decidable) test of membership of the element x in the set A, informally denoted by $x \in A$. Similarly, A \subset B denotes formally the (decidable) test of inclusion of the set A in the set B, which tests whether the true values of the mask defining A are also true values in the mask defining B. Informally we denote this test by $A \subset B$. The expression A :|: B (resp. A :|: B) denotes the intersection (resp. union) of two sets over the same finite carrier. The corresponding informal notation is $A \cap B$ (resp. $A \cup B$).

The expression #|A| denotes the cardinal of a finite set, which is the number of $true$ values in the mask. The notation f @: A is used for the image set of A by the function f from a finite type to an other finite type [3], which we denote informally by $f(A)$. The notation f @^-1: A is used for the preimage set of A by the function f from a finite type to an arbitrary type, which we denote informally by $f^{-1}(A)$. We will also use in section 3.1 the possibility of defining a set by comprehension: the expression [set x | P x] formally denotes the set of elements satisfying the (decidable) property P, and we denote this set informally by $\{x \mid P(x)\}$.

In all what follows, by set we mean a finite set of elements in a finite type. The reader can safely forget about the implementation described in the present section to apprehend the rest of this paper and rely on his or her classical intuition of sets.

Libraries. The corresponding files to this subsection are choice.v, fintype.v, finfun.v and finset.v.

3 Elementary Notions of Finite Group Theory

In this section we describe the datastructures adopted in the libraries about the elementary concepts of finite group theory. The design choices evolved in time and are now different from their earliest published description [12]. Garillot's PhD thesis [8] provides a more recent and accurate account of these choices, targeted at an audience expert in proof assistants.

The datastructures representing formally the operations defining finite groups of interest and the operations combining finite groups are shaped by two important remarks. First, we model groups as certain subsets of an ambient, larger group, which fixes the data all its subgroups share: the type of the elements, the group operation, the identity element. Hence groups are not types but objects, namely some sets of a finite type. This choice is motivated by the observation that finite group theory is not about the properties of the elements of a given group, but mostly about the study of how finite *subgroups* (of a larger finite

---

[3] Since we define an image set we need the codomain type to be also a finite type.

group) can combine. The second remark is that it is possible to revisit the standard definitions of the literature, so that they apply to arbitrary subsets of an ambient group, and not only to the special subsets that are also groups. The motivation for this generalization is to make the related constructions total and the statements of the related results less constrained and hence more usable.

3.1 Finite Groups

We reproduce below excerpts borrowed from the preliminary results of Aschbacher's book [2].

Définition 1 (Group, subgroup). A group is a set G together with an associative binary operation which possesses an identity and such that each element of G possesses an inverse. In the remainder of this section G is a group written multiplicatively. (...) A subgroup of G is a nonempty subset H of G such that for each $x, y \in H$, xy and x^{-1} are in H. This insures that the binary operation on G restricts to a binary operation on H which makes H into a group with the same identity as G and the same inverse.

Définition 2 (Product). For $X, Y \subseteq G$ define $XY = \{xy;\ x \in X, y \in Y\}$. The set XY is the product of X with Y.

In définition 2, we can observe that the group G is only here to fix the group operation and identity shared by the two sets X and Y and is not otherwise part of the definition. Moreover, these standard definitions and notational conventions are an instance of the standard practice which consists in using product notations both for points and sets: a similar convention apply for the inverse $X^{-1} = \{x^{-1};\ x \in X\}$ of a set $X \subseteq G$ and the constant 1 denotes both the identity of the group and the singleton $\{1\}$.

The library for elementary finite group theory defines two main structures. A first structure packages a finite type with a monoid operation and an involutive antimorphism. This structure is formally called baseFinGroupType[4] and all its instances share three common notations: the infix notation * denotes the monoid operation, the postfix $^{-1}$ notation denotes the involution and 1 denotes the neutral element. A second structure enriches the previous one to obtain all of the group axioms, hence describes what we call *group types* in the sequel. This second structure is formally called finGroupType and its instances inherit from the notations for the group operation, for the inverse and for the identity.

Let G be a group type. Both G and the type of sets of G are instances of the baseFinGroupType structure. For G, this holds by construction of a group type. For the type of sets of G, this comes from the properties of set product and set inverse. We can therefore utilize the notations *, $^{-1}$ and 1 for both point-wise and set-wise operations of G. Informally, we use a multiplicative convention and denote by xy the product of the element x by the element y of a group type.

---

[4] We do not use an informal name for this concept which we use only once in the rest of this text.

Similarly, we denote by AB the set product of two sets A and B of a group type.

In order to avoid useless rigid type constraints in the formalized statements of finite group theory, we generalize as much as possible the standard concepts of finite group theory to sets of group type. In all that follows a *mere set*, sometimes even abbreviated in a *set*, refers to an arbitrary set of a group type G. Elements of a mere set A can be multiplied by the group operation defined by G, although this product does not necessarily belong to A, nor the identity of G.

For instance if A is an *arbitrary mere set* of a group type G and x an arbitrary element of G, we define the *conjugate* of A by x as the set of conjugates $x^{-1}ax$ of elements $a \in A$ by x:

$$A^x := \{x^{-1}ax \mid a \in A\}$$

Note that we use here the group operation of G to describe the elements of A^x. Formally this set is defined as the image set of the set A by the function $y \mapsto x^{-1}yx$. Similarly, we define the conjugate of the set A by the set B as A^B, the image of the set B by the function $x \mapsto A^x$. The *normalizer of a mere set* A is defined as:

$$N(A) := \{x \mid A^x \subseteq A\}$$

The definition of $N(A)$ is formalized using the comprehension-style construction mentioned in section 2.2. The *centralizer of a mere set* A is the intersection of the normalizers of all the singleton sets of its elements:

$$C(A) := \bigcup_{x \in A} N(\{x\})$$

The formal definition of $C(A)$ uses a library about iterated operators [5], which provides a modular infrastructure for notations, theory and computation of indexed constructions like $\bigcap_{x \in A}$. We also introduce a notation for the localization of the normalizer and centralizer of a set A to a set B: we denote informally by $N_B(A)$ (resp. $C_B(A)$) the intersection $N(A) \cap B$ (resp. $C(A) \cap B$).

Finally, for any group type G, a *group* of G (or just a *group*) is a mere set G of G which contains the identity element $(1 \in G)$ and is closed under the product $(GG \subseteq G)$. Note that what we call a group here is necessarily a finite group. The singleton set $\{1\}$ of the identity element of G is a group of G as well as the total set containing all the elements of G. This total set actually plays the role of the ambient group G postulated in many statements, like for instance in definition 2. A subgroup H of a group G is a group whose underlying set is a subset of the one underlying G. Formally, {group gT} denotes the type of groups of a group type gT. For any set A, the group *generated* by A is obtained as the intersection of all the sets that are groups and contain A, formalized by the means of aforementioned library about iterated operators [5]. As usual, the group generated by A is formally defined as a set, later equipped with a canonical group structure of group. We denote informally by $\langle A \rangle$ and formally by <<A>> the group generated by a mere set A. We also prove that some of the above constructions on mere sets preserve the property of being a group: the intersection of two groups is a group, the normalizer of a group is a group.... Note that the set-level product of two groups is not necessarily a group.

Lemma 1 (Product group [2]). *Let* X, Y *be subgroups of a group* G. *Then* XY *is a subgroup of* G *if and only if* $XY = YX$.

In our formal library, we model this fact by defining an alternate product operation on sets of a group type that always produces a group: the join product of two sets A and B is simply the group generated by A and B, which coincides with AB when $AB = BA$.

As a conclusion of this subsection, let us summarize two differences in flavor between the usual paper versions of statements and definitions in finite group theory and their formal versions. First, the standard constructions of new groups from known groups like normalizer, centralizer, etc. are defined as the construction of new sets from known sets. The resulting sets are later equipped with a group structure under the suitable assumption on their components. Second, every formal statement features one more universal quantification or parameter, for the parameter group type. For instance, the original statement of the Odd Order Theorem is the following:

Theorem 1 (Odd Order theorem [7]). *Every finite group of odd order is soluble.*

And its formal statement in Coq is:

```
Theorem Feit_Thompson : forall (gT : finGroupType),
forall (G : {group gT}), odd #|G| -> solvable G.
```

This formal version is not less general than the original one: it can be read as "every subgroup of odd order of a group is solvable", where the group and the subgroup can for instance be the same.

Libraries. The corresponding file to this subsection is fingroup.v.

3.2 Group Morphisms, Isomorphisms

We again quote Aschbacher's definition [2] of the homomorphisms associated with the structure of group:

Definition 3 (Group homomorphism). *A group homomorphism from a group* G *into a group* H *is a function* $\alpha : G \to H$ *of the set* G *into the set* H *which preserves the group operations: that is for all* x, y *in* G, $(xy)\alpha = x\alpha\, y\alpha$. *Notice that I usually write my maps on the right, particularly those that are homomorphisms. The homomorphism* α *is an isomorphism* α *is a bijection* (..) H *is said to be a homomorphic image of* G *if there is a surjective homomorphism of* G *onto* H.

In all that follows, we slightly depart from Aschbacher's choices: we use the words *group morphism* instead of *group homomorphism* and we write these maps applicatively ($\alpha(x)$) instead of on the right ($x\alpha$). Definition 3 describes a group morphism as a function whose domain is a specified group. In type theory, a function is defined as an object (f : A -> B) whose type A -> B specifies the domain type A of its arguments and the codomain type B of its values. Such a function f is necessarily total on its domain type A. However we argued in section 3.1 that groups are better represented not as types but as objects, namely as sets of an ambient group type. Hence in our formalization, two groups G and H are modeled as sets of two (group) types G_1 and G_2 respectively and definition 3 only specifies a morphism $\phi : G \leftarrow H$ as a function and its morphism properties for certain G_1, the ones in G. We are hence left to choosing one of the several standard ways of dealing with partiality issues in type theory: assigning a clever default value outside the domain, restricting the type of the domain, using a monadic style....

Several approaches have been successively considered for the formal definition of group morphisms, leading to different versions of the related libraries. We eventually reverted the choice described in our earliest publication [12]. Garillot [8] has discussed the motivations for the change to the datastructure we describe hereafter. The current structure of group morphism, formally denoted by {morphism D >-> rT}, has three parameters:

- a group type aT called the domain type;
- a group type rT called the codomain type;
- a mere set D of the group type aT called the domain.

The domain type is not displayed to the user in the type {morphism D >-> rT} because it is implicit: it can be inferred from the type of the parameter D. This interface describes functions of type aT -> rT which distribute over the product of two elements if they both belong to the set D.

Since this definition ensures the distributive property only on the domain of a morphism, it becomes natural to consider alternative definitions of images and preimages for group morphisms. More precisely, consider f a group morphism and denote D the domain of f. Let A be a set of the domain type of f. We define the *morphic image* of the set A by the group morphism f as the image by f of the intersection of A with the domain D:

$$f_*(A) := f(A \cap D)$$

where $f(A \cap D)$ refers to the image set by f (see section 2.2). Formally, this morphic image is denoted by f @* A. Similarly, we define the *morphic preimage* of the set R by the group morphism f as the intersection of the domain D with the preimage by f of R:

$$f^{-1*}(R) := f^{-1}(R) \cap D$$

where $f^{-1}(R)$ is the preimage set of R by f. In Coq, this morphic preimage is denoted by f @*^-1 R. Note that the image and the morphic preimage domain D coincide. When both the domain D of f and the set A (resp. the set

R) are groups, the morphic image (resp. the morphic preimage) of A (resp. of R) is a group. For instance, the morphic preimage of the singleton set $\{1\}$ of the identity is a group, called the *kernel* of the group morphism. Informally we denote by *Ker* f the kernel of a group morphism and by *Im* f its image. We denote by $Ker_A\, f$ the intersection of the kernel of a morphism with a set A.

As a consequence of this formalization choice, each new definition of a morphism should come with the explicit mention of the domain it is a morphism on. However, we use the facilities offered by Coq's type inference mechanism to compute automatically a non-trivial domain for morphisms resulting from standard operations. For instance if f and g are two group morphisms, under the obvious compatibility condition on their domain and codomain types, Coq can infer that the composition $g \circ f$ is a morphism with domain (at least) $f^{-1}_*(H)$ where H is the domain of the morphism g. This means that if we dispose of (f : {morphism G >-> hT}) and (g : {morphism H >-> rT}) two morphisms, Coq infers the type (g \o f : {morphism f @*^-1 H >-> rT}) automatically. Similarly, the inverse of an injective morphism f with domain D is canonically a morphism with domain $f(D)$. However, it might sometimes be necessary to restrict by hand the domain of a morphism. We hence provide an operator which constructs a new morphism g with domain the set A from a known morphism f and a proof that $A \subseteq D$. Let us mention a last example of useful operation on group morphisms. Let f_1 and f_2 be two group morphisms with possibly different codomain types but with the same domain type. Let D be the domain of f_1 and G be the domain of f_2. We moreover assume that G is a group. Under this assumption we can construct the natural *factor morphism* $f_1^*(G)$ to $f_2^*(G)$ provided that G is included in D and that the kernel of f_1 is included in the kernel of f_2. The kernel of this morphism is the morphic image $f_1^*(\mathrm{Ker}\, f_2)$ of the kernel of f_2 by f_1.

When there exists a group morphism such that a set B is the morphic image of the set A by this morphism, we say that B is the *homomorphic image* of A. In Coq, this is denoted by (B \homg A). We say that a group morphism f with domain D maps a set A *isomorphically* to the set B when both A is included in D and the image of the complement of A (in the domain type) by f is *equal* to the complement of B (in the codomain type). This seemingly contrived definition is a concise way of ensuring both that the morphism is injective and that B is the image of A. In Coq, we denote by (B \isog A) the existence of a group morphism which maps A to B isomorphically. Note that once again we have posed these definitions at the level of sets of the group type.

Let us conclude this subsection by mentioning that we adopt a completely different datastructure for the set $Aut(G)$ of automorphisms of a group G, which is the set of endomorphisms of G that are also isomorphisms. The set of automorphisms of a group G of a group type $\mathbb{G}$ is defined as the set of permutations of $\mathbb{G}$ that are the identity function outside G and distribute over the group product inside G. It is easy to interpret canonically such a permutation as a group morphism but defining automorphisms as permutations actually allows us to transpose the (group) theory already developed for the permutations of a set to the $Aut(G)$ group.

Libraries. The corresponding files to this subsection are `morphism.v` and `automorphism.v`

4   Cosets, Normal Subgroups and Quotients

In this section again we quote Aschbacher's definitions [2].

Definition 4 (Normal subgroup). *A subgroup H of G is normal if $g^{-1}hg \in H$ for each $g \in G$ and $h \in H$. Write $H \trianglelefteq G$ to indicate H is a normal subgroup.*

Definition 5 (Cosets, coset space). *Let H be a subgroup of G. For $x \in G$ write $Hx = \{hx : h \in H\}$ and $xH = \{xh : h \in H\}$. Hx and xH are cosets of H in G. Hx is a right coset and xH is a left coset. To be consistent I'll work with right cosets Hx in this section. G/H denotes the set of all (right) cosets of H in G. G/H is the* coset space *of H in G.*

Both the definition of the "normal" predicate and the one of right and left cosets are literally formalized in our formal development, except that they are defined as usual for mere sets instead of groups. However we depart from Aschbacher's choice when it comes to the definition of coset spaces. Instead, we somehow take backward the following definition [2]:

Definition 6 (Factor group). *If $H \trianglelefteq G$ the coset space G/H is made into a group by defining multiplication via*

$$(Hx)(Hy) = Hxy \quad x, y \in G$$

Moreover there is a natural surjective homomorphism $\pi : G \to G/H$ defined by $\pi : x \mapsto Hx$. Notice $ker(\pi) = H$. Conversely if $\alpha : G \to L$ is a surjective homomorphism with $ker(\alpha) = H$ then the map $\beta : Hx \mapsto x\alpha$ is an isomorphism of G/H with L such that $\pi\beta = \alpha$. G/H is called the factor group *of G by H. Therefore the factor groups of G over its various normal subgroups are, up to isomorphism, precisely the homomorphic images to G.*

Instead of studying the entire coset space in the sense of definition 5, we formally define the coset space of a mere set A as the type whose elements are the right cosets Ax, with x spanning the normalizer $N(A)$ of A (see section 3.1). Informally, we denote by $\mathcal{C}_A$ the coset space of a set A according to this alternate definition. An element of $\mathcal{C}_A$ is a *bilateral coset*, since for $x \in N(A)$, we have $Ax = xA$. Note that when H is a group, $N(H)$ happens to be the largest group (for inclusion) in which H is normal.

Coset spaces as defined in 5 are actually of little use when H is not normal in G, assumption which make the coset space a group. Hence the theory of coset spaces described in the literature actually boils down to the theory of quotients of a group by one of its normal subgroup. In our setting, we are able to define a better theory of coset spaces which allows us to simplify greatly the manipulation of quotients by erasing these normality assumptions. Without this effort, the requirements put in the type constraints are soon too demanding for the lemmas to be usable as such. We explain this formalization in the rest of this subsection and illustrate its impact on the three isomorphism theorems of group theory.

Consider a group H of a group type $\mathcal{G}$. Its coset space $\mathcal{C}_H$ is an instance of group type: the group product operation and the inverse operation are respectively the set product and the set inverse operations, and the identity element $1 \in \mathcal{C}_H$ is $1H = H$. In CoQ, this new instance $\mathcal{C}_H$ of group type is named (coset_groupType H). For $x, y \in N(H)$, the product $xHyH$ of two bilateral cosets is the coset xyH of the product $xy \in N(H)$. But interestingly, if we extend this correspondence by associating each element $x \notin N(H)$ with the identity value $1 \in \mathcal{C}_H$, we obtain a total function on $\mathcal{G}$, which is an instance of group morphism, in the sense discussed in section 3.2. The domain type of this morphism is $\mathcal{G}$, its codomain type is $\mathcal{C}_H$ and its domain is the normalizer $N(H)$. Informally, we denote this morphism by $./H$. Formally, we denote this morphism by (coset H). The quotient of a mere set A by the group H, denoted A/H, is defined as the *morphic image* of the set A by this morphism $./H$.

When H is a normal subgroup of A, our definition coincide with the one of the standard literature. It is however more general for it provides a precise meaning to the quotient of a mere set A by H which requires A neither to be a group nor to be included in $N(H)$. Using the standard definition 6, what we define as A/H would be indeed described as $N_A(H)H/H$. In fact we make the definition of quotient even more general: if A and B are sets of $\mathcal{G}$, then A/B denotes the quotient of the set A by the group $\langle B \rangle$ generated by B: A/B is a set of the group type $\mathcal{C}_{\langle B \rangle}$.

We conclude this subsection by commenting three elementary but crucial results of finite group theory, called isomorphism theorems [18]. The first one is a rephrasing of the result contained in Aschbacher's definition 6.

Theorem 2 (First isomorphism theorem). *Let f be a group morphism from G to K. Then*

$$G/(Ker\ f) \to H \quad with \quad (Ker\ f)x \mapsto xf$$

is an injective group morphism. In particular

$$G/(Ker\ f) \text{ is isomorphic to } (Im\ f).$$

In our setting, starting from a group morphism f with domain a group G, we construct a new morphism g with domain $G/(Ker\ f)$ which is injective and such that for any set A, the morphic image $f^*(A)$ of A by f is equal to $g^*(A/(Ker\ f))$. This is a slight generalization of the above statement, which corresponds to the case where we take $A = G$. This morphism g can be obtained as the factor morphism mentioned in section 3.2, taking f as f_2 and $./(Ker\ f)$ as f_1: the kernel of $./(Ker\ f)$ is $(Ker\ f)$, hence included in (in fact equal to) the one of f. Moreover the domain of a group morphism is included in the normalizer of its kernel, thus the domain of f is included in the one of $./(Ker\ f)$. We can therefore form this factor morphism g, which sends $A/(Ker\ f)$ to $f^*(A)$ for any set A. The kernel of this morphism is $(Ker\ f)/(Ker\ f)$ which is trivial, hence the morphism is injective.

An easy corollary follows from this first isomorphism theorem. For an extra subgroup H of G, we can construct a morphism g with domain $H/(Ker_H\ f)$, which is injective and such that for any subset A of H, the morphic image $f^*(A)$ of A by f is equal to $g^*(A/(Ker_H\ f))$.

The second isomorphism theorem is a central ingredient in the butterfly argument of the proof of Jordan-Hölder theorem (see section 5).

Theorem 3 (Second isomorphism theorem). *Let G be a group and H and K two subgroups of G such that $H \subset N(K)$. Then HK is a subgroup of G and thus K is a normal subgroup of HK and:*

$$\phi : H \to HK/K \quad with \quad u \mapsto uK$$

is an injective group morphism with $Ker\ \phi = H \cap K$ and

$$H/H \cap K \text{ is isomorphic to } HK/K.$$

In this theorem, we can observe an instance of the partiality issues raised by the standard definition of quotients in the literature. The statement of the theorem has two parts: the first one establishes the conditions under which some objects are well defined. The second one is an isomorphism involving these objects. The situation is quite easier using the generalized definitions set up in the previous subsections.

In our setting, an ambient group type $\mathcal{G}$ plays the role of the above G, and we consider two groups H and K of $\mathcal{G}$, such that $H \subset N(K)$. We prove the second isomorphism theorem by constructing a morphism g with domain $H/H \cap K$, which is injective and such that $g^*(A/H \cap K) = A/K$ for any subset A of H. Noticing that $H \cap K$ is $Ker_H\ (./K)$, the existence of g is a direct application of the above corollary of the first isomorphism theorem, applied to the morphism $./K$. The usual version of the second isomorphism theorem, as stated in theorem 3, follows from this construction: in our setting HK/K, which is the morphic image of HK by $./K$, is *equal* to the morphic image H/K of H by $./K$.

The conclusion of the third isomorphism theorem uses three distinct quotient operations, each of which deserves a side condition of well-formedness.

Theorem 4 (Third isomorphism theorem). *Let H and K be normal subgroups of G such that H is a subgroup of K. Then*

$$\phi : G/H \to G/K \quad with \quad Hx \mapsto Kx$$

is an injective morphism with $Ker\ \phi = K/H$, and

$$(G/H)/(K/H) \text{ is isomorphic to } G/K.$$

We consider three groups G, H, and K of a group type $\mathcal{G}$. We suppose that H and K are normal subgroups of G and that H is a subgroup of K. Again, we slightly generalize the above statement by constructing a morphism g with domain $(G/H)/(K/H)$ such that for any subset A of G, $g^*(A/H/(K/H)) = A/K$. We proceed in three steps. First, we consider the morphism $./K$, whose domain is $N(K)$. We restrict it to G, using the restriction operation mentioned in section 3.2. This is possible since $G \subset N(K)$ by hypothesis. Then, we factor the morphism $./H$ by this restriction. This means we apply the factor operation described in section 3.2 with f_1 equal to the morphism $./H$ and f_2 equal to the restriction of $./K$ to G. We hence obtain a morphism g' which maps any subset A of G/H to A/K and has kernel K/H. Finally, we apply the corollary of the first isomorphism theorem to g', which constructs the announced morphism g.

Libraries. The corresponding file to this section is `quotient.v`.

5   The Jordan Hölder Theorem(s)

In this section, we sketch the well-known proof of the Jordan-Hölder theorem for finite groups [17, 15] as formalized in CoQ on top of the infrastructure presented in section 3.

5.1   Simple Groups, Composition Series

A normal series is a sequence of successive quotients of a group.

Definition 7 (Normal series, factors). *A normal series for a group G is a sequence $1 = G_0 \trianglelefteq G_1 \ldots \trianglelefteq G_n = G$, and the successive quotients $(G_{k+1}/G_k)_{0 \le k < n}$ are called the factors of the series.*

Formalizing normal series poses no particular problem: it is a sequence of groups where the sets underlying two consecutive elements are related by the relation $\trianglelefteq$. The corresponding formal definition is actually obtained from a more general pattern, that we call subgroup series. Subgroup series are defined as sequences of groups for which the sets of two consecutive elements related by a binary relation (on sets). This simple definition suits the formalization of several notions like normal series, ascending series, descending series, chief series....

A formal definition of the sequence of factors of a normal series is however slightly more uneasy at first sight. Let G be a group of the group type $\mathcal{G}$, and $(G_k)_{0 \le k \le n}$ a normal series for G. All the elements of the series are groups of $\mathcal{G}$. By contrast, each factor G_{k+1}/G_k is a group of the group type $\mathcal{C}_{G_k}$. Since the elements of the sequence of factors have pairwise distinct types a formal definition of this sequence would be very intricate. Instead, we represent a factor (G_{k+1}/G_k) of a normal series by a pair of groups (G'_{k+1}, G_k) where (G'_{k+1}/G_k) is a canonical representative of the isomorphism class of (G_{k+1}/G_k) inside $\mathcal{C}_{G_k}$. The sequence of factors can hence be represented as a homogeneous sequence, whose elements are pairs of group of $\mathcal{G}$. The use of the isomorphism representative is motivated by the proof described in section 5.2. The formalization of this definition uses the choice operator mentioned in section 2.2.

Definition 8 (Simple group). *A group G is simple when its only proper normal subgroup is the trivial group 1.*

Definition 9 (Composition series). *A normal series whose factors are all simple groups is called a composition series.*

Definitions 8 and 9 are translated literally in the libraries, using the material presented so far. Simple groups are exactly groups with composition series of length 1 (containing only the group itself and 1). Similarly trivial groups are are exactly groups with empty composition series (containing only the group itself).

Lemma 2 (Existence of a composition series). *Every finite group has a composition series.*

The proof of lemma 2 is an induction on the cardinal of a group G, which is either simple, or trivial, or has a non-trivial proper normal subgroup H, maximal for inclusion. In the last case, the quotient G/H is simple and we conclude by applying the induction hypothesis to H.

Libraries. The corresponding files to this subsection are `gseries.v` and `jordanholder.v`.

5.2   Uniqueness of Composition Series

The Jordan-Hölder theorem states that the (simple) factors of a composition series play a role analogous to the prime factors of a number. They however do not control completely the structure of a group: unlike natural numbers non-isomorphic groups may have composition series with isomorphic factors.

Theorem 5 (Jordan-Hölder Uniqueness). *Two composition series of a same group have the same length and the same factors up to permutation and isomorphism.*

Let G be a group of a group type $\mathcal{G}$. We prove that for any two composition series of G, the corresponding sequences of factors are equal up to permutation, since we have already picked canonical isomorphism representatives for the factors. Again we proceed by induction on the cardinal of the group G. In the inductive case, we can assume that G is neither trivial nor simple, and we consider two non empty composition series of G, $(N_i)_{0 \le i \le r+1}$ and $(M_j)_{0 \le j \le s+1}$. Note that $G = N_{r+1} = M_{s+1}$. We call N (resp. M) the group N_r (resp. M_s): $(N_i)_{0 \le i \le r}$ is a composition series of N and $(M_j)_{0 \le j \le s}$ is a composition series of M. Both N and M are normal subgroups of G. If N and M are equal, then the theorem is proved from the induction hypothesis. Otherwise we pose $I = M \cap N$, which is normal in both N and M. Now comes the crux of the demonstration: G/N is isomorphic to M/I and G/M is isomorphic to N/I. This step is called the butterfly lemma, or also Zassenhaus lemma [23] and both these isomorphisms are easy consequences of the second isomorphism theorem 3.

To finish the proof, we use lemma 2 to construct a composition series $(I_k)_{0 \le k \le t}$ for I. The butterfly lemma ensures that the quotient N/I is simple. Therefore $(I_k)_{0 \le k \le t}$ extends to a composition series for N by taking $I_{t+1} := N$. We dispose of two composition series $(I_k)_{0 \le k \le t+1}$ and $(N_i)_{0 \le i \le r}$ for the group N whose cardinal is smaller that the one of G: the induction hypothesis applies and these series have the same length and the same factors. Similarly we apply the induction hypothesis to the two composition series we dispose of for M. Hence up to isomorphism the set of factors associated with $(N_i)_{0 \le i \le r+1}$ is G/N, N/I and a set F_N of other factors, the set of factors associated with $(M_i)_{0 \le i \le r+1}$ is G/M, M/I and a set F_M of other factors, such that F_N and F_M are the same up to isomorphism and permutation. The isomorphisms established by the butterfly lemma conclude the proof.

Libraries. The corresponding file to the subsection is `jordanholder.v`.

5.3   More Butterflies

The more general version of the Jordan-Hölder theorem for finite groups deals with a more general kind of composition series: given a set A which acts on a group G, an A-composition series is an increasing sequence $(G_k)_{0 \le k \le n}$ of subgroups of G, with $G = G_n$ and such that for each k, G_k is a maximal subgroup of G_{k+1} invariant by the action of A. A finite group G has an A-composition series as soon as A acts on G and the uniqueness theorem transposes to the factors of A-composition series of a same group, with a little more work, in particular for establishing the butterfly lemma. We have also formalized this more general version [2], which we do not detail here by lack of space.

The library also features the analogue Jordan-Hölder theorem for the theory of representations of finite groups [16], whose proof is again analogue in shape. However the algebraic structures at stake in that case are much more sophisticated than the ones of finite groups, and their formalization is based on a significant reworking of the standard mathematical presentations of elementary linear algebra [9].

As a final remark we would like to mention that these butterfly lemmas are quite typical, although rather simple, examples where two objects play a symmetrical role, which is broken *without loss of generality* at the beginning of the proof. The version of the Jordan-Hölder theorem we detailed in section 5.2 is so simple that no additional support is really needed in that proof. However the code formalizing the two more advanced versions we mentioned above are using a specific feature of the proof shell [13] used to develop these libraries, called the `wlog` tactic. This command is a key ingredient in order to avoid extremely painful redundancy in the script describing these mathematical arguments based on symmetries. This quite elementary feature of the tactic language has actually been instrumental at several places of the libraries, including advanced group theory for the proof of the Odd Order Theorem [10].

Libraries. The corresponding files to this subsection are `jordanholder.v` and `mxrepresentation.v`. See for instance `PFsection9.v` for an instance of `wlog` tactic, on a tricky chain of circular inequalities with equality conditions.

6   Conclusion

The structure of the paper reflects in miniature the one of the whole set of libraries of the formal proof of the Odd Order Theorem. Libraries on elementary concepts, like types with decidable equality or finite sets, are tightly related to the type system underlying the COQ proof assistant. They provide an infrastructure which allows us to ignore the details of their implementation when it comes to formalizing finite groups as finite sets of a group type. Here again part of the basic libraries about finite groups, morphisms, and quotients are devoted to the infrastructure work which aims at providing the same flavor of mathematical notations and packaging as in the standard literature of finite group theory. As a result, there is not much left to say when it comes to describing the formalized proof of the Jordan-Hölder, and this was precisely the purpose of the upstream effort. The elementary examples from finite group theory presented here also illustrate the fact that textbook presentations of abstract algebra are not necessarily sufficient references in order to design the appropriate abstractions for formal libraries to scale. Future formalizations will show whether the techniques employed in the present libraries are general enough to apply to more mathematical structures. The design of these patterns will for sure be impacted by improvements in the implementation of proof assistants [19] but also possibly by evolutions of the type theory they implement [21].

The difference in purpose of the different layers of libraries affect their de Bruijn factor [22], a criterion measuring the difference in size between the code describing a formal proof and the code of the typeset description of a paper proof. Lower level libraries feature by far the highest de Bruijn factor because they describe a lot of material which addresses the implicit content of paper mathematics. This implicit content is not only about datastructures, but also about how to recompute the implicit content of notational conventions, or abuse thereof, without which a paper text appears as extremely pedantic and soon unreadable. By contrast, for advanced libraries like the ones corresponding to the final chapters of the proof of the Odd Order theorem, it is possible to obtain a one to one correspondence, and even sometimes a shorter formal proof, which illustrates the benefits of the re-factoring of the mathematics.

Libraries. An example of infrastructure file with a very large de Bruijn factor is `bigop.v`. By contrast, file `BGappendixC.v` has a very small de Bruijn factor (3 pages for 170 lines of script according to G. Gonthier, author of the script). In file `PFsection3.v`, the local definition of some appropriate boilerplate [10] significantly shortens a pedestrian computational proof.

Acknowledgments. The author wishes to thank Jacques Carette, William Farmer, Laurence Rideau and Enrico Tassi for their proofreading.

References

[1] The Coq Proof Assistant, http://coq.inria.fr
[2] Aschbacher, M.: Finite Group Theory. Cambridge Studies in Advanced Mathematics. Cambridge University Press (2000)
[3] Bender, H., Glauberman, G.: Local analysis for the Odd Order Theorem. London Mathematical Society, LNS, vol. 188. Cambridge University Press (1994)
[4] Bertot, Y., Castéran, P.: Interactive theorem proving and program development: Coq'Art: The calculus of inductive constructions. Springer, Berlin (2004)
[5] Bertot, Y., Gonthier, G., Ould Biha, S., Pasca, I.: Canonical big operators. In: Mohamed, O.A., Muñoz, C., Tahar, S. (eds.) TPHOLs 2008. LNCS, vol. 5170, pp. 86–101. Springer, Heidelberg (2008)
[6] de Bruijn, N.G.: The mathematical language AUTOMATH, its usage, and some of its extensions. In: Laudet, M., Lacombe, D., Nolin, L., Schützenberger, M. (eds.) Symposium on Automatic Demonstration. Lecture Notes in Mathematics, vol. 125, pp. 29–61. Springer, Heidelberg (1970)
[7] Feit, W., Thompson, J.G.: Solvability of groups of odd order. Pacific Journal of Mathematics 13(3), 775–1029 (1963)
[8] Garillot, F.: Generic Proof Tools and Finite Group Theory. PhD thesis, École polytechnique (2011)
[9] Gonthier, G.: Point-free, set-free concrete linear algebra. In: van Eekelen, M., Geuvers, H., Schmaltz, J., Wiedijk, F. (eds.) ITP 2011. LNCS, vol. 6898, pp. 103–118. Springer, Heidelberg (2011)
[10] Gonthier, G., Asperti, A., Avigad, J., Bertot, Y., Cohen, C., Garillot, F., Roux, S.L., Mahboubi, A., O'Connor, R., Biha, S.O., Pasca, I., Rideau, L., Solovyev, A., Tassi, E., Théry, L.: A machine-checked proof of the odd order theorem. To appear in the Proceedings of the ITP 2013 Conference (2013)
[11] Gonthier, G., Mahboubi, A.: An introduction to small scale reflection in Coq. Journal of Formalized Reasoning 3(2), 95–152 (2010)
[12] Gonthier, G., Mahboubi, A., Rideau, L., Tassi, E., Théry, L.: A Modular Formalisation of Finite Group Theory. In: Schneider, K., Brandt, J. (eds.) TPHOLs 2007. LNCS, vol. 4732, pp. 86–101. Springer, Heidelberg (2007)
[13] Gonthier, G., Mahboubi, A., Tassi, E.: A Small Scale Reflection Extension for the Coq system. Rapport de recherche RR-6455, INRIA (2012)
[14] Hedberg, M.: A coherence theorem for Martin-Löf's type theory. Journal of Functional Programming 8(4), 413–436 (1998)
[15] Hölder, O.: Zurückführung einer beliebigen algebraischen Gleichung auf eine Kette von Gleichungen. Mathematische Annalen 34(1), 26–56 (1889)
[16] Isaacs, I.: Character Theory of Finite Groups. AMS Chelsea Pub. Series (1976)
[17] Jordan, C.: Traité des substitutions et des équations algébriques. Gauthier-Villars, Paris (1870)
[18] Kurzweil, H., Stellmacher, B.: The Theory of Finite Groups: An Introduction. Universitext Series. Springer (2010)
[19] Mahboubi, A., Tassi, E.: Canonical structures for the working coq user. To appear in the Proceedings of the ITP 2013 Conference (2013)
[20] Peterfalvi, T.: Character Theory for the Odd Order Theorem. London Mathematical Society, LNS, vol. 272. Cambridge University Press (2000)
[21] The Univalent Foundations Program. Homotopy type theory: Univalent foundations of mathematics. Technical report, Institute for Advanced Study (2013)
[22] Wiedijk, F.: The "de Bruijn factor", http://www.cs.ru.nl/~freek/factor/
[23] Zassenhaus, H.: Zum satz von Jordan-Hölder-Schreier. Abhandlungen aus dem Mathematischen Seminar der Universität Hamburg 10(1), 106–108 (1934)

Optimising Problem Formulation
for Cylindrical Algebraic Decomposition

Russell Bradford, James H. Davenport, Matthew England, and David Wilson

University of Bath, Bath, BA2 7AY, U.K.
{R.J.Bradford,J.H.Davenport,M.England,D.J.Wilson}@bath.ac.uk
http://people.bath.ac.uk/masjhd/Triangular/

Abstract. Cylindrical algebraic decomposition (CAD) is an important
tool for the study of real algebraic geometry with many applications both
within mathematics and elsewhere. It is known to have doubly exponential complexity in the number of variables in the worst case, but the
actual computation time can vary greatly. It is possible to offer different
formulations for a given problem leading to great differences in tractability. In this paper we suggest a new measure for CAD complexity which
takes into account the real geometry of the problem. This leads to new
heuristics for choosing: the variable ordering for a CAD problem, a designated equational constraint, and formulations for truth-table invariant
CADs (TTICADs). We then consider the possibility of using Gröbner
bases to precondition TTICAD and when such formulations constitute
the creation of a new problem.

Keywords: cylindrical algebraic decomposition, problem formulation,
Gröbner bases, symbolic computation.

1 Introduction

Cylindrical algebraic decomposition (CAD) is a key tool in real algebraic geometry both for its original motivation, quantifier elimination (QE) problems [10,
etc.], but also in other applications ranging from robot motion planning [25, etc.]
to programming with complex functions [13, etc.] and branch cut analysis [17,
etc.]. Decision methods for real closed fields are used in theorem proving [15], so
CAD has much potential here. In particular MetiTarski employs QEPCAD [4] to
decide statements in special functions using polynomial bounds [1, 2, 23]. Work
is ongoing to implement a verified CAD procedure in CoQ [9, 22].

Since its inception there has been much research on CAD. New types of CAD
and new algorithms have been developed, offering improved performance and
functionality. The thesis of this paper is that more attention should now be
given to how problems are presented to these algorithms.

How a problem is formulated can be of fundamental importance to algorithms,
rendering simple problems infeasible and vice versa. In this paper we take some
steps towards better formulation by introducing a new measure of CAD complexity and new heuristics for many of the choices required by CAD algorithms.
We also further explore preconditioning the input via Gröbner bases.

J. Carette et al. (Eds.): CICM 2013, LNAI 7961, pp. 19–34, 2013.

1.1 Background on CAD

A CAD is a decomposition of $\mathbb{R}^n$ into cells arranged cylindrically (meaning their projections are equal or disjoint) and described by semi-algebraic sets. Traditionally CADs are produced sign-invariant to a given set of polynomials in n variables **x**, meaning the sign of the polynomials does not vary on the cells. This definition was provided by Collins in [10] along with an algorithm which proceeded in two main phases. The first, *projection*, applies a projection operator repeatedly to a set of polynomials, each time producing another set of polynomials in one fewer variables. Together these sets provide the *projection polynomials*. The second phase, *lifting*, then builds the CAD incrementally from these polynomials. First $\mathbb{R}$ is decomposed into cells which are points and intervals corresponding to the real roots of the univariate polynomials. Then $\mathbb{R}^2$ is decomposed by repeating the process over each cell using the bivariate polynomials at a sample point of the cell. The output for each cell consists of *sections* of polynomials (where a polynomial vanishes) and *sectors* (the regions between these). Together these form the *stack* over the cell, and taking the union of these stacks gives the CAD of $\mathbb{R}^2$. This process is repeated until a CAD of $\mathbb{R}^n$ is produced. This final CAD will have cells ranging in dimension from 0 (single points) to n (full dimensional portions of space). The cells of dimension d are referred to as *d-cells*.

It has often been noted that such decompositions actually do much more work than is required for most applications, motivating theory which considers not just polynomials but their origin. For example, partial CAD [12, etc.] avoids unnecessary lifting over a cell if the solution to the QE problem on a cell is already apparent. Another example is the use of CAD with equational constraints [21, etc.] where sign-invariance is only ensured over the sections of a designated equation, thus reducing the number of projection polynomials required. It is worth noting that while the lifting stage takes far more resources that the projection, improvements of the projection operator have offered great benefits.

Applications often analyse formulae (boolean combinations of polynomial equations, inequations and inequalities) by constructing a sign invariant CAD for the polynomials involved. However this analyses not only the given problem, but any formula built from these polynomials. In [3] the authors note that it would be preferable to build CADs directly from the formulae and so define a Truth Table Invariant CAD (TTICAD) as one which is has invariant truth values of various quantifier-free formulae (QFFs) in each cell. In [3] an algorithm was produced which efficiently constructed such objects for a wide class of problems by utilising the theory of equational constraints.

1.2 Formulating Problems for CAD Algorithms

The TTICAD algorithm in [3] takes as input a sequence of QFFs, each of which is a formula with a designated equational constraint (an equation logically implied by the formula). It outputs a CAD such that on each cell of the decomposition each QFF has constant truth value. The algorithm is more efficient than constructing a full sign-invariant CAD for the polynomials in the QFFs, since it uses the theory of equational constraints for each QFF to reduce the projection polynomials used and hence the number of cells required. Its benefit over equational constraints alone is that it may be used for formulae which do not have an overall explicit equational constraint (and to greater advantage than the use of implicit equational constraints). Many applications present problems in a suitable form for TTICAD, such as problems from branch cut analysis [17].

However, it is possible to envision problems where although separate QFFs are not imposed they could still lead to more economical CADs, (see Example 6). Further, we may consider splitting up individual QFFs if more than one equational constraint is present. This leads to the question of how best to formulate the input to TTICAD, a question which motivated this paper and is answered in Section 4. Some of this analysis could equally be applied to the theory of equational constraints alone and so this is considered in Section 3.

In devising heuristics to guide this process we realised that the existing measures for predicting CAD complexity could be misled. An important use for these is choosing a variable ordering for a CAD; a choice which can make a substantial difference to the tractability of problems. We use $x \prec y$ to indicate x is less than y in an ordering. In [14] the authors presented measures for CAD complexity but none of these consider aspects of the problem sensitive to the domain we work in (namely real geometry rather than complex). In Section 2 we suggest a simple new measure (the number of zero cells in the induced CAD of $\mathbb{R}^1$) leading to a new heuristic for use in conjunction with [14]. We demonstrate in general it does well at discriminating between variable choices, and for certain problems is more accurate than existing heuristics.

These three topics are all examples of choices for the formulation of problems for CAD algorithms. They are presented in the opposite order to which they were considered above, as it is more natural for presenting the theory. Problem formulation was considered in this conference series last year [27] where the idea of preconditioning CAD using Gröbner bases was examined. This work is continued in Section 5 where we now consider preconditioning TTICAD.

The tools developed for the formulation of input here lead to the question of whether their use is merely an addition to the algorithm or leads to the creation of a new problem. This question also arose in [26] where a project collecting together a repository of examples for CAD is described. In Section 6 we give our thoughts on this along with our conclusions and ideas for future work.

2   Choosing a Variable Ordering for CAD

2.1   Effects of Variable Ordering on CAD

It is well documented [14, etc.] that the variable ordering used to construct a CAD can have a large impact on the number of cells and computation time. Example 1 gives a simple illustration. Note that the effect of the variable ordering can be far greater than the numbers presented here and can change the feasibility of a given problem. In [5] the authors prove there are problems where one variable ordering will lead to a CAD with a constant number of cells while another will give a number of cells doubly exponential in the number of variables.

Example 1. Consider the polynomial $f := (x-1)(y^2+1)-1$ whose graph is the solid curve in Figure 1. We have two choices of variable ordering, which lead to the two different CADs visualised. Each cell is indicated by a sample point (the solid circles). Setting $y \prec x$ we obtain a CAD with 3 cells; the curve itself and the portions of the plane either side. However, setting $x \prec y$ leads to a CAD with 11 cells; five 2-cells, five 1-cells and one 0-cell. The dotted lines indicate the stacks over the 0-cells in the induced CAD of $\mathbb{R}^1$. With $y \prec x$ the CAD of $\mathbb{R}^1$ had just one cell (the entire real line) while with $x \prec y$ there are five cells.

We note that these numbers occur using various CAD algorithms. Indeed, for this simple example it is clear that these CADs are both minimal for their respective variable orderings, (i.e. there is no other decomposition which could have less cells whilst maintaining cylindricity.)

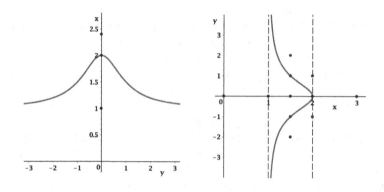

Fig. 1. Plots visualising the CADs described in Example 1

2.2   Heuristics for Choosing Variable Ordering

In [14] the authors considered the problem of choosing a variable ordering for CAD and QE via CAD. They identified a measure of CAD complexity that was correlated to the computation time, number of cells in the CAD and number of leaves in a partial CAD. They identified the *sum of total degrees of all monomials of all projection polynomials*, known as sotd and proposed the heuristic of picking the ordering with the lowest sotd. Although the best known heuristic, sotd does not always pick the ideal ordering as demonstrated by some experiments in [14] and sometimes cannot distinguish between orderings as shown in Example 2.

Example 2. Consider again the problem from Example 1. Applying any known valid projection operator to f gives, with respect to y, the set of projection factors $\{x - 1, x - 2\}$, (arising from the coefficients and discriminant of f). Similarly, applying a projection operator with respect to x gives $\{y^2 + 1\}$. Hence in this case both variable orderings have the same `sotd`.

We consider why `sotd` cannot differentiate between the orderings in this case. Algebraically, the only visible difference is that one ordering offers two factors of degree one while the other offers a single factor of degree two. From Figure 1 we see that one noticeable difference between the variable orderings is the number of 0-cells in the CAD of $\mathbb{R}^1$ (the dotted lines). This is a feature of the real geometry of the problem as opposed to properties of the algebraic closure, measured by `sotd`. Investigating examples of this sort we devised a new measure `ndrr` defined to be the *number of distinct real roots of the univariate projection polynomials* and created the associated heuristic of picking the variable ordering with lowest `ndrr`. Considering again the projection factors from Example 2 we see that this new heuristic will correctly identify the ordering with the least cells.

The number of real roots can be identified, for example, using the theory of Sturm chains. This extra calculation will likely take more computation time than the measuring of degrees required for `sotd`. However, both costs are usually negligible compared to the cost of lifting in the CAD algorithm.

2.3   Relative Merits of the Heuristics

We do not propose `ndrr` as a replacement for `sotd` but suggest they are used together since both have relative merits. We have already noted that the strength of `ndrr` is its ability to give information on the real geometry of the CAD. Its weakness is that it only gives information on the complexity of the univariate polynomials, compared to `sotd` which measures at all levels. If the key differences between orderings are not apparent in the univariate polynomials then `ndrr` is of little use, as in Example 3.

Example 3. Consider the problem of finding necessary and sufficient conditions on the coefficients of a quartic polynomial so that it is positive semidefinite: eliminate the quantifier in, $\forall x (px^2 + qx + r + x^2 \geq 0)$. This classic QE problem was first proposed in [18] and was a test case in [14]. There are six admissible variable orderings (since x must always be projected first). In all of these orderings the univariate projection factor set will consist of just the single variable of lowest order, (either p, q or r) and hence all orderings will have an `ndrr` of one. However, the `sotd` can distinguish between the orderings as reported in [14].

Despite the shortcoming of only considering the first level, `ndrr` should not be dismissed as effects at the bottom level can be magnified. We suggest using the heuristics in tandem, either using one to break ties between orderings which the other cannot discriminate or by taking a combination of the two measures.

In [14] the authors suggested a second heuristic, a greedy algorithm based on sotd. This approach avoided the need to calculate the projection polynomials for all orderings, instead choosing one variable at a time using the sum of total degree of the monomials from those projection polynomials obtained so far. Unfortunately there is not an obvious greedy approach to using ndrr. For problems involving many variables (so that calculating the full set of projection polynomials for each ordering is infeasible) we should revert to the sotd greedy algorithm, perhaps making use of ndrr to break ties.

2.4 Coupled Variables

It has been noted in [24] that a class of problems particularly unsuitable for sotd is choosing between coupled variables (two variables which are the real and imaginary parts of a complex variable). These are used, for example, when analysing complex functions by constructing a CAD to decompose the domain according to their branch cuts. The ordering of the coupled variables for the CAD can affect the efficiency of the algorithm, as in Example 4.

Example 4. Consider $f = \sqrt{z^2 + 1}$ where $z \in \mathbb{C}$. The square root function has a branch cut along the negative real axis and so f has branch cuts when

$$\Re(z^2 + 1) = x^2 - y^2 + 1 < 0 \quad \text{and} \quad \Im(z^2 + 1) = 2xy = 0,$$

where x, y are coupled real variables such that $z = x + iy$. With variable ordering $x \prec y$ we have $\text{sotd} = 8, \text{ndrr} = 4$ and a CAD with 21 cells while with variable ordering $y \prec x$ we have $\text{sotd} = 8, \text{ndrr} = 5$ and a CAD with 29 cells. The CADs are visualised in Figure 2 using the same techniques as described for Figure 1.

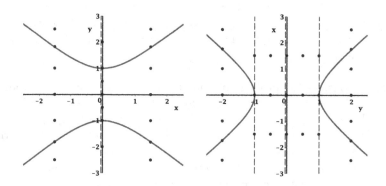

Fig. 2. Plots visualising the CADs described in Example 4

3   Designating Equational Constraints

An **equational constraint** is an equation logically implied by a formula. The theory of equational constraints is based on the observation that the formula will be false for any cell in the CAD where the equation is not satisfied. Hence the polynomials forming any other constraints need only be sign invariant over the sections of the equational constraint. The observation was first made in [11] with McCallum providing the first detailed approach in [21]. Given a problem with an equational constraint McCallum suggested a reduced projection operator, which will usually result in far fewer projection factors and a simpler CAD.

This approach has been implemented in QEPCAD, a command line interface for quantifier elimination through partial CAD [4]. It can also be induced in any implementation of TTICAD as discussed in Section 4. The use of equational constraints can offer increased choice for problem formulation beyond that of picking a variable order. If a problem has more than one equational constraint then one must be *designated* for use in the algorithm. We propose simple heuristics for making this choice based on sotd and ndrr.

Let P be the McCallum projection operator which, informally, is applied to a set of polynomials to produce the coefficients, discriminant and cross resultants. The full technical details are available in [19] and a validated algorithm was given in [20]. Note that implementations usually make some trivial simplifications such as removal of constants, exclusion of polynomials that are identical to a previous entry (up to constant multiple), and only including those coefficients which are really necessary for the theory to hold.

Next, for some equational constraint f let P_f be the reduced projection operator relative to f described in [21]. Informally, this consists of the coefficients and discriminant of f together with the resultant of f taken with each of the other polynomials. This is used for the first projection, reverting to P for subsequent projections. We can then apply the sotd and ndrr measures to the sets of projection polynomials as a measure of the complexity of the CADs that would be produced. We denote these values by S and N respectively and our heuristics are then to choose the equational constraint that minimises these values.

We ran experiments to test the effectiveness of these heuristics using problems from the CAD repository described in [26][1]. We selected those problems with more than one equational constraint, for which at least one of the choices is tractable. The experiments were run in MAPLE using the `ProjectionCAD` package [16] and the results are displayed in Table 1 with the cell count, computation time and heuristic values given for each problem and choice of equational constraint.

The full details on the problems can be found in the repository. The examples each contain two or three equational constraints and the numbering of the choices in the table refers to the order the equational constraints are listed in the repository. The variable orderings used were those suggested in the repository.

---

[1] Freely available at `http://opus.bath.ac.uk/29503`

Table 1. Comparing the choice of equational constraint for a selection of problems. The lowest cell count for each problem is highlighted and the minimal values of the heuristics emboldened.

Problem	EC Choice 1				EC Choice 2				EC Choice 3			
	Cells	Time	S	N	Cells	Time	S	N	Cells	Time	S	N
Intersection A	657	5.6	61	7	463	5.1	64	8	269	1.3	**42**	**4**
Intersection B	711	6.3	66	6	471	5.4	71	6	303	1.1	**40**	**5**
Random A	375	2.7	81	9	435	3.6	**73**	**8**	425	2.8	80	8
Random B	1295	21.4	140	13	477	3.8	**84**	**9**	1437	23.9	158	14
Sphere-Catastrophe	285	2.0	61	7	169	1.0	**59**	**5**				
Arnon84-2	39	0.1	54	5	9	0.0	**47**	**1**				
Hong-90	F	-	14	0	F	-	14	0	27	0.1	**14**	**0**
Cyclic-3	57	0.3	**32**	**3**	117	0.7	35	3	119	0.6	36	4

The time taken to calculate S and N for each problem was always less than 0.05 seconds and so insignificant to the overall timings.

For each problem the equational constraint choice resulting in the lowest cell count and timing has been highlighted and the minimal values of the heuristics emboldened. We can see that for almost all cases both the heuristics point to the best choice. However, there is an example (Random A) where both point to an incorrect choice. The heuristic based on sotd is more sensitive (because it measures at all levels) and as a result is sometimes more effective. For example, it picks the appropriate choice for the Cyclic-3 example while the other does not.

Although the sotd heuristic is superior for all these examples it can be misled by examples where the real geometry differs, as in Example 5.

Example 5. Consider the polynomials

$$f := y^5 - 2y^3x + yx^2 + y = y(y^2 - (x + \mathrm{i}))(y^2 - (x - \mathrm{i}))$$
$$g := y^5 - 2y^3x + yx^2 - y = y(y^2 - (x + 1))(y^2 - (x - 1))$$

along with the formula $f = 0 \wedge g = 0$ and variable ordering $x \prec y$. We could use either f or g as an equational constraint when constructing a CAD. We have

$$\mathrm{discrim}(f) = 256(x^2 + 1)^3, \qquad \mathrm{discrim}(g) = 256(x - 1)^3(x + 1)^3$$

and so both the projection sets have the same sotd. However, with f as an equational constraint the projection set has ndrr$= 0$ while with g it is 2. The CADs of $\mathbb{R}^2$ have 3 and 31 cells respectively.

4   Formulating Input for TTICAD

Let Φ represent a set of QFFs, $\{\phi_i\}$. In [3] the authors define a Truth-Table Invariant CAD (TTICAD) as a CAD such that the boolean value of each ϕ_i is constant (either true or false) on each cell. Clearly such a CAD is sufficient for solving many problems involving the formulae.

A sign-invariant CAD is also a TTICAD, however, in [3] the authors present an algorithm to construct TTICADs more efficiently for the case where each ϕ_i has a designated equational constraint f_i (an equation logically implied by ϕ_i). They adapt the theory of equational constraints to define a TTICAD projection operator and prove a key theorem explaining when it is valid. Informally, the TTICAD projection operator produces the union of the application of the equational constraints projection operator to each ϕ_i along with the cross resultants of all the designated equational constraints, (see [3] for the full technical details). As noted in the introduction, TTICAD is more efficient than equational constraints alone.

If there is more than one equational constraint present within a single ϕ_i then a choice must be made as to which is designated for use in the algorithm, (the others would then be treated as any other constraint). As with choosing equational constraints in Section 3 the two different projection sets could be calculated and the measures sotd and ndrr taken and used as heuristics, picking the choice that leads to the lowest values.

However, this situation actually offers further choice for problem formulation than the designation. If ϕ_i had two equational constraints then it would be admissible to split this into two QFFs $\phi_{i,1}, \phi_{i,2}$ with one equational constraint assigned to each and the other constraints partitioned between them in any manner. (Admissible because any TTICAD for $\phi_{i,1}, \phi_{i,2}$ is also a TTICAD for ϕ_i.) This is a generalisation of the following observation: given a formula ϕ with two equational constraints a CAD could be constructed using either the traditional theory of equational constraints or the TTICAD algorithm applied to two QFFs. On the surface it is not clear why the latter option would ever be chosen since it would certainly lead to more projection polynomials after the first projection. However, a specific equational constraint may have a comparatively large number of intersections with another constraint, in which case, while separating these into different QFFs would likely increase the number of projection polynomials it may still reduce the number of cells in the CAD, (since the resultants taken would be less complicated leading to fewer projection factors at subsequent steps). Example 6 describes a simple problem which could be tackled using the theory of equational constraints alone, but for which it is beneficial to split into two QFFs and tackle with TTICAD.

Example 6. Let $x \prec y$ and consider the polynomials

$$f_1 := (y - 1) - x^3 + x^2 + x, \qquad g_1 := y - \tfrac{x}{4} + \tfrac{1}{2},$$
$$f_2 := (-y - 1) - x^3 + x^2 + x, \qquad g_2 := -y - \tfrac{x}{4} + \tfrac{1}{2},$$

and the formula $\phi := f_1 = 0 \wedge g_1 > 0 \wedge f_2 = 0 \wedge g_2 < 0$.

The polynomials are plotted in Figure 3 where the solid curve is f_1, the solid line g_1, the dashed curve f_2 and the dashed line g_2. The three figures also contain dotted lines indicating the stacks over the 0-cells of the CAD of $\mathbb{R}^1$ arising from the decomposition of the real line using various CAD algorithms.

First, if we use the theory of equational constraints (with either f_1 or f_2 as the designated equational constraint) then a CAD is constructed which identifies all the roots and intersection between the four polynomials except for the intersection of g_1 and g_2. (Note that this would be identified by a full sign-invariant CAD). This is visualised by the plot on the left while the plot on the right relates to a TTICAD with two QFFs. In this case only three 0-cells are identified, with the intersections of g_2 with f_1 and g_1 with f_2 ignored.

The TTICAD has 31 cells while the CADs produced using equational constraints both have 39 cells. The TTICAD projection set has an `sotd` of 26 and an `ndrr` of 3 while each of the CADs produced using equational constraints have projection sets with values of 30 and 6 for `sotd` and `ndrr`.

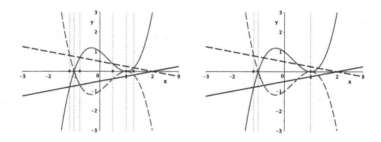

Fig. 3. Plots visualising the induced CADs of $\mathbb{R}^1$ described in Example 6

As suggested by Example 6 we propose using the measures `sotd` and `ndrr` applied to the set of projection polynomials as heuristics for picking an approach. We can apply these with the TTICAD projection operator for deciding if it would be beneficial to split QFFs. This can also be used for choosing whether to use TTICAD instead of equational constraints alone, since applying the TTICAD algorithm from [3] on a single QFF is equivalent to creating a CAD invariant with respect to an equational constraint.

We may also consider whether it is possible to combine any QFFs. If the formulae were joined by conjunction then it would be permitted and probably beneficial but we would then need to choose which equational constraint to designate. Formulae joined by disjunction could also be combined if they share an equational constraint, (with that becoming the designated choice in the combined formula). Such a situation is common for the application to branch cut analysis since many branch cuts come in pairs which lie on different portions of the same curve. However, upon inspection of the projection operators, we see that such a merger would not change the set of projection factors in the case where the shared equational constraint is the designated one for each formula. Note, if the shared equational constraint is not designated in both then the only way to merge would be by changing designation.

When considering whether to split and which equational constraint to designate the number of possible formulations increases quickly. Hence we propose a method for TTICAD QFF formulation, making the choices one QFF at a time. Given a list $\hat{\Phi}$ of QFFs (quantifier free formulae):

(1) Take the disjunction of the QFFs and put that formula into disjunctive normal form, $\bigvee \hat{\phi}_i$ so that each $\hat{\phi}_i$ is a conjunction of atomic formulae.
(2) Consider each $\hat{\phi}_i$ in turn and let m_i be the number of equational constraints.
    – If $m_i = 0$ then $\hat{\Phi}$ is not suitable for the TTICAD algorithm of [3], (although we anticipate that it could be adapted to include such cases).
    – If $m_i = 1$ then the sole equational constraint is designated trivially.
    – If $m_i > 1$ then we consider all the possible partitions of the formula in $\hat{\phi}_i$ into sub QFFs with at least one equational constraint each, and all the different designations of equational constraint within those sub-QFFs with more than one. Choose a partition and designation for this clause according to the heuristics based on sotd and ndrr applied to the projections polynomials from the clause.
(3) Let Φ be the list of new QFFs, ϕ_i, and the input to TTICAD.

5   Using Gröbner Bases to Precondition TTICAD QFFs

Recall that for an ideal, $I \subset \mathbb{R}[\mathbf{x}]$, a *Gröbner basis* (for a given monomial ordering) is a polynomial basis of I such that $\{lm(g) \mid g \in G\}$ is also a basis for $\{lm(f) \mid f \in I\}$. In [7] experiments were conducted to see if Gröbner basis techniques could precondition problems effectively for CAD. Given a problem:

$$\varphi := \bigwedge_{i=1}^{s} f_i(\mathbf{x}) = 0,$$

a purely lexicographical Gröbner basis $\{\hat{f}_i\}_{i=1}^{t}$ for the f_i, (taken with respect to the same variable ordering as the CAD), could take their place to form an equivalent sentence:

$$\hat{\varphi} := \bigwedge_{i=1}^{t} \hat{f}_i(\mathbf{x}) = 0.$$

Initial results suggested that this preconditioning can be hugely beneficial in certain cases, but may be disadvantageous in others.

In [27] this idea was considered in greater depth. A larger base of problems was tested and the idea extended to include Gröbner reduction. Given a problem:

$$\psi := \left(\bigwedge_{i=1}^{s_1} f_i(\mathbf{x}) = 0\right) \wedge \left(\bigwedge_{i=1}^{s_2} g_i(\mathbf{x}) *_i 0\right), \qquad *_i \in \{=, \neq, >, <\},$$

you can first compute $\{\hat{f}_i\}_{i=1}^{t_1}$ followed by reducing the g_i with respect to the $\hat{f}_i$ to obtain $\{\hat{g}_i\}_{i=1}^{t_2}$. Then the following sentence will be equivalent to ψ:

$$\hat{\psi} := \left(\bigwedge_{i=1}^{t_1} \hat{f}_i(\mathbf{x}) = 0\right) \wedge \left(\bigwedge_{i=1}^{t_2} \hat{g}_i(\mathbf{x}) *_i 0\right).$$

Experimentation showed that this Gröbner preconditioning can be highly beneficial with respect to both computation time and cell count, however the effect

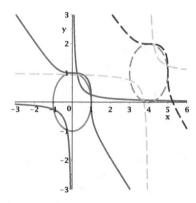

Fig. 4. Plot of the functions described in Example 7

is not universal. To identify when preconditioning is beneficial a simple metric was posited and shown to be a good indicator. The quantity TNoI (*total number of indeterminates*) for a set of polynomials F is simply defined to be the sum of the number of variables present in each polynomial in F. In all testing carried out (both for [27] and henceforth) if the produced Gröbner basis has a lower TNoI than the original set of polynomials then preconditioning is beneficial for sign-invariant CAD (the converse is not always true).

A natural question is whether Gröbner preconditioning can be adapted for TTICAD. This is possible by performing the Gröbner preconditioning on the individual QFFs. There is a necessity, however, for a problem to be suitably complicated for this preconditioning to work: each QFF must have multiple equational constraints amenable to the creation of a Gröbner Basis. This required complexity means there are few examples in the literature which are suitable and tractable for experimentation. We demonstrate the power of combining these two techniques through a worked example.

Example 7. Consider the polynomials

$$
\begin{array}{ll}
f_{1,1} := x^2 + y^2 - 1, & f_{2,1} := (x - 4)^2 + (y - 1)^2 - 1, \\
f_{1,2} := x^3 + y^3 - 1, & f_{2,2} := (x - 4)^3 + (y - 1)^3 - 1, \\
g_1 := xy - \frac{1}{4}, & g_2 := (x - 4)(y - 1) - \frac{1}{4}
\end{array}
$$

and the formula $[f_{1,1} = 0 \wedge f_{1,2} = 0 \wedge g_1 > 0] \vee [f_{2,1} = 0 \wedge f_{2,2} = 0 \wedge g_2 > 0]$.

The polynomials are plotted in Figure 4 where the solid curves represent $f_{1,1}, f_{1,2}, g_1$, and the dashed curves $f_{2,1}, f_{2,2}, g_2$.

We will consider both variable orderings: $y \prec x$ and $x \prec y$. We can compute full CADs for this problem, with 725 and 657 cells for the respective orderings. If we use TTICAD to tackle the problem then there are four possible two-QFF formulations, (splitting QFFs is not beneficial for this problem). The four formulations are described in the second column of Table 2.

Table 2. Experimental results relating to Example 7. The lowest cell counts are highlighted and the minimal values of the heuristics emboldened.

Order	Full CAD Cells	Full CAD Time	TTI CAD Eq Const	TTI CAD Cells	TTI CAD Time	S	N	TTI+Grö CAD Eq Const	TTI+Grö CAD Cells	TTI+Grö CAD Time	S	N
$y \prec x$	725	22.802	$f_{1,1}, f_{2,1}$	153	0.818	**62**	12	$\hat{f}_{1,1}, \hat{f}_{2,1}$	27	0.095	**37**	**3**
			$f_{1,1}, f_{2,2}$	111	0.752	94	10	$\hat{f}_{1,1}, \hat{f}_{2,2}$	47	0.361	50	5
			$f_{1,2}, f_{2,1}$	121	0.732	85	9	$\hat{f}_{1,1}, \hat{f}_{2,3}$	93	0.257	50	9
			$f_{1,2}, f_{2,2}$	75	0.840	99	**7**	$\hat{f}_{1,2}, \hat{f}_{2,1}$	47	0.151	47	5
								$\hat{f}_{1,2}, \hat{f}_{2,2}$	83	0.329	63	7
								$\hat{f}_{1,2}, \hat{f}_{2,3}$	145	0.768	81	11
								$\hat{f}_{1,3}, \hat{f}_{2,1}$	95	0.263	46	10
								$\hat{f}_{1,3}, \hat{f}_{2,2}$	151	0.712	80	12
								$\hat{f}_{1,3}, \hat{f}_{2,3}$	209	0.980	62	16
$x \prec y$	657	22.029	$f_{1,1}, f_{2,1}$	125	0.676	**65**	14	$\hat{f}_{1,1}, \hat{f}_{2,1}$	29	0.085	**39**	**4**
			$f_{1,1}, f_{2,2}$	117	0.792	96	11	$\hat{f}_{1,1}, \hat{f}_{2,2}$	53	0.144	52	6
			$f_{1,2}, f_{2,1}$	117	0.728	88	11	$\hat{f}_{1,1}, \hat{f}_{2,3}$	97	0.307	53	97
			$f_{1,2}, f_{2,2}$	85	0.650	101	**8**	$\hat{f}_{1,2}, \hat{f}_{2,1}$	53	0.146	49	6
								$\hat{f}_{1,2}, \hat{f}_{2,2}$	93	0.332	65	8
								$\hat{f}_{1,2}, \hat{f}_{2,3}$	149	0.782	81	13
								$\hat{f}_{1,3}, \hat{f}_{2,1}$	97	0.248	48	11
								$\hat{f}_{1,3}, \hat{f}_{2,2}$	149	0.798	82	13
								$\hat{f}_{1,3}, \hat{f}_{2,3}$	165	1.061	65	18

We can apply Gröbner preconditioning to both QFFs separately, computing a Gröbner basis, with respect to the compatible ordering, of $\{f_{i,1}, f_{i,2}\}$. For both QFFs and both variable orderings three polynomials are produced. We denote them by $\{\hat{f}_{i,1}, \hat{f}_{i,2}, \hat{f}_{i,3}\}$ (note the polynomials differ depending on the variable ordering). The algorithm used to compute these bases gives the polynomials in decreasing order of leading monomials with respect to the order used to compute the basis (purely lexicographical).

Table 2 shows that the addition of Gröbner techniques to TTICAD can produce significant reductions: a drop from 153 cells in 0.8s to 27 cells in under 0.1s (including the time required to compute the Gröbner bases). As discussed in [27], preconditioning is not always beneficial, as evident from the handful of cases that produce more cells than TTICAD alone. As with Table 1 we have highlighted the examples with lowest cell count and emboldened the lowest heuristic. Looking at the values of S and N we see that for this example ndrr is the best measure to use.

In [27] TNoI was used to predict whether preconditioning by Gröbner Basis would be beneficial. In this example TNoI is increased in both orderings by taking a basis, which correctly predicts a bigger full CAD after preconditioning. However, TNoI does not take into account the added subtlety of TTICAD (as shown by the huge benefit above).

6   Conclusions and Future Work

In this paper we have considered various issues based around the formulation of input for CAD algorithms. We have revisited the classic question of choosing the variable ordering, proposing a new measure of CAD complexity ndrr to complement the existing sotd measure. We then used these measures as heuristics for the problem of designating equational constraints and QFF formulation for TTICAD. Finally we considered the effect of preconditioning by Gröbner bases.

It is important to note that these are just heuristics and, as such, can be misleading for certain examples. Although the experimental results in Section 3 suggest sotd is a finer heuristic than ndrr we have demonstrated that there are examples when ndrr performs better, not just Example 5 which was contrived for the purpose but also Example 7 introduced for the work on Gröbner bases.

These issues have been treated individually but of course they intersect. For example it is also necessary to pick a variable ordering for TTICAD. This choice will need to made before employing the method for choosing QFF formulation described in Section 4. However, the optimal choice of variable ordering for one QFF formulation may not be optimal for another! For example, the TTICAD formulation with two QFFs was the best choice in Example 6 where the variable ordering was stated as $x \prec y$ but if we had $y \prec x$ then a single QFF is superior.

The idea of combining TTICAD with Gröbner preconditioning (discussed in [7], [27]) is shown, by a worked example, to have the potential of being a very strong tool. However, this adds even more complication in choosing a formulation for the problem. Taken together, all these choices of formulation can become combinatorially overwhelming and so methods to reduce this, such as the greedy algorithm in [14] or the method at the end of Section 4, are of importance.

All these options for problem formulation motivate the somewhat philosophical question of when a reformulation results in a new problem. When a variable ordering is imposed by an application (such as projecting quantified variables first when using CAD for quantifier elimination) then violating this would clearly lead to a new problem while changing the ordering within quantifier blocks could be seen to be a optimisation of the algorithm. Similar distinctions could be drawn for other issues of formulation.

Given the significant gains available from problem reformulation it would seem that the existing technology could benefit from a redesign to maximise the possibility of its use. For example, CAD algorithms could allow the user to input the variables is quantifier blocks so that the technology can choose the most appropriate ordering that still solves the problem.

We finish with some ideas for future work on these topics.

- All the work in this paper has been stated with reference to CAD algorithms based on projection and lifting. A quite different approach, CAD via Triangular Decomposition, has been developed in [8] and implemented as part of the core MAPLE distribution. This constructs a (sometimes quite different) sign-invariant CAD by transferring the problem to complex space for solving. A key question is how much of the work here transfers to this approach?

- Can the heuristics for choosing equational constraints also be used for choosing pivots when using the theory of bi-equational constraints in [6]?
- Can the `ndrr` measure be adapted to consider also the real roots of those projection polynomials with more than one variable?

We finish by discussing one of the initial motivations for engaging in work on problem formulation: a quantifier elimination problem proving a property of Joukowski's transformation. This is the transformation $z \mapsto \frac{1}{2}(z + \frac{1}{z})$ which is used in aerodynamics to create an aerofoil from the unit circle. The fact it is bijective on the upper half plane is relatively simple to prove analytically but we found the state of the art CAD technology was incapable of producing an answer in reasonable time. Then, in a personal communication, Chris Brown described how reformulating the problem with a succession of simple logical steps makes it amenable to QEPCAD, allowing for a solution in a matter of seconds. These steps included splitting a disjunction to form two separate problems and the (counter-intuitive) removal of quantifiers which block QEPCAD's use of equational constraints. Further details are given in [13, Sec. III] and in the future we aim to extend our work on problem formulation to develop techniques to automatically render this problem feasible.

Acknowledgements. This work was supported by the EPSRC grant: EP/J003247/1. The authors would like to thank Scott McCallum for many useful conversations on TTICAD and Chris Brown for sharing his work on the Joukowski transformation.

References

1. Akbarpour, B., Paulson, L.C.: MetiTarski: An Automatic Prover for the Elementary Functions. In: Autexier, S., Campbell, J., Rubio, J., Sorge, V., Suzuki, M., Wiedijk, F. (eds.) AISC/Calculemus/MKM 2008. LNCS (LNAI), vol. 5144, pp. 217–231. Springer, Heidelberg (2008)
2. Akbarpour, B., Paulson, L.C.: MetiTarski: An automatic theorem prover for real-valued special functions. Journal of Automated Reasoning 44(3), 175–205 (2010)
3. Bradford, R., Davenport, J.H., England, M., McCallum, S., Wilson, D.: Cylindrical algebraic decompositions for boolean combinations. In Press: Proc. ISSAC 2013 (2013), Preprint at http://opus.bath.ac.uk/33926/
4. Brown, C.W.: QEPCAD B: A program for computing with semi-algebraic sets using CADs. ACM SIGSAM Bulletin 37(4), 97–108 (2003)
5. Brown, C.W., Davenport, J.H.: The complexity of quantifier elimination and cylindrical algebraic decomposition. In: Proc. ISSAC 2007, pp. 54–60. ACM (2007)
6. Brown, C.W., McCallum, S.: On using bi-equational constraints in CAD construction. In: Proc. ISSAC 2005, pp. 76–83. ACM (2005)
7. Buchberger, B., Hong, H.: Speeding up quantifier elimination by Gröbner bases. Technical report, 91-06. RISC, Johannes Kepler University (1991)
8. Chen, C., Moreno Maza, M., Xia, B., Yang, L.: Computing cylindrical algebraic decomposition via triangular decomposition. In: Proc. ISSAC 2009, pp. 95–102. ACM (2009)
9. Cohen, C., Mahboubi, A.: Formal proofs in real algebraic geometry: from ordered fields to quantifier elimination. LMCS 8(1:02), 1–40 (2012)

10. Collins, G.E.: Quantifier elimination for real closed fields by cylindrical algebraic decomposition. In: Brakhage, H. (ed.) GI-Fachtagung 1975. LNCS, vol. 33, pp. 134–183. Springer, Heidelberg (1975)
11. Collins, G.E.: Quantifier elimination by cylindrical algebraic decomposition – 20 years of progress. In: Caviness, B., Johnson, J. (eds.) Quantifier Elimination and Cylindrical Algebraic Decomposition. Texts & Monographs in Symbolic Computation, pp. 8–23. Springer (1998)
12. Collins, G.E., Hong, H.: Partial cylindrical algebraic decomposition for quantifier elimination. J. Symb. Comput. 12, 299–328 (1991)
13. Davenport, J.H., Bradford, R., England, M., Wilson, D.: Program verification in the presence of complex numbers, functions with branch cuts etc. In: Proc. SYNASC 2012 (2012)
14. Dolzmann, A., Seidl, A., Sturm, T.: Efficient projection orders for CAD. In: Proc. ISSAC 2004, pp. 111–118. ACM (2004)
15. Dolzmann, A., Sturm, T., Weispfenning, V.: A New Approach for Automatic Theorem Proving in Real Geometry. Journal of Automated Reasoning 21(3), 357–380 (1998)
16. England, M.: An implementation of CAD in Maple utilising McCallum projection. Department of Computer Science Technical Report series 2013-02, University of Bath (2013), http://opus.bath.ac.uk/33180/
17. England, M., Bradford, R., Davenport, J.H., Wilson, D.: Understanding branch cuts of expressions. In: Carette, J., Aspinall, D., Lange, C., Sojka, P., Windsteiger, W. (eds.) CICM 2013. LNCS (LNAI), vol. 7961, pp. 136–151. Springer, Heidelberg (2013)
18. Lazard, D.: Quantifier elimination: Optimal solution for two classical examples. J. Symb. Comput. 5(1-2), 261–266 (1988)
19. McCallum, S.: An improved projection operation for cylindrical algebraic decomposition of three-dimensional space. J. Symb. Comput. 5(1-2), 141–161 (1988)
20. McCallum, S.: An improved projection operation for cylindrical algebraic decomposition. In: Caviness, B., Johnson, J. (eds.) Quantifier Elimination and Cylindrical Algebraic Decomposition. Texts & Monographs in Symbolic Computation, pp. 242–268. Springer (1998)
21. McCallum, S.: On projection in CAD-based quantifier elimination with equational constraint. In: Proc. ISSAC 1999, pp. 145–149. ACM (1999)
22. Mahboubi, A.: Implementing the cylindrical algebraic decomposition within the Coq system. Math. Struct. in Comp. Science 17(1), 99–127 (2007)
23. Passmore, G.O., Paulson, L.C., de Moura, L.: Real Algebraic Strategies for Meti-Tarski Proofs. In: Jeuring, J., Campbell, J.A., Carette, J., Dos Reis, G., Sojka, P., Wenzel, M., Sorge, V. (eds.) CICM 2012. LNCS, vol. 7362, pp. 358–370. Springer, Heidelberg (2012)
24. Phisanbut, N.: Practical Simplification of Elementary Functions using Cylindrical Algebraic Decomposition. PhD thesis, University of Bath (2011)
25. Schwartz, J.T., Sharir, M.: On the "Piano-Movers" Problem: II. General techniques for computing topological properties of real algebraic manifolds. Adv. Appl. Math. 4, 298–351 (1983)
26. Wilson, D.J., Bradford, R.J., Davenport, J.H.: A repository for CAD examples. ACM Communications in Computer Algebra 46(3), 67–69 (2012)
27. Wilson, D.J., Bradford, R.J., Davenport, J.H.: Speeding up cylindrical algebraic decomposition by Gröbner bases. In: Jeuring, J., Campbell, J.A., Carette, J., Dos Reis, G., Sojka, P., Wenzel, M., Sorge, V. (eds.) CICM 2012. LNCS, vol. 7362, pp. 280–294. Springer, Heidelberg (2012)

The Formalization of Syntax-Based Mathematical Algorithms Using Quotation and Evaluation*

William M. Farmer

Department of Computing and Software
McMaster University
Hamilton, Ontario, Canada
wmfarmer@mcmaster.ca

Abstract. Algorithms like those for differentiating functional expressions manipulate the syntactic structure of mathematical expressions in a mathematically meaningful way. A formalization of such an algorithm should include a specification of its computational behavior, a specification of its mathematical meaning, and a mechanism for applying the algorithm to actual expressions. Achieving these goals requires the ability to integrate reasoning about the syntax of the expressions with reasoning about what the expressions mean. A *syntax framework* is a mathematical structure that is an abstract model for a syntax reasoning system. It contains a mapping of expressions to *syntactic values* that represent the syntactic structures of the expressions; a language for reasoning about syntactic values; a *quotation* mechanism to refer to the syntactic value of an expression; and an *evaluation* mechanism to refer to the value of the expression represented by a syntactic value. We present and compare two approaches, based on instances of a syntax framework, to formalize a syntax-based mathematical algorithm in a formal theory T. In the first approach the syntactic values for the expressions manipulated by the algorithm are members of an inductive type in T, but quotation and evaluation are functions defined in the metatheory of T. In the second approach every expression in T is represented by a syntactic value, and quotation and evaluation are operators in T itself.

1 Introduction

A great many of the algorithms employed in mathematics work by manipulating the syntactic structure of mathematical expressions in a mathematically meaningful way. Here are some examples:

1. Arithmetic operations applied to numerals.
2. Operations such as factorization applied to polynomials.
3. Simplification of algebraic expressions.

---

* This research was supported by NSERC.

4. Operations such as transposition performed on matrices.
5. Symbolic differentiation and antidifferentiation of expressions with variables.

The study and application of these kinds of algorithms is called *symbolic compu-tation*. For centuries symbolic computation was performed almost entirely using pencil and paper. However, today symbolic computation can be performed by computer, and algorithms that manipulate mathematical expressions are the main fare of computer algebra systems.

In this paper we are interested in the problem of how to formalize syntax-based mathematical algorithms. These algorithms manipulate members of a formal language in a computer algebra system, but their behavior and meaning are usually not formally expressed in a computer algebra system. However, we want to use these algorithms in formal theories and formally understand what they do. We are interested in employing existing external implementations of these algorithms in formal theories as well as implementing these algorithms directly in formal theories.

As an illustration, consider an algorithm, say named RatPlus, that adds ra-tional number numerals, which are represented in memory in some suitable way. (An important issue, that we will not address, is how the numerals are repre-sented to optimize the efficiency of RatPlus.) For example, if the numerals $\frac{2}{5}$ and $\frac{3}{8}$ are given to RatPlus as input, the numeral $\frac{31}{40}$ is returned by RatPlus as output. What would we need to do to use RatPlus to add rational numbers in a formal theory T and be confident that the results are correct? First, we would have to introduce values in T to represent rational number numerals as syntactic structures, and then define a binary operator O over these values that has the same input-output relation as RatPlus. Second, we would have to prove in T that, if $O(a, b) = c$, then the sum of the rational numbers represented by a and b is the rational number represented by c. And third, we would have to devise a mechanism for using the definition of O to add rational numbers in T.

The second task is the most challenging. The operator O, like RatPlus, ma-nipulates numerals as syntactic structures. To state and then prove that these manipulations are mathematically meaningful requires the ability to express the interplay of how the numerals are manipulated and what the manipulations mean with respect to rational numbers. This is a formidable task in a traditional logic in which there is no mechanism for directly referring to the syntax of the expres-sions in the logic. We need to reason about a rational number numeral $\frac{2}{5}$ both as a syntactic structure that can be deconstructed into the integer numerals 2 and 5 and as an expression that denotes the rational number $2/5$.

Let us try to make the problem of how to formalize syntax-based mathemat-ical algorithms like RatPlus more precise. Let T be a theory in a traditional logic like first-order logic or simple type theory, and let A be an algorithm that manipulates certain expressions of T. To formalize A in T we need to do three things:

1. *Define an operator O_A in T that represents A*: Introduce values in T that represent the expressions manipulated by A. Introduce an operator O_A in T that maps the values that represent the input expressions taken by A

to the values that represent the output expressions produced by A. Write a sentence named CompBehavior in T that specifies the *computational behavior* of O_A to be the same as that of A. That is, if A takes an input expression e and produces an output expression e', then CompBehavior should say that O_A maps the value that represents e to the value that represents e'.

2. *Prove in T that O_A is mathematically correct*: Write a sentence named MathMeaning in T that specifies the *mathematical meaning* of O_A to be the same as that of A. That is, if the value of an input expression e given to A is related to the value of the corresponding output expression e' produced by A in a particular way, then MathMeaning should say that the value of the expression representing e should be related to the value of the expression representing e' in the same way. Finally, prove MathMeaning from CompBehavior in T.

3. *Devise a mechanism for using O_A in T*: An application $O_A(a_1, \ldots, a_n)$ of O_A to the values $a_1, \ldots, a_n$ can be computed in T by instantiating CompBehavior with $a_1, \ldots, a_n$ and then simplifying the resulting formula to obtain the value of $O_A(a_1, \ldots, a_n)$. For the sake of convenience or efficiency, we might want to use A itself to compute $O_A(a_1, \ldots, a_n)$. We will know that results produced by A are correct provided A and O_A have the same computational behavior.

If we believe that A works correctly and we are happy to do our computations with A outside of T, we can skip the writing of CompBehavior and use MathMeaning as an axiom that asserts A has the mathematical meaning specified by MathMeaning for O_A. The idea of treating specifications of external algorithms as axioms is a key component of the notion of a *biform theory* [2,6].

So to use A in T we need to formalize A in T, and to do this, we need a system that integrates reasoning about the syntax of the expressions with reasoning about what the expressions mean. A *syntax framework* [9] is a mathematical structure that is an abstract model for a syntax reasoning system. It contains a mapping of expressions to *syntactic values* that represent the syntactic structures of the expressions; a language for reasoning about syntactic values; a quotation mechanism to refer to the syntactic value of an expression; and an evaluation mechanism to refer to the value of the expression represented by a syntactic value. A syntax framework provides the tools needed to reason about the interplay of syntax and semantics. It is just what we need to formalize syntax-based mathematical algorithms.

Reflection is a technique to embed reasoning about a reasoning system (i.e., metareasoning) in the reasoning system itself. Reflection has been employed in logic [13], theorem proving [12], and programming [5]. Since metareasoning very often involves the syntactic manipulation of expressions, a syntax framework is a natural subcomponent of a reflection mechanism.

This paper attacks the problem of formalizing a syntax-based mathematical algorithm A in a formal theory T using syntax frameworks. Two approaches are presented and compared. The first approach is local in nature. It employs a syntax framework in which there are syntactic values only for the expressions manipulated by A. The second approach is global in nature. It employs a syntax framework in which there are syntactic values for all the expressions of T. We will see that these two approaches have contrasting strengths and weaknesses. The local approach offers an incomplete solution at a low cost, while the global approach offers a complete solution at a high cost.

The two approaches will be illustrated using the familiar example of polynomial differentiation. In particular, we will discuss how the two approaches can be employed to formalize an algorithm that differentiates expressions with variables that denote real-valued polynomial functions. We will show that algorithms like differentiation that manipulate expressions with variables are more challenging to formalize than algorithms like symbolic arithmetic that manipulate numerals without variables.

The following is the outline of the paper. The next section, Section 2, presents the paper's principal example, polynomial differentiation. The notion of a syntax framework is defined in Section 3. Sections 4 and 5 present the local and global approaches to formalizing syntax-based mathematical algorithms. And the paper concludes with Section 6.

2   Example: Polynomial Differentiation

We examine in this section the problem of how to formalize a symbolic differentiation algorithm and then prove that the algorithm actually computes derivatives. We start by defining what a derivative is.

Let $f : \mathbb{R} \to \mathbb{R}$ be a function over the real numbers and $a \in \mathbb{R}$. The *derivative of f at a*, written $\mathsf{deriv}(f, a)$, is

$$\lim_{h \to 0} \frac{f(a + h) - f(a)}{h}$$

if this limit exists. The *derivative of f*, written $\mathsf{deriv}(f)$, is the function

$$\lambda x : \mathbb{R} . \mathsf{deriv}(f, x).$$

Notice that we are using the traditional definition of a derivative in which a derivative of a function is defined pointwise.

Differentiation is in general the process of finding derivatives which ultimately reduces to finding limits. *Symbolic differentiation* is the process of mechanically transforming an expression with variables that represents a function over the real numbers into an expression with variables that represents the derivative of the function. For example, the result of symbolically differentiating the expression $\mathsf{sin}(x^2)$ which represents the function $\lambda x : \mathbb{R} . \mathsf{sin}(x^2)$ is the expression $2 \cdot x \cdot \mathsf{cos}(x^2)$ which represents the function $\lambda x : \mathbb{R} . 2 \cdot x \cdot \mathsf{cos}(x^2)$. Symbolic differentiation is performed by applying certain *differentiation rules* and *simplification rules* to a starting expression until no rule is applicable.

Let us look at how symbolic differentiation works on polynomials. A *polynomial* is an expression constructed from real-valued constants and variables by applying addition, subtraction, multiplication, and natural number exponentiation. For example, $x \cdot (x^2 + y)$ is a polynomial. The symbolic differentiation of polynomials is performed using the following well-known differentiation rules:

Constant Rule

$$\frac{d}{dx}(c) = 0 \quad \text{where } c \text{ is a constant or a variable different from } x.$$

Variable Rule

$$\frac{d}{dx}(x) = 1.$$

Sum and Difference Rule

$$\frac{d}{dx}(u \pm v) = \frac{d}{dx}(u) \pm \frac{d}{dx}(v).$$

Product Rule

$$\frac{d}{dx}(u \cdot v) = \frac{d}{dx}(u) \cdot v + u \cdot \frac{d}{dx}(v).$$

Power Rule

$$\frac{d}{dx}(u^n) = \begin{cases} 0 & \text{if } n = 0. \\ n \cdot u^{n-1} \cdot \frac{d}{dx}(u) & \text{if } n > 0. \end{cases}$$

Written using traditional Leibniz notation, the rules specify how symbolic differentiation is performed with respect to the variable x. The symbols u and v range over polynomials that may contain x as well as other variables, and the symbol n ranges over natural numbers. Notice that these rules are not meaning preserving in the usual way; for example, the rule $\frac{d}{dx}(c) = 0$ is not meaning preserving if we view c as a value and not as an expression.

Let PolyDiff be the algorithm that, given a polynomial u and variable x, applies the five differentiation rules above to the starting expression $\frac{d}{dx}(u)$ until there are no longer any expressions starting with $\frac{d}{dx}$ and then simplifies the resulting expression using the rules $0 + u = u + 0 = 0$ and $1 \cdot u = u \cdot 1 = u$ and collecting like terms. Applied to $x \cdot (x^2 + y)$, PolyDiff would perform the following steps:

$$\frac{d}{dx}(x \cdot (x^2 + y)) = \frac{d}{dx}(x) \cdot (x^2 + y) + x \cdot \frac{d}{dx}(x^2 + y) \tag{1}$$

$$= 1 \cdot (x^2 + y) + x \cdot \left(\frac{d}{dx}(x^2) + \frac{d}{dx}(y) \right) \tag{2}$$

$$= 1 \cdot (x^2 + y) + x \cdot \left(2 \cdot x^1 \cdot \frac{d}{dx}(x) + 0 \right) \tag{3}$$

$$= 1 \cdot (x^2 + y) + x \cdot (2 \cdot x^1 \cdot 1 + 0) \tag{4}$$

$$= 3 \cdot x^2 + y \tag{5}$$

Line (1) is by the Product Rule; (2) is by the Variable and Sum and Difference Rules; (3) is by the Power and Constant Rules; (4) is by the Variable Rule; and (5) is by the simplification rules. Thus, given the function

$$f = \lambda x : \mathbb{R} . x \cdot (x^2 + y),$$

using PolyDiff we are able to obtain the derivative

$$\lambda\, x : \mathbb{R} \,.\, 3 \cdot x^2 + y$$

of f via mechanical manipulation of the expression $x \cdot (x^2 + y)$.

Algorithms similar to PolyDiff are commonly employed in informal mathematics. In fact, they are learned and applied by every calculus student. They should be as available and useful in formal mathematics as they are in informal mathematics. We thus need to formalize them as described in the Introduction.

The main objective of this paper is to show how syntax-based mathematical algorithms can be formalized using PolyDiff as example. We will begin by making the task of formalizing PolyDiff precise.

Let a *theory* be a pair $T = (L, \Gamma)$ where L is a formal language and Γ is a set of sentences in L that serve as the axioms of the theory. Define $T_R = (L_R, \Gamma_R)$ to be a theory of the real numbers in (many-sorted) simple type theory. We assume that L_R is a set of expressions over a signature that includes a type $\mathbb{R}$ of the real numbers, constants for each natural number, and constants for addition, subtraction, multiplication, natural number exponentiation, and the unary and binary deriv operators defined above. We assume that Γ_R contains the axioms of a complete ordered field as well as the definitions of all the defined constants in L_R (see [8] for further details).

Let $L_{\mathrm{var}} \subseteq L_R$ be the set of variables of type $\mathbb{R}$ and $L_{\mathrm{poly}} \subseteq L_R$ be the set of expressions constructed from members of L_{var}, constants of type $\mathbb{R}$, addition, subtraction, multiplication, and natural number exponentiation. Finally, assume that $\mathsf{PolyDiff} : L_{\mathrm{poly}} \times L_{\mathrm{var}} \to L_{\mathrm{poly}}$ is the algorithm described in the previous section adapted to operate on expressions of L_R.

Thus to formalize PolyDiff we need to:

1. Define an operator O_{pd} in T_R that represents PolyDiff.
2. Prove in T_R that O_{pd} is mathematically correct.
3. Devise a mechanism for using O_{pd} in T.

Formalizing PolyDiff should be much easier than formalizing differentiation algorithms for larger sets of expressions that include, for example, rational expressions and transcendental functions. Polynomial functions are total (i.e., they are defined at all points on the real line) and their derivatives are also total. As a result, issues of undefinedness do not arise when specifying the mathematical meaning of PolyDiff.

However, functions more general than polynomial functions as well as their derivatives may be undefined at some points. Thus using a differentiation algorithm to compute the derivative of one of these more general functions requires care in determining the precise domain of the derivative. For example, differentiating the rational expression x/x using the well-known Quotient Rule yields the expression 0, but the derivative of $\lambda\, x : \mathbb{R} \,.\, x/x$ is not $\lambda\, x : \mathbb{R} \,.\, 0$. The derivative is actually the partial function

$$\lambda\, x : \mathbb{R} \,.\, \text{if } x \neq 0 \text{ then } 0 \text{ else } \bot.$$

We restrict our attention to differentiating polynomial functions so that we can focus on reasoning about syntax without being concerned about issues of undefinedness.

3 Syntax Frameworks

A syntax framework [9] is a mathematical structure that is intended to be an abstract model of a system for reasoning about the syntax of an interpreted language (i.e., a formal language with a semantics). It will take several definitions from [9] to present this structure.

Definition 1 (Interpreted Language). An *interpreted language* is a triple $I = (L, D_{sem}, V_{sem})$ where:

1. L is a formal language, i.e, a set of expressions.[1]
2. D_{sem} is a nonempty domain (set) of *semantic values*.
3. $V_{sem} : L \to D_{sem}$ is a total function, called a *semantic valuation function*, that assigns each expression $e \in L$ a semantic value $V_{sem}(e) \in D_{sem}$. □

A syntax representation of a formal language is an assignment of syntactic values to the expressions of the language:

Definition 2 (Syntax Representation). Let L be a formal language. A *syntax representation* of L is a pair $R = (D_{syn}, V_{syn})$ where:

1. D_{syn} is a nonempty domain (set) of *syntactic values*. Each member of D_{syn} represents a syntactic structure.
2. $V_{syn} : L \to D_{syn}$ is an injective, total function, called a *syntactic valuation function*, that assigns each expression $e \in L$ a syntactic value $V_{syn}(e) \in D_{syn}$ such that $V_{syn}(e)$ represents the syntactic structure of e. □

A syntax language for a syntax representation is a language of expressions that denote syntactic values in the syntax representation:

Definition 3 (Syntax Language). Let $R = (D_{syn}, V_{syn})$ be a syntax representation of a formal language L_{obj}. A *syntax language* for R is a pair (L_{syn}, I) where:

1. $I = (L, D_{sem}, V_{sem})$ in an interpreted language.
2. $L_{obj} \subseteq L$, $L_{syn} \subseteq L$, and $D_{syn} \subseteq D_{sem}$.
3. V_{sem} restricted to L_{syn} is a total function $V'_{sem} : L_{syn} \to D_{syn}$. □

Finally, we are now ready to define a syntax framework:

---

[1] No distinction is made between how expressions are constructed in this definition as well as in subsequent definitions. In particular, expressions constructed by binding variables are not treated in any special way.

Definition 4 (Syntax Framework in an Interpreted Language).

Let $I = (L, D_{\mathrm{sem}}, V_{\mathrm{sem}})$ be an interpreted language and L_{obj} be a sublanguage of L. A *syntax framework* for (L_{obj}, I) is a tuple $F = (D_{\mathrm{syn}}, V_{\mathrm{syn}}, L_{\mathrm{syn}}, Q, E)$ where:

1. $R = (D_{\mathrm{syn}}, V_{\mathrm{syn}})$ is a syntax representation of L_{obj}.
2. (L_{syn}, I) is a syntax language for R.
3. $Q : L_{\mathrm{obj}} \to L_{\mathrm{syn}}$ is an injective, total function, called a *quotation function*, such that:
   Quotation Axiom. For all $e \in L_{\mathrm{obj}}$,

   $$V_{\mathrm{sem}}(Q(e)) = V_{\mathrm{syn}}(e).$$

4. $E : L_{\mathrm{syn}} \to L_{\mathrm{obj}}$ is a (possibly partial) function, called an *evaluation function*, such that:
   Evaluation Axiom. For all $e \in L_{\mathrm{syn}}$,

   $$V_{\mathrm{sem}}(E(e)) = V_{\mathrm{sem}}(V_{\mathrm{syn}}^{-1}(V_{\mathrm{sem}}(e)))$$

   whenever $E(e)$ is defined.   □

A syntax framework is depicted in Figure 1. For $e \in L_{\mathrm{obj}}$, $Q(e)$ is called the *quotation* of e. $Q(e)$ denotes a value in D_{syn} that represents the syntactic structure of e. For $e \in L_{\mathrm{syn}}$, $E(e)$ is called the *evaluation* of e. If it is defined, $E(e)$ denotes the same value in D_{sem} that the expression represented by the value of e denotes. Since there will usually be different $e_1, e_2 \in L_{\mathrm{syn}}$ that denote the same syntactic value, E will usually not be injective. Q and E correspond to the quote and eval operators in Lisp and other languages.

Common examples of syntax frameworks are based on representing the syntax of expressions by Gödel numbers, strings, and members of an inductive type. Programming languages that support metaprogramming — such as Lisp, F# [10], MetaML [18], MetaOCaml [15], reFLect [11], and Template Haskell [16] — are instances of a syntax framework if mutable variables are disallowed. See [9] for these and other examples of syntax frameworks.

The notion of a syntax framework can be easily lifted from an interpreted language to an interpreted theory. This is the version of a syntax framework that we will use in this paper.

Definition 5 (Model). Let $T = (L, \Gamma)$ be a theory. A *model* of T is a pair $M = (D_{\mathrm{sem}}^M, V_{\mathrm{sem}}^M)$ such that D_{sem}^M is a nonempty set of semantic values that includes the truth values T (true) and F (false) and $V_{\mathrm{sem}}^M : L \to D_{\mathrm{sem}}^M$ is a total function such that, for all sentences $A \in \Gamma$, $V_{\mathrm{sem}}^M(A) = \mathrm{T}$. □

Definition 6 (Interpreted Theory). An *interpreted theory* is a pair $I = (T, \mathcal{M})$ where T is a theory and $\mathcal{M}$ is a set of models of T. (If $T = (L, \Gamma)$, $(L, D_{\mathrm{sem}}^M, V_{\mathrm{sem}}^M)$ is obviously an interpreted language for each $M \in \mathcal{M}$.) □

Definition 7 (Syntax Framework in an Interpreted Theory).
Let $I = (T, \mathcal{M})$ be an interpreted theory where $T = (L, \Gamma)$ and $L_{\mathrm{obj}} \subseteq L$. A *syntax framework* for (L_{obj}, I) is a triple $F = (L_{\mathrm{syn}}, Q, E)$ where:

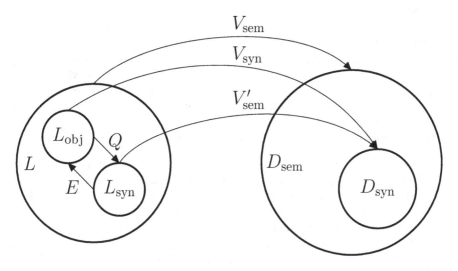

Fig. 1. A Syntax Framework

1. $L_{\text{syn}} \subseteq L$.
2. $Q : L_{\text{obj}} \to L_{\text{syn}}$ is an injective, total function.
3. $E : L_{\text{syn}} \to L_{\text{obj}}$ is a (possibly partial) function.
4. For all $M = (D_{\text{sem}}^M, V_{\text{sem}}^M) \in \mathcal{M}$, $F^M = (D_{\text{syn}}^M, V_{\text{syn}}^M, L_{\text{syn}}, Q, E)$ is a syntax framework for $(L_{\text{obj}}, (L, D_{\text{sem}}^M, V_{\text{sem}}^M))$ where D_{syn}^M is the range of V_{sem}^M restricted to L_{syn} and $V_{\text{syn}}^M = V_{\text{sem}}^M \circ Q$.

Let $I = (L, D, V)$ be an interpreted language, $L_{\text{obj}} \subseteq L$, and $F = (D_{\text{syn}}, V_{\text{syn}}, L_{\text{syn}}, Q, E)$ be a syntax framework for (L_{obj}, I). F has *built-in quotation* if there is an operator (which we will denote as quote) such that, for all $e \in L_{\text{obj}}$, $Q(e)$ is the syntactic result of applying the operator to e (which we will denote as quote(e)). F has *built-in evaluation* if there is an operator (which we will denote as eval) such that, for all $e \in L_{\text{syn}}$, $E(e)$ is the syntactic result of applying the operator to e (which we will denote as eval(e)) whenever $E(e)$ is defined. There are similar definitions as those above when F is a syntax framework in an interpreted theory.

A syntax framework F for (L_{obj}, I), where I is either an interpreted language or an interpreted theory, is *replete* if $L_{\text{obj}} = L$ and F has both built-in quotation and evaluation. If F is replete, it has the facility to reason about the syntax of all of L within L itself. Examples of a replete syntax framework are rare. The programming language Lisp with a simplified semantics is the best known example of a replete syntax framework [9]. T. Æ. Mogensen's self-interpretation of lambda calculus [14] and the logic Chiron [7], derived from classical NBG set theory, are two other examples of replete syntax frameworks [9].

4   Local Approach

In order to formalize PolyDiff in T_R we need the ability to reason about the polynomials in L_{poly} as syntactic structures (i.e., as syntax trees). This can be achieved by constructing a syntax framework for (L_{poly}, I'_R) where $I'_R = (T'_R, \mathcal{M}')$ is an interpreted theory such that T'_R is a conservative extension of T_R. Since we seek to reason about just the syntax of L_{poly} instead of a larger language, we call this the *local approach*.

The construction of the syntax framework requires the following steps:

1. Define in T_R an inductive type whose members are the syntax trees of the polynomials in L_{poly}. The inductive type should include a new type symbol $\mathbb{S}$ and appropriate constants for constructing and deconstructing expressions of type $\mathbb{S}$. Let L_{syn} be the set of expressions of type $\mathbb{S}$. For example, if $x + 3$ is a polynomial in L_{poly}, then an expression like $\mathsf{plus}(\mathsf{var}(\mathsf{s}_x), \mathsf{con}(\mathsf{s}_3))$ could be the expression in L_{syn} that denotes the syntax tree of $x + 3$. Next add an unspecified "binary" constant O_{pd} of type $\mathbb{S} \to (\mathbb{S} \to \mathbb{S})$ to L_R (that is intended to represent PolyDiff). Let $T'_R = (L'_R, \Gamma'_R)$ be the resulting extension of T_R. T'_R is clearly a conservative extension of T_R.
2. In the metatheory of T'_R define an injective, total function $Q : L_{\text{poly}} \to L_{\text{syn}}$ such that, for each polynomial $u \in L_{\text{poly}}$, $Q(u)$ is an expression e that denotes the syntax tree of u. For example, $Q(x + 3)$ could be $\mathsf{plus}(\mathsf{var}(\mathsf{s}_x), \mathsf{con}(\mathsf{s}_3))$.
3. In the metatheory of T'_R define an injective, total mapping $E : L_{\text{syn}} \to L_{\text{poly}}$ such that, for each expression $e \in L_{\text{syn}}$, $E(e)$ is the polynomial whose syntax tree is denoted by e. For example, $E(\mathsf{plus}(\mathsf{var}(\mathsf{s}_x), \mathsf{con}(\mathsf{s}_3)))$ would be $x + 3$.

Let (L_{poly}, I'_R) where $I'_R = (T'_R, \mathcal{M}')$ and $\mathcal{M}'$ is the set of standard models of T'_R in simple type theory (see [8]). It is easy to check that $F = (L_{\text{syn}}, Q, E)$ is a syntax framework for (L_{poly}, I'_R). Notice that E is the left inverse of Q and hence the *law of disquotation* holds: For all $u \in L_{\text{poly}}$, $E(Q(u)) = u$.

We are now ready to formalize PolyDiff in T'_R. First, we need to define an operator in T'_R to represent PolyDiff. We will use O_{pd} for this purpose. We write a sentence CompBehavior

$$\lambda\, a, b : \mathbb{S}\, .\, \mathsf{is\text{-}var}(b) \Rightarrow B(a, b, O_{\text{pd}}(a)(b))$$

in T'_R where, for all $u \in L_{\text{poly}}$ and $x \in L_{\text{var}}$,

$$B(Q(u), Q(x), O_{\text{pd}}(Q(u))(Q(x)))$$

holds iff

$$\mathsf{PolyDiff}(u, x) = E(O_{\text{pd}}(Q(u))(Q(x))).$$

That is, we specify the computational behavior of O_{pd} to be the same as that of PolyDiff.

Second, we need to prove that O_{pd} is mathematically correct. We write the sentence MathMeaning

$$\text{for all } u \in L_{\text{poly}}, \mathsf{deriv}(\lambda\, x : \mathbb{R}\, .\, u) = \lambda\, x : \mathbb{R}\, .\, E(O_{\text{pd}}(Q(u))(Q(x)))$$

in the metatheory of T'_R that says O_{pd} computes derivatives (with respect to the variable x). That is, we specify the mathematical meaning of O_{pd} to be the same as that of deriv. And then we prove in T'_R that MathMeaning follows from CompBehavior. The proof requires showing that the value of $E(O_{pd}(Q(u))(Q(x)))$ equals $\mathsf{deriv}((\lambda x : \mathbb{R} . u), x)$, the derivative of $(\lambda x : \mathbb{R} . u)$ at x, which is

$$\lim_{h \to 0} \frac{(\lambda x : \mathbb{R} . u)(x + h) - (\lambda x : \mathbb{R} . u)(x)}{h}.$$

The details of the proof are found in any good calculus textbook such as [17].

Third, we need to show how PolyDiff can be used to compute the derivative of a function $\lambda x : \mathbb{R} . u$ in T'_R. There are two ways. The first way is to instantiate the sentence CompBehavior with $Q(u)$ and $Q(x)$ and then simplify the resulting expression (e.g., by beta-reduction). The second way is to replace $E(O_{pd}(Q(u))(Q(x)))$ in MathMeaning with the result of applying PolyDiff to u and x. The first way requires that PolyDiff is implemented in T'_R as O_{pd}. The second way does not require that PolyDiff is implemented in T'_R, but only that its meaning is specified in T'_R.

The local approach is commonly used to reason about the syntax of expressions in a formal theory. It embodies a *deep embedding* [1] of the object language (e.g., L_{poly}) into the underlying formal language (e.g., L_R). The local approach to reason about syntax can be employed in almost any proof assistant in which it is possible to define an inductive type (e.g., see [1,4,20]).

The local approach has both strengths and weaknesses. These are the strengths of the local approach:

1. *Indirect Reasoning about the syntax of L_{poly} in the Theory.* In T'_R using L_{syn}, we can indirectly reason about the syntax of the polynomials in L_{poly}. This thus enables us to specify the computational behavior of PolyDiff via O_{pd}.
2. *Direct Reasoning about the syntax of L_{poly} in the Metatheory.* In the metatheory of T'_R using the formula

$$\text{for all } u \in L_{poly}, x \in L_{var}, \mathsf{PolyDiff}(u, x) = E(O_{pd}(Q(u))(Q(x))),$$

we can directly reason about the syntax of the polynomials in L_{poly} and specify the mathematical meaning of PolyDiff.

And these are the weaknesses:

1. *Syntax Problem.* We cannot refer in T'_R to the syntax of polynomials. Also the variable x is free in $x + 3$ but not in $Q(x + 3) = \mathsf{plus}(\mathsf{var}(\mathsf{s}_x), \mathsf{con}(\mathsf{s}_3))$. As a result, Q and E cannot be defined in T'_R and thus PolyDiff cannot be fully formalized in T'_R. In short, we can reason about syntax in T'_R but not about the interplay of syntax and semantics in T'_R.
2. *Coverage Problem.* The syntax framework F can only be used for reasoning about the syntax of polynomials. It cannot be used for reasoning, for example, about rational expressions. To do that a new syntax framework must be constructed.

3. *Extension Problem.* L_{poly}, L_{syn}, Q, and E must be extended each time a new constant of type $\mathbb{R}$ is defined in T'_R .

In summary, the local approach only gives us indirect access to the syntax of polynomials and must be modified to cover new or enlarged contexts.

If L_{obj} (which is L_{poly} in our example) does not contain variables, then we can define E to be a total operator in the theory. (If the theory is over a traditional logic, we will still not be able to define Q in the theory.) This variant of the local approach is used, for example, in the Agda reflection mechanism [19].

5   Global Approach

The *global approach* described in this section utilizes a replete syntax framework. Assume that we have modified T_R and simple type theory so that there is a replete syntax framework $F = (L_{\mathrm{syn}}, Q, E)$ for (L_R, I_R) where $I_R = (T_R, \mathcal{M})$ and $\mathcal{M}$ is the set of standard models of T_R in the modified simple type theory. Let us also assume that L_{syn} is the set of expressions of type $\mathbb{S}$ and L_R includes a constant O_{pd} of type $\mathbb{S} \to (\mathbb{S} \to \mathbb{S})$. By virtue of F being replete, F embodies a deep embedding of L_R into itself.

As far as we know, no one has ever worked out the details of how to modify simple type theory so that it admits built-in quotation and evaluation for the full language of a theory. However, we have shown how NBG set theory can be modified to admit built-in quotation and evaluation for its entire language [7]. Simple type theory can be modified in a similar way. We plan to present a version of simple type theory with a replete syntax framework in a future paper.

We can formalize PolyDiff in T_R as follows. We will write $\mathsf{quote}(e)$ and $\mathsf{eval}(e)$ as $\ulcorner e \urcorner$ and $[\![e]\!]$, respectively. First, we define the operator O_{pd} in T_R to represent PolyDiff. We write a sentence CompBehavior

$$\lambda\, a, b : \mathbb{S}\ .\ \mathsf{is\text{-}poly}(a) \wedge \mathsf{is\text{-}var}(b) \Rightarrow B(a, b, O_{\mathrm{pd}}(a)(b))$$

in T_R where, for all $u \in L_{\mathrm{poly}}$ and $x \in L_{\mathrm{var}}$,

$$B(\ulcorner u \urcorner, \ulcorner x \urcorner, O_{\mathrm{pd}}(\ulcorner u \urcorner)(\ulcorner x \urcorner))$$

holds iff

$$\mathsf{PolyDiff}(u, x) = [\![O_{\mathrm{pd}}(\ulcorner u \urcorner)(\ulcorner x \urcorner)]\!].$$

That is, we specify the computational behavior of O_{pd} to be the same as that of PolyDiff.

Second, we prove in T_R that O_{pd} is mathematically correct. We write the sentence MathMeaning

$$\forall\, a : \mathbb{S}\ .\ \mathsf{is\text{-}poly}(a) \Rightarrow \mathsf{deriv}(\lambda\, x : \mathbb{R}\ .\ [\![a]\!]) = \lambda\, x : \mathbb{R}\ .\ [\![O_{\mathrm{pd}}(a)(\ulcorner x \urcorner)]\!]$$

in T_R that says O_{pd} computes derivatives (with respect to the variable x). That is, we specify the mathematical meaning of O_{pd} to be the same as that of deriv. And then we prove in T'_R that MathMeaning follows from CompBehavior.

Third, we use PolyDiff to compute the derivative of a function $\lambda x : \mathbb{R} \,.\, u$ in T_R' in either of the two ways described for the local approach.

The strengths of the global approach are:

1. *Direct Reasoning about the syntax of L_{poly} in the Theory.* In T_R using L_{syn}, quote, and eval, we can directly reason about the syntax of the polynomials in L_{poly}. As a result, we can formalize PolyDiff in T_R as described in the Introduction.

2. *Direct Reasoning about the syntax of L_R in the Theory.* In T_R using L_{syn}, quote, and eval, we can directly reason about the syntax of the expressions in the entire language L_R. As a result, the syntax framework F can cover all current and future syntax reasoning needs. Moreover, we can express such things as syntactic side conditions, formula schemas, and substitution for a variable directly in T_R (see [7] for details).

In short, not only does the global approach enable us to formalize PolyDiff in T_R, it provides us with the facility to move syntax-based reasoning from the metatheory of T_R to T_R itself. This seems to be a wonderful result that solves the problem of formalizing syntax-based mathematical algorithms. Unfortunately, the global approach has the following serious weaknesses that temper the enthusiasm one might have for its strengths:

1. *Evaluation Problem.*

   Claim: eval cannot be defined on all expressions in L_R.

   Proof: Suppose eval is indeed total. T_R is sufficiently expressive, in the sense of Gödel's incomplete theorem, to apply the the diagonalization lemma [3], to obtain a formula LIAR such that

   $$\mathsf{LIAR} = \ulcorner \neg [\![\mathsf{LIAR}]\!] \urcorner.$$

   Then

   $$[\![\mathsf{LIAR}]\!] = [\![\ulcorner \neg [\![\mathsf{LIAR}]\!] \urcorner]\!] = \neg [\![\mathsf{LIAR}]\!],$$

   which is a contradiction. $\square$

   This means that the liar paradox limits the use of eval and, in particular, the law of disquotation does not hold universally, i.e., there are expressions e in L_R such that $[\![\ulcorner e \urcorner]\!] \neq e$.

2. *Variable Problem.* The variable x is not free in the expression $\ulcorner x + 3 \urcorner$ (or in any quotation). However, x is free in $[\![\ulcorner x + 3 \urcorner]\!]$ because $[\![\ulcorner x + 3 \urcorner]\!] = x + 3$. If the value of the variable e is $\ulcorner x + 3 \urcorner$, then both e and x are free in $[\![e]\!]$ because $[\![e]\!] = [\![\ulcorner x + 3 \urcorner]\!] = x + 3$.

   This example shows that the notions of a free variable, substitution for a variable, etc. are significantly more complex when expressions contain eval.

3. *Extension Problem.* We can define $L_{\mathrm{con}} \subseteq L_{\mathrm{syn}}$ in T_R as the language of expressions denoting the syntactic values of constants in L_R.

Claim: Assume the set of constants in L is finite and $T'_R = (L'_R, \Gamma'_R)$ is an extension of T_R such that there is a constant in L'_R but not in L_R. Then T'_R is not a conservative extension of T_R.

Proof: Let $\{c_1, \ldots, c_n\}$ be the set of constants in L. Then

$$L_{\text{con}} = \{\ulcorner c_1 \urcorner, \ldots, \ulcorner c_n \urcorner\}$$

is valid in T_R but not in T'_R. $\square$.

This shows that in the global approach the development of a theory via definitions requires that the notion of a conservative extension be weakened.

4. *Interpretation Problem.*

Let $T = (L, \Gamma)$ and $T' = (L', \Gamma')$ in be two theories in a simple type theory that has been modified to admit built-in quotation and evaluation for the entire language of a theory.

Claim: Let Φ be an interpretation of T in T' such that Φ is a homomorphism with respect to the logical operators of the underlying logic. Then Φ must be injective on the constants of L.

Proof: Assume that Φ is not injective on constants. Then there are two different constants a, b such that $\Phi(a) = \Phi(b)$. $\ulcorner a \urcorner \neq \ulcorner b \urcorner$ is valid in T. Hence

$$\Phi(\ulcorner a \urcorner \neq \ulcorner b \urcorner) = \ulcorner \Phi(a) \urcorner \neq \ulcorner \Phi(b) \urcorner$$

since Φ is a homomorphism, and the latter inequality must be valid in T' since Φ is an interpretation (which maps valid formulas of T to valid formulas of T'). However, since Q is injective, our hypothesis $\Phi(a) = \Phi(b)$ implies $\ulcorner \Phi(a) \urcorner = \ulcorner \Phi(b) \urcorner$, which is a contradiction. $\square$

This shows that the use of interpretations is more cumbersome in a logic that admits quotation than one that does not.

6 Conclusion

Syntax-based mathematical algorithms are employed throughout mathematics and are one of the main offerings of computer algebra systems. They are difficult, however, to formalize since they manipulate the syntactic structure of expressions in mathematically meaningful ways. We have presented two approaches to formalizing syntax-based mathematical algorithms in a formal theory, one called the *local approach* and the other the *global approach*. Both are based on the notion of a *syntax framework* which provides a foundation for integrating reasoning about the syntax of expressions with reasoning about what the expressions mean. Syntax frameworks include a syntax representation, a syntax language for reasoning about the representation, and quotation and evaluation mechanisms. Common syntax reasoning systems are instances of a syntax framework.

The local approach and close variants are commonly used for formalizing syntax-based mathematical algorithms. Its major strength is that it provides the means to formally reason about the syntactic structure of expressions, while its major weakness is that the mathematical meaning of a syntax-based mathematical algorithm cannot be expressed in the formal theory. Another weakness is that an application of the local approach cannot be easily extended to cover new or enlarged contexts.

The global approach enables one to reason in a formal theory T directly about the syntactic structure of the expressions in T as well as about the interplay of syntax and semantics in T. As a result, it is possible to fully formalize syntax-based algorithms like PolyDiff and move syntax-based reasoning, like the use of syntactic side conditions, from the metatheory of T to T itself. Unfortunately, these highly desirable results come with a high cost: Significant change must be made to the underlying logic as illustrated by the Evaluation, Variable, Extension, and Interpretation Problems given in the previous section.

One of the main goals of the MathScheme project [2], led by J. Carette and the author, is to see if the global approach can be used as a basis to integrate axiomatic and algorithmic mathematics. The logic Chiron [7] demonstrates that it is possible to modify a traditional logic to support the global approach. Although we have begun an implementation of Chiron, it remains an open question whether a logic modified in this way can be effectively implemented. As part of the MathScheme project, we are now pursuing this problem as well as developing the techniques needed to employ the global approach.

Acknowledgments. The author would like to thank Jacques Carette and Pouya Larjani for many fruitful discussions on ideas related to this paper. The author is also grateful to the referees for their comments and careful review of the paper.

References

1. Boulton, R., Gordon, A., Gordon, M., Harrison, J., Herbert, J., Van Tassel, J.: Experience with embedding hardware description languages in HOL. In: Stavridou, V., Melham, T.F., Boute, R.T. (eds.) Proceedings of the IFIP TC10/WG 10.2 International Conference on Theorem Provers in Circuit Design: Theory, Practice and Experience. IFIP Transactions A: Computer Science and Technology, vol. A-10, pp. 129–156. North-Holland (1993)
2. Carette, J., Farmer, W.M., O'Connor, R.: Mathscheme: Project description. In: Davenport, J.H., Farmer, W.M., Urban, J., Rabe, F. (eds.) Calculemus/MKM 2011. LNCS, vol. 6824, pp. 287–288. Springer, Heidelberg (2011)
3. Carnap, R.: Die Logische Syntax der Sprache. Springer (1934)
4. Contejean, E., Courtieu, P., Forest, J., Pons, O., Urbain, X.: Certification of automated termination proofs. In: Konev, B., Wolter, F. (eds.) FroCos 2007. LNCS (LNAI), vol. 4720, pp. 148–162. Springer, Heidelberg (2007)
5. Demers, F.-N., Malenfant, J.: Reflection in logic, functional and object-oriented programming: A short comparative study. In: IJCAI 1995 Workshop on Reflection and Metalevel Architectures and their Applications in AI, pp. 29–38 (1995)
6. Farmer, W.M.: Biform theories in Chiron. In: Kauers, M., Kerber, M., Miner, R., Windsteiger, W. (eds.) MKM/CALCULEMUS 2007. LNCS (LNAI), vol. 4573, pp. 66–79. Springer, Heidelberg (2007)

7. Farmer, W.M.: Chiron: A set theory with types, undefinedness, quotation, and evaluation. SQRL Report No. 38, McMaster University (2007), http://imps.mcmaster.ca/doc/chiron-tr.pdf (revised 2012)
8. Farmer, W.M.: The seven virtues of simple type theory. Journal of Applied Logic 6, 267–286 (2008)
9. Farmer, W.M., Larjani, P.: Frameworks for reasoning about syntax that utilize quotation and evaluation. McSCert Report No. 9, McMaster University (2013), http://imps.mcmaster.ca/doc/syntax.pdf
10. The F# Software Foundation
11. Grundy, J., Melham, T., O'Leary, J.: A reflective functional language for hardware design and theorem proving. Journal of Functional Programming 16 (2006)
12. Harrison, J.: Metatheory and reflection in theorem proving: A survey and critique. Technical Report CRC-053, SRI Cambridge (1995), http://www.cl.cam.ac.uk/~jrh13/papers/reflect.ps.gz
13. Koellner, P.: On reflection principles. Annals of Pure and Applied Logic 157, 206–219 (2009)
14. Mogensen, T.Æ.: Efficient self-interpretation in lambda calculus. Journal of Functional Programming 2, 345–364 (1994)
15. Rice University Programming Languages Team. Metaocaml: A compiled, type-safe, multi-stage programming language (2011), http://www.metaocaml.org/
16. Sheard, T., Jones, S.P.: Template meta-programming for Haskell. ACM SIGPLAN Notices 37, 60–75 (2002)
17. Spivak, M.: Calculus, 4th edn. Publish or Perish (2008)
18. Taha, W., Sheard, T.: MetaML and multi-stage programming with explicit annotations. Theoretical Computer Science 248, 211–242 (2000)
19. van der Walt, P.: Reflection in Agda. Master's thesis, Universiteit Utrecht (2012)
20. Wildmoser, M., Nipkow, T.: Certifying machine code safety: Shallow versus deep embedding. In: Slind, K., Bunker, A., Gopalakrishnan, G.C. (eds.) TPHOLs 2004. LNCS, vol. 3223, pp. 305–320. Springer, Heidelberg (2004)

Certification of Bounds of Non-linear Functions: The Templates Method

Xavier Allamigeon[1], Stéphane Gaubert[1],
Victor Magron[2], and Benjamin Werner[2]

[1] INRIA and CMAP, École Polytechnique, Palaiseau, France
{Xavier.Allamigeon,Stephane.Gaubert}@inria.fr
[2] INRIA and LIX, École Polytechnique, Palaiseau, France
magron@lix.polytechnique.fr, benjamin.werner@polytechnique.edu

Abstract. The aim of this work is to certify lower bounds for real-valued multivariate functions, defined by semialgebraic or transcendental expressions. The certificate must be, eventually, formally provable in a proof system such as Coq. The application range for such a tool is widespread; for instance Hales' proof of Kepler's conjecture yields thousands of inequalities. We introduce an approximation algorithm, which combines ideas of the max-plus basis method (in optimal control) and of the linear templates method developed by Manna et al. (in static analysis). This algorithm consists in bounding some of the constituents of the function by suprema of quadratic forms with a well chosen curvature. This leads to semialgebraic optimization problems, solved by sum-of-squares relaxations. Templates limit the blow up of these relaxations at the price of coarsening the approximation. We illustrate the efficiency of our framework with various examples from the literature and discuss the interfacing with Coq.

Keywords: Polynomial Optimization Problems, Hybrid Symbolic-numeric Certification, Semidefinite Programming, Transcendental Functions, Semialgebraic Relaxations, Flyspeck Project, Quadratic Cuts, Max-plus Approximation, Templates Method, Proof Assistant.

1 Introduction

Numerous problems coming from various fields boil down to the computation of a certified lower bound for a real-valued multivariate function $f : \mathbb{R}^n \to \mathbb{R}$ over a compact semialgebraic set $K \subset \mathbb{R}^n$.

Our aim is to automatically provide lower bounds for the following global optimization problem:

$$f^* := \inf_{\mathbf{x} \in K} f(\mathbf{x}) \ , \tag{1.1}$$

We want these bounds to be certifiable, meaning that their correctness must be, eventually, formally provable in a proof system such as Coq. One among many applications is the set of several thousands of non-linear inequalities which occur

J. Carette et al. (Eds.): CICM 2013, LNAI 7961, pp. 51–65, 2013.
© Springer-Verlag Berlin Heidelberg 2013

in Thomas Hales' proof of Kepler's conjecture, which is formalized in the Flyspeck project [1,2]. Several inequalities issued from Flyspeck actually deal with special cases of Problem (1.1). For instance, f may be a multivariate polynomial (polynomial optimization problems (POP)), or belong to the algebra $\mathcal{A}$ of semialgebraic functions which extends multivariate polynomials with arbitrary compositions of $(\cdot)^p, (\cdot)^{\frac{1}{p}} (p \in \mathbb{N}_0), |\cdot|, +, -, \times, /, \sup(\cdot, \cdot), \inf(\cdot, \cdot)$ (semialgebraic optimization problems), or involve transcendental functions (sin, arctan, etc).

Formal methods that produce precise bounds are mandatory because of the tightness of these inequalities. However, we also need to tackle scalability issues, which arise when one wants to provide coarser lower bounds for optimization problems with a larger number of variables or polynomial inequalities of a higher degree, etc. A common idea to handle Problem (1.1) is to first approximate f by multivariate polynomials through a semialgebraic relaxation and then obtain a lower bound of the resulting POP with a specialized software. This implies being able to also certify the approximation error in order to conclude. Such techniques rely on hybrid symbolic-numeric certification methods, see Peyrl and Parrilo [3] and Kaltofen et al. [4]. They allow one to produce positivity $certificates$ for such POP which can be checked in proof assistants such as Coq [5,6], HOL-light [7] or MetiTarski [8]. Recent efforts have been made to perform a formal verification of several Flyspeck inequalities with Taylor interval approximations [9]. We also mention procedures that solve SMT problems over the real numbers, using interval constraint propagation [10].

Solving POP is already a hard problem, which has been extensively studied. Semidefinite programming (SDP) relaxations based methods have been developed by Lasserre [11] and Parrilo [12]. A sparse refinement of the hierarchy of SDP relaxations by Kojima [13] has been implemented in the SparsePOP solver. Other approaches are based on Bernstein polynomials [14], global optimization by interval methods (see e.g. [15]), branch and bound methods with Taylor models [16].

Inequalities involving transcendental functions are typically difficult to solve with interval arithmetic, in particular due to the correlation between arguments of unary functions (e.g. sin) or binary operations (e.g. $+, -, \times, /$). For illustration purpose, we consider the following running example coming from the global optimization literature:

Example 1 (Modified Schwefel Problem 43 from Appendix B in [17]).

$$\min_{\mathbf{x} \in [1,500]^n} f(\mathbf{x}) = -\sum_{i=1}^{n} (x_i + \epsilon x_{i+1}) \sin(\sqrt{x_i}),$$

where $x_{n+1} = x_1$, and ϵ is a fixed parameter in $\{0,1\}$. In the original problem, $\epsilon = 0$, *i.e.* the objective function f is the sum of independent functions involving a single variable. This property may be exploited by a global optimization solver by reducing it to the problem $\min_{x \in [1,500]} x \sin(\sqrt{x})$. Hence, we also consider a modified version of this problem with $\epsilon = 1$.

Contributions. In this paper, we present an exact certification method, aiming at handling the approximation of transcendental functions and increasing the size of certifiable instances. It consists in combining SDP relaxations à la Lasserre / Parrilo, with an abstraction or approximation method. The latter is inspired by the linear template method of Sankaranarayanan, Sipma and Manna in static analysis [18], its nonlinear extension by Adjé et al. [19], and the maxplus basis method in optimal control introduced by Fleming and McEneaney [20], and developed by several authors [21–24].

The non-linear template method is a refinement of polyhedral based methods in static analysis. It allows one to determine invariants of programs by considering a parametric family of sets, $S(\alpha) = \{x \mid w_i(x) \leqslant \alpha_i, 1 \leqslant i \leqslant p\}$, where the vector $\alpha \in \mathbb{R}^p$ is the parameter, and $w_1, \ldots, w_p$ (the template) are fixed possibly non-linear functions, tailored to the program characteristics. The max-plus basis method is equivalent to the approximation of the epigraph of a function by a set $S(\alpha)$. In most basic examples, the functions w_i of the template are linear or quadratic functions.

In the present application, templates are used both to approximate transcendental functions, and to produce coarser but still tractable relaxations when the standard SDP relaxation of the semialgebraic problem is too complex to be handled. Indeed, SDP relaxations are a powerful tool to get tight certified lower bound for semialgebraic optimization problems, but their applicability is so far limited to small or medium size problems: their execution time grows exponentially with the relaxation order, which itself grows with the degree of the polynomials to be handled. Templates allow one to reduce these degrees, by approximating certain projections of the feasible set by a moderate number of nonconvex quadratic inequalities.

Note that by taking a trivial template (bound constraints, i.e., functions of the form $w_i(x) = \pm x_i$), the template method specializes to a version of interval calculus, in which bounds are derived by SDP techniques. By comparison, templates allow one to get tighter bounds, taking into account the correlations between the different variables. They are also useful as a replacement of standard Taylor approximations of transcendental functions: instead of increasing the degree of the approximation, one increases the number of functions in the template. A geometrical way to interpret the method is to think of it in terms of "quadratic cuts": quadratic inequalities are successively added to approximate the graph of a transcendental function.

The present paper is a followup of [25], in which the idea of max-plus approximation of transcendental function was applied to formal proof. By comparison, the new ingredient is the introduction of the template technique (approximating projections of the feasible sets), leading to an increase in scalability.

The paper is organized as follows. In Section 2, we recall the definition and properties of Lasserre relaxations of polynomial problems (Section 2.1), together with reformulations by Lasserre and Putinar of semialgebraic problems classes. In Section 2.2, we outline the conversion of the numerical SOS produced by the SDP solvers into an exact rational certificate. Then we explain how to verify this certificate in Coq. The max-plus approximation, and the main algorithm based on the non-linear templates method are presented in Section 3. Numerical results are presented in Section 4. We demonstrate the scalability of our approach by certifying bounds of non-linear problems involving up to 10^3 variables, as well as non trivial inequalities issued from the Flyspeck project.

2 Notation and Preliminary Results

Let $\mathbb{R}_d[\mathbf{x}]$ be the vector space of multivariate polynomials in n variables of degree d and $\mathbb{R}[\mathbf{x}]$ the set of multivariate polynomials in n variables. We also define the cone of sums of squares of degree at most $2d$:

$$\Sigma_d[\mathbf{x}] = \left\{ \sum_i q_i^2, \text{ with } q_i \in \mathbb{R}_d[\mathbf{x}] \right\}. \tag{2.1}$$

The set $\Sigma_d[\mathbf{x}]$ is a closed, fully dimensional convex cone in $\mathbb{R}_{2d}[\mathbf{x}]$. We denote by $\Sigma[\mathbf{x}]$ the cone of sums of squares of polynomials in n variables.

2.1 Constrained Polynomial Optimization Problems and SDP

We consider the general constrained polynomial optimization problem (POP):

$$f_{\text{pop}}^* := \inf_{\mathbf{x} \in K_{\text{pop}}} f_{\text{pop}}(\mathbf{x}), \tag{2.2}$$

where $f_{\text{pop}} : \mathbb{R}^n \to \mathbb{R}$ is a d-degree multivariate polynomial, K_{pop} is a compact set defined by inequalities $g_1(\mathbf{x}) \geq 0, \ldots, g_m(\mathbf{x}) \geq 0$, where $g_j(\mathbf{x}) : \mathbb{R}^n \to \mathbb{R}$ is a real-valued polynomial of degree ω_j, for $j = 1, \ldots, m$. Recall that the *set of feasible points* of an optimization problem is simply the domain over which the optimum is taken, i.e., here, K_{pop}.

Lasserre's Hierarchy of Semidefinite Relaxations. We set $g_0 := 1$ and take $k \geq k_0 := \max(\lceil d/2 \rceil, \max_{1 \leq j \leq m} \lceil \omega_j/2 \rceil)$. We consider the following hierarchy of semidefinite relaxations for Problem (2.2), consisting of the optimization problems Q_k, $k \geq k_0$,

$$Q_k : \begin{cases} \sup_{\mu, \sigma_j} \mu \\ \text{s.t.} \quad f_{\text{pop}}(\mathbf{x}) - \mu = \sum_{j=0}^m \sigma_j(\mathbf{x}) g_j(\mathbf{x}), \\ \quad \mu \in \mathbb{R}, \quad \sigma_j \in \Sigma_{k - \lceil \omega_j/2 \rceil}[\mathbf{x}], j = 0, \cdots, m. \end{cases}$$

We denote by $\sup(Q_k)$ the optimal value of Q_k. A feasible point $(\mu, \sigma_0, \ldots, \sigma_m)$ of Problem Q_k is said to be a *SOS certificate*, showing the implication $g_1(\mathbf{x}) \geq 0, \ldots, g_m(\mathbf{x}) \geq 0 \implies f_{\text{pop}}(\mathbf{x}) \geq \mu$.

The sequence of optimal values $(\sup(Q_k))_{k \geq k_0}$ is non-decreasing. Lasserre showed [11] that it does converge to f_{pop}^* under certain assumptions on the polynomials g_j. Here, we will consider sets K_{pop} included in a box of $\mathbb{R}^n$, so that Lasserre's assumptions are automatically satisfied.

Application to Semialgebraic Optimization. Given a semialgebraic function f_{sa}, we consider the problem $f_{\mathrm{sa}}^* = \inf_{\mathbf{x} \in K_{\mathrm{sa}}} f_{\mathrm{sa}}(\mathbf{x})$, where K_{sa} is a basic semialgebraic set. Moreover, we assume that f_{sa} has a basic semialgebraic lifting (for more details, see e.g. [26]). This implies that we can add auxiliary variables $z_1, \ldots, z_p$ (lifting variables), and construct polynomials $h_1, \ldots, h_s \in \mathbb{R}[\mathbf{x}, z_1, \ldots, z_p]$ defining the semialgebraic set $K_{\mathrm{pop}} := \{(\mathbf{x}, z_1, \ldots, z_p) \in \mathbb{R}^{n+p} : \mathbf{x} \in K_{\mathrm{sa}}, h_1(\mathbf{x}, \mathbf{z}) \geqslant 0, \ldots, h_s(\mathbf{x}, \mathbf{z}) \geqslant 0\}$, such that $f_{\mathrm{pop}}^* := \inf_{(\mathbf{x}, \mathbf{z}) \in K_{\mathrm{pop}}} z_p$ is a lower bound of f_{sa}^*.

2.2 Hybrid Symbolic-Numeric Certification and Formalization

The previous relaxation Q_k can be solved with several semidefinite programming solvers (e.g. SDPA [27]). These solvers are implemented using floating-point arithmetics. In order to build formal proofs, we currently rely on exact rational certificates which are needed to make formal proofs: Coq, being built on a computational formalism, is well equipped for checking the correctness of such certificates.

Such rational certificates can be obtained by a rounding and projection algorithm of Peyrl and Parillo [3], with an improvement of Kaltofen et al. [4]. Note that if the SDP formulation of Q_k is not strictly feasible, then the rounding and projection algorithm fails. However, Monniaux and Corbineau proposed a partial workaround for this issue [5]. In this way, except in degenerate situations, we arrive at a candidate SOS certificate with rational coefficients, $(\mu, \sigma_0, \ldots, \sigma_m)$. This certificate can straightforwardly be translated to Coq; the verification then boils down to formally checking that this SOS certificate does satisfy the equality constraint in Q_k with Coq's `field` tactic, which implies that $f_{\mathrm{pop}}^* \geqslant \mu$. This checking is typically handled by generating Coq scripts from the OCaml framework, when the lower bound μ obtained at the relaxation Q_k is accurate enough.

Future improvements could build, for instance, on future Coq libraries handling algebraic numbers or future tools to better handle floating point approximations inside Coq.

3 Max-plus Approximations And Non-linear Templates

3.1 Max-plus Approximations and Non-linear Templates

The max-plus basis method in optimal control [20, 21, 23] involves the approximation from below of a function f in n variables by a supremum

$$f \gtrsim g := \sup_{1 \leqslant i \leqslant p} \lambda_i + w_i \ . \tag{3.1}$$

The functions w_i are fixed in advance, or dynamically adapted by exploiting the problem structure. The parameters λ_i are degrees of freedom.

This method is closely related to the non-linear extension [19] of the template method [18]. This extension deals with parametric families of subsets of $\mathbb{R}^n$ of the form $S(\alpha) = \{x \mid w_i(x) \leqslant \alpha_i, 1 \leqslant i \leqslant p\}$. The template method consists in propagating approximations of the set of reachables values of the variables of a program by sets of the form $S(\alpha)$. The non-linear template and max-plus approximation methods are somehow equivalent. Indeed, the 0-level set of g, $\{x \mid g(x) \leqslant 0\}$, is nothing but $S(-\lambda)$, so templates can be recovered from max-plus approximations, and vice versa.

The functions w_i are usually required to be quadratic forms,

$$w_i(x) = p_i^\top x + \frac{1}{2}x^\top A_i x ,$$

where $p_i \in \mathbb{R}^n$ and A_i is a symmetric matrix. A basic choice is $A_i = -cI$, where c is a fixed constant, and I the identity matrix. Then, the parameters p remain the only degrees of freedom.

The consistency of the approximation follows from results of Legendre-Fenchel duality. Recall that a function f is said to be c-semiconvex if $x \mapsto f(x) + c\|x\|^2$ is convex. Then, if f is c-semiconvex and lowersemicontinuous, as the number of basis functions r grows, the best approximation $g \lesssim f$ by a supremum of functions of type (3.1), with $A_i = -cI$, is known to converge to f [20]. The same is true without semiconvexity assumptions if one allows A_i to vary [28].

A basic question is to estimate the number of basis functions needed to attain a prescribed accuracy. A typical result is proved in [24, Theorem 3.2], as a corollary of techniques of Grüber concerning the approximation of convex bodies by circumscribed polytopes. This theorem shows that if f is $c - \epsilon$ semiconvex, for $\epsilon > 0$, twice continuously differentiable, and if X is a full dimensional compact convex subset of $\mathbb{R}^n$, then, the best approximation g of f as a supremum or r functions as in (3.1), with $w_i(x) = p_i^\top x - c\|x\|^2/2$, satisfies

$$\|f - g\|_{L_\infty(X)} \simeq \frac{C(f)}{r^{2/n}} \tag{3.2}$$

where the constant $C(f)$ is explicit (it depends of $\det(f'' + cI)$ and is bounded away from 0 when ϵ is fixed). This estimate indicates that some curse of dimensionality is unavoidable: to get a uniform error of order ϵ, one needs a number of basis functions of order $1/\epsilon^{n/2}$. However, in what follows, we shall always apply the approximation to small dimensional constituents of the optimization problems ($n = 1$ when one needs to approximate transcendental functions in a single variable). We shall also apply the approximation by templates to certain relevant small dimensional projections of the set of lifted variables, leading to a smaller effective n. Note also that for optimization purposes, a uniform approximation is not needed (one only needs an approximation tight enough near the optimum, for which fewer basis functions are enough).

3.2   A Templates Method Based on Max-plus Approximations

We now consider an instance of Problem (1.1). We assume that K is a box and we identify the objective function f with its abstract syntax tree t_f. We suppose that the leaves of t_f are semialgebraic functions, and that the other nodes are either basic binary operations ($+$, $\times$, $-$, $/$), or unary transcendental functions (sin, etc).

Our main algorithm `template_optim` (Figure 1) is based on a previous method of the authors [25], in which the objective function is bounded by means of semialgebraic functions. For the sake of completeness, we first recall the basic principles of this method.

Bounding the objective function by semialgebraic estimators. Given a function represented by an abstract tree t, semialgebraic lower and upper estimators t^- and t^+ are computed by induction. If the tree is reduced to a leaf, *i.e.* $t \in \mathcal{A}$, it suffices to set $t^- = t^+ := t$. If the root of the tree corresponds to a binary operation `bop` with children c_1 and c_2, then the semialgebraic estimators c_1^-, c_1^+ and c_2^-, c_2^+ are composed using a function `compose_bop` to provide bounding estimators of t. Finally, if t corresponds to the composition of a transcendental (unary) function ϕ with a child c, we first bound c with semialgebraic functions c^+ and c^-. We compute a lower bound c_m of c^- as well as an upper bound c_M of c^+ to obtain an interval $I := [c_m, c_M]$ enclosing c. Then, we bound ϕ from above and below by computing parabola at given control points (function `build_par`), thanks to the semiconvexity properties of ϕ on the interval I. These parabola are composed with c^+ and c^-, thanks to a function denoted by `compose`.

These steps correspond to the part of the algorithm `template_optim` from Lines 1 to 10.

Reducing the complexity of semialgebraic estimators using templates. The semialgebraic estimators previously computed are used to determine lower and upper bounds of the function associated with the tree t, at each step of the induction. The bounds are obtained by calling the functions `min_sa` and `max_sa` respectively, which reduce the semialgebraic optimization problems to polynomial optimization problems by introducing extra lifting variables (see Section 2).

However, the complexity of solving the POPs can grow significantly because of the number n_{lifting} of lifting variables. If k denotes the relaxation order, the corresponding SDP problem Q_k indeed involve linear matrix inequalities of size $O((n + n_{\text{lifting}})^k)$ over $O((n + n_{\text{lifting}})^{2k})$ variables.

Consequently, this is crucial to control the number of lifting variables, or equivalently, the complexity of the semialgebraic estimators. For this purpose, we introduce the function `build_template`. It allows to compute approximations of the tree t by means of suprema/infima of quadratic functions, when the number of lifting variables exceeds a user-defined threshold value $n_{\text{lifting}}^{\max}$. The algorithm is depicted in Figure 2. Using a heuristics, it first builds candidate quadratic forms q_j^- and q_j^+ approximating t at each control point $\mathbf{x}_j$ (function `build_quadratic_form`, described below). Since each q_j^- does not necessarily underestimate the function t, we then determine the lower bound m_j^- of the semialgebraic function $t^- - q_j^-$, which ensures that $q_j^- + m_j^-$ is a quadratic lower-approximation of t. Similarly, the function $q_j^+ + M_j^+$ is an upper-approximation of t. The returned semialgebraic expressions $\max_{1 \leqslant j \leqslant r} \{q_j^- + m_j^-\}$ and $\min_{1 \leqslant j \leqslant r} \{q_j^+ + M_j^+\}$ now generate only one lifting variable (representing max or min).

Input: tree t, box K, SDP relaxation order k, control points sequence $s = \{x_1, \ldots, x_r\} \subset K$

Output: lower bound m, upper bound M, lower semialgebraic estimator t_2^-, upper semialgebraic estimator t_2^+

1: **if** $t \in \mathcal{A}$ **then**
2:    $t^- := t$, $t^+ := t$
3: **else if** $bop := \text{root }(t)$ is a binary operation with children c_1 and c_2 **then**
4:    $m_{c_i}, M_{c_i}, c_i^-, c_i^+ := \texttt{template\_optim}(c_i, K, k, s)$ for $i \in \{1, 2\}$
5:    $t^-, t^+ := \texttt{compose\_bop}(c_1^-, c_1^+, c_2^-, c_2^+)$
6: **else if** $r := \text{root}(t) \in \mathcal{T}$ with child c **then**
7:    $m_c, M_c, c^-, c^+ := \texttt{template\_optim}(c, K, k, s)$
8:    $par^-, par^+ := \texttt{build\_par}(r, m_c, M_c, s)$
9:    $t^-, t^+ := \texttt{compose}(par^-, par^+, c^-, c^+)$
10: **end**
11: $t_2^-, t_2^+ := \texttt{build\_template}(t, K, k, s, t^-, t^+)$
12: **return** $\texttt{min\_sa}(t_2^-, k)$, $\texttt{max\_sa}(t_2^+, k)$, t_2^-, t_2^+

Fig. 1. `template_optim`

Quadratic functions returned by $\texttt{build\_quadratic\_form}(t, x_j)$ are of the form:

$$q_{x_j, \lambda} : x \mapsto t(x_j) + \mathcal{D}(t)(x_j)\,(x - x_j) + \frac{1}{2}(x - x_j)^T \mathcal{D}^2(t)(x_j)\,(x - x_j) + \frac{1}{2}\lambda(x - x_j)^2$$

(we assume that t is twice differentiable) where λ is computed as follows. We sample the Hessian matrix difference $\mathcal{D}^2(t)(x) - \mathcal{D}^2(t)(x_j)$ over a finite set of random points $R \subset K$, and construct a matrix interval D enclosing all the entries of $(\mathcal{D}^2(t)(x) - \mathcal{D}^2(t)(x_j))$ for $x \in R$. A lower bound λ^- of the minimal eigenvalue of D is obtained by applying a robust SDP method on interval matrix described by Calafiore and Dabbene in [29]. Similarly, we get an upper bound λ^+ of the maximal eigenvalue of D. The function $\texttt{build\_quadratic\_form}(t, x_j)$ then returns the two quadratic forms $q^- := q_{x_j, \lambda^-}$ and $q^+ := q_{x_j, \lambda^+}$.

Example 2 (Modified Schwefel Problem). We illustrate our method with the function f from Example 1 and the finite set of three control points $\{135, 251, 500\}$. For each $i = 1, \ldots, n$, consider the sub-tree $\sin(\sqrt{x_i})$. First, we represent each sub-tree $\sqrt{x_i}$ by a lifting variable y_i and compute $a_1 := \sqrt{135}$, $a_2 := \sqrt{251}$, $a_3 := \sqrt{500}$. Then, we get the equations of $par_{a_1}^-$, $par_{a_2}^-$ and $par_{a_3}^-$ with $\texttt{build}_{par}$, which are three underestimators of the function $\sin$ on the real interval $I := [1, \sqrt{500}]$. Similarly we obtain three overestimators $par_{a_1}^+$, $par_{a_2}^+$ and $par_{a_3}^+$. Finally, we obtain the underestimator $t_{1,i}^- := \max_{j \in \{1,2,3\}}\{par_{a_j}^-(y_i)\}$ and the overestimator $t_{1,i}^+ := \min_{j \in \{1,2,3\}}\{par_{a_j}^+(y_i)\}$. To solve the modified Schwefel problem, we consider the following POP:

$$\begin{cases} \min_{x \in [1,500]^n, y \in [1,\sqrt{500}]^n, z \in [-1,1]^n} & -\sum_{i=1}^{n}(x_i + \epsilon x_{i+1})z_i \\ \text{s.t.} & z_i \leqslant par_{a_j}^+(y_i), j \in \{1,2,3\}, i = 1, \cdots, n \\ & y_i^2 = x_i, i = 1, \cdots, n \end{cases}$$

Input: tree t, box K, SDP relaxation order k, control points sequence $s = \{x_1, \ldots, x_r\} \subset K$, lower/upper semialgebraic estimator t^-, t^+
1: **if** the number of lifting variables exceeds $n_{\text{lifting}}^{\max}$ **then**
2:     **for** $x_j \in s$ **do**
3:         $q_j^-, q_j^+ := \texttt{build\_quadratic\_form}(t, x_j)$
4:         $m_j^- := \texttt{min\_sa}(t_1^- - q_j^-, k)$ $\triangleright\ q_j^- + m_j^- \leqslant t^- \leqslant t$
5:         $M_j^+ := \texttt{max\_sa}(q_j^+ - t_1^+, k)$ $\triangleright\ q_j^+ + M_j^+ \geqslant t^+ \geqslant t$
6:     **done**
7:     **return** $\max_{1 \leqslant j \leqslant r}\{q_j^- + m_j^-\}$, $\min_{1 \leqslant j \leqslant r}\{q_j^+ + M_j^+\}$
8: **else**
9:     **return** t^-, t^+
10: **end**

Fig. 2. build_template

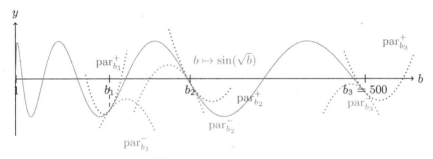

Fig. 3. Templates based on Max-plus Semialgebraic Estimators for $b \mapsto \sin(\sqrt{b})$:
$t_{2,i}^- := \max_{j \in \{1,2,3\}}\{\text{par}_{b_j}^-(x_i)\} \leqslant \sin\sqrt{x_i} \leqslant t_{2,i}^+ := \min_{j \in \{1,2,3\}}\{\text{par}_{b_j}^+(x_i)\}$

Notice that the number of lifting variables is $2n$ and the number of equality constraints is n, thus we can obtain coarser semialgebraic approximations of f by considering the function $b \mapsto \sin(\sqrt{b})$ (see Figure 3). We get new estimators $t_{2,i}^-$ and $t_{2,i}^+$ of each sub-tree $\sin(\sqrt{x_i})$ with the functions $\texttt{build\_quadratic\_form}$, $\texttt{min\_sa}$ and $\texttt{max\_sa}$. The resulting POP involves only n lifting variables. Besides, it does not contain equality constraints anymore, which improves in practice the numerical stability of the POP solver.

Dynamic choice of the control points. As in [25], the sequence s of control points is computed iteratively. We initialize the set s to a single point of K, chosen so as to be a minimizer candidate for t (e.g. with a local optimization solver). Calling the algorithm $\texttt{template\_optim}$ on the main objective function t_f yields an underestimator t_f^-. Then, we compute a minimizer candidate x_{opt} of the underestimator tree t_f^-. It is obtained by projecting a solution x_{sdp} of the SDP relaxation of Section 2.1 on the coordinates representing the first order moments, following [11, Theorem 4.2]. We add x_{opt} to the set of control points s. Consequently, we can refine dynamically our templates based max-plus approximations by iterating the previous procedure to get tighter lower bounds. This procedure can be stopped as soon as the requested lower bound is attained.

Remark 1 (Exploiting the system properties). Several properties of the POP can be exploited to decrease the size of the SDP relaxations such as symmetries [30] or sparsity [31]. Consider Problem (1.1) with f having some sparsity pattern or being invariant under the action of a finite subgroup symmetries. Then the same properties hold for the resulting semialgebraic relaxations that we build with our non-linear templates method.

4   Results

Comparing Three Certification Methods. We next present numerical results obtained by applying the present template method to examples from the global optimization literature, as well as inequalities from the Flyspeck project. Our tool is implemented in OCaml and interfaced with the SparsePOP solver [31].

In each example, our aim is to certify a lower bound m of a function f on a box K. We use the algorithm `template_optim`, keeping the SOS relaxation order k sufficiently small to ensure the fast computation of the lower bounds. The algorithm `template_optim` returns more precise bounds by successive updates of the control points sequence s. However, in some examples, the relaxation gap is too high to certify the requested bound. Then, we perform a domain subdivision in order to reduce this gap: we divide the maximal width interval of K in two halves to get two sub-boxes K_1 and K_2 such that $K = K_1 \cup K_2$. We repeat this subdivision procedure, by applying `template_optim` on a finite set of sub-boxes, until we succeed to certify that m is a lower bound of f. We denote by #boxes the total number of sub-boxes generated by the algorithm.

For the sake of comparison, we have implemented a template-free SOS method `ia_sos`, which coincides with the particular case of `template_optim` in which $\#s = 0$ and $n_{\text{lifting}} = 0$. It computes the bounds of semialgebraic functions with standard SOS relaxations and bounds the univariate transcendental functions by interval arithmetic. We also tested the MATLAB toolbox algorithm `intsolver` [32], which is based on the Newton interval method [33]. Experiments are performed on an Intel Core i5 CPU (2.40 GHz).

Global Optimization Problems. The following test examples are taken from Appendix B in [17]. Some of these examples depend on numerical constants, the values of which can be found there.

- *Hartman 3 (H3):* $\displaystyle \min_{\mathbf{x} \in [0,1]^3} f(\mathbf{x}) = -\sum_{i=1}^{4} c_i \exp\left[-\sum_{j=1}^{3} a_{ij}(x_j - p_{ij})^2\right]$
- *Mc Cormick (MC),* with $K = [-1.5, 4] \times [-3, 3]$:
  $\displaystyle \min_{\mathbf{x} \in K} f(\mathbf{x}) = \sin(x_1 + x_2) + (x_1 - x_2)^2 - 0.5x_1 + 2.5x_2 + 1$
- *Modified Langerman (ML):*
  $\displaystyle \min_{\mathbf{x} \in [0,10]^n} f(\mathbf{x}) = \sum_{j=1}^{5} c_j \cos(d_j/\pi) \exp(-\pi d_j), \text{ with } d_j = \sum_{i=1}^{n} (x_i - a_{ji})^2$

- *Paviani Problem (PP)*, with $K = [2.01, 9.99]^{10}$:

$$\min_{\mathbf{x} \in K} f(\mathbf{x}) = \sum_{i=1}^{10} \left[(\log(x_i - 2))^2 - \log(10 - x_i))^2 \right] - \left(\prod_{i=1}^{10} x_i \right)^{0.2}$$

- *Shubert (SBT)*: $\min_{\mathbf{x} \in [-10,10]^n} f(\mathbf{x}) = \prod_{i=1}^{n} \left(\sum_{j=1}^{5} j \cos((j+1)x_i + j) \right)$

- *Modified Schwefel (SWF)*: see Example 1

Informal certification of lower bounds of non-linear problems. In Table 1, the *time* column indicates the total informal verification time, *i.e.* without the exact certification of the lower bound m with Coq. Each occurrence of the symbol "−" means that m could not be determined within one day of computation by the corresponding solver. We see that ia_sos already outperforms the interval arithmetic solver intsolver on these examples. However, it can only be used for problems with a moderate number of variables. The algorithm template_optim allows us to overcome this restriction, while keeping a similar performance (or occasionally improving this performance) on moderate size examples.

Notice that reducing the number of lifting variables allows us to provide more quickly coarse bounds for large-scale instances of *SWF*. We discuss the results appearing in the two last lines of Table 1. Without any box subdivision, we can certify a better lower bound $m = -967n$ with $n_{\text{lifting}} = 2n$ since our semialgebraic estimator is more precise. However the last lower bound $m = -968n$ can be computed twice faster by considering only n lifting variables, thus reducing the size of the POP described in Example 2. This indicates that the method is able to avoid the blow up for certain hard sub-classes of problems where a standard (template free) POP formulation would involve a large number of lifting variables.

Formal certification of lower bounds of POP. For some small size instances of POP, our tool can prove the correctness of lower bounds. Our solver is interfaced with the framework mentioned in [5] to provide exact rational certificates, which can be formally checked with Coq. This formal verification is much slower. As an example, for the *MC* problem, it is 36 times slower to generate exact SOS certificates and 13 times slower to prove its correctness in Coq. Note that the interface with Coq still needs some streamlining.

High-degree polynomial approximations. An alternative approach consists in approximating the transcendental functions by polynomial functions of sufficiently high degree, and then applying sums of squares approach to the polynomial problems. Given $d \in \mathbb{N}$ and a floating-point interval I, we can approximate an univariate transcendental function on I by the best uniform degree-d polynomial approximation and obtain an upper bound of the approximation error. This technique, based on Remez algorithm, is implemented in the Sollya tool (for further details, see e.g. [34]).

We interfaced our tool with Sollya and performed some numerical tests. The minimax approximation based method is eventually faster than the templates

Table 1. Comparison results for global optimization examples

Problem	n	m	template_optim					ia_sos		intsolver
			k	#s	n_{lifting}	#boxes	time	#boxes	time	time
H3	3	-3.863	2	3	4	99	101 s	1096	247 s	3.73 h
H6	6	-3.33	2	1	6	113	102 s	113	45 s	$> 4\,h$
MC	2	-1.92	1	2	1	17	1.8 s	92	7.6 s	4.4 s
ML	10	-0.966	1	1	6	8	8.2 s	8	6.6 s	$> 4\,h$
PP	10	-46	1	3	2	135	89 s	3133	115 s	56 min
SBT	2	-190	2	3	2	150	36 s	258	0.6 s	57 s
SWF ($\epsilon = 0$)	10	$-430n$	2	6	$2n$	16	40 s	3830	129 s	18.5 min
	100	$-440n$	2	6	$2n$	274	1.9 h	> 20000	$> 10\,h$	$-$
	1000	$-486n$	2	4	$2n$	1	450 s	$-$	$-$	$-$
	1000	$-488n$	2	4	n	1	250 s	$-$	$-$	$-$
SWF ($\epsilon = 1$)	1000	$-967n$	3	2	$2n$	1	543 s	$-$	$-$	$-$
	1000	$-968n$	3	2	n	1	272 s	$-$	$-$	$-$

Table 2. Results for Flyspeck inequalities using template_optim with $n = 6$, $k = 2$ and $m = 0$

Inequality id	$n_{\mathcal{T}}$	#s	n_{lifting}	#boxes	time
9922699028	1	4	9	47	241 s
9922699028	1	4	3	39	190 s
3318775219	1	2	9	338	26 min
7726998381	3	4	15	70	43 min
7394240696	3	2	15	351	1.8 h
4652969746_1	6	4	15	81	1.3 h
OXLZLEZ 6346351218_2_0	6	4	24	200	5.7 h

method for moderate instances. For the examples *H3* and *H6*, the speed-up factor is 8 when the function exp is approximated by a quartic minimax polynomial.

However, this approach is much slower to compute lower bounds of problems involving a large number of variables. It requires 57 times more CPU time to solve *SWF* ($\epsilon = 1$) with $n = 10$ by considering a cubic minimax polynomial approximation of the function $b \mapsto \sin(\sqrt{b})$ on a floating-point interval $I \supseteq [1, \sqrt{500}]$. These experiments indicate that a high-degree polynomial approximation is not suitable for large-scale problems.

Certification of Various Flyspeck Inequalities. In Table 2, we present some test results for several non-linear Flyspeck inequalities. The information in the columns *time*, #boxes, and n_{lifting} is the same as above. The integer $n_{\mathcal{T}}$ represents the number of transcendental univariate nodes in the corresponding abstract syntax trees. These inequalities are known to be tight and involve sum of arctan of correlated functions in many variables, whence we keep high the number of lifting variables to get precise max-plus estimators. However, some inequalities (e.g. 9922699028) are easier to solve by using coarser semialgebraic estimators. For instance, the first line ($n_{\text{lifting}} = 9$) corresponds to the algorithm described in [25] and the second one ($n_{\text{lifting}} = 3$) illustrates our improved templates method. For the latter, we do not use any lifting variables to represent square roots of univariate functions.

5 Conclusion

The present quadratic templates method computes certified lower bounds for global optimization problems. It can provide tight max-plus semialgebraic estimators to certify non-linear inequalities involving transcendental multivariate functions (e.g. for Flyspeck inequalities). It also allows one to limit the growth of the number of lifting variables as well as of polynomial constraints to be handled in the POP relaxations, at the price of a coarser approximation. Thus, our method is helpful when the size of optimization problems increases. Indeed, the coarse lower bounds obtained (even with a low SDP relaxation order) are better than those obtained with interval arithmetic or high-degree polynomial approximation. For future work, we plan to study how to obtain more accurate non-linear templates by constructing a sequence of semialgebraic estimators, which converges to the "best" max-plus estimators (following the idea of [35]).

Furthermore, the formal part of our implementation, currently can only handle small size POP certificates. We plan to address this issue by a more careful implementation on the Coq side, but also by exploiting system properties of the problem (sparsity, symmetries) in order to reduce the size of the rational SOS certificates. Finally, it remains to complete the formal verification procedure by additionally proving in Coq the correctness of our semialgebraic estimators.

Acknowledgements. The authors thank the anonymous referees for helpful comments and suggestions to improve this paper.

References

1. Hales, T.C., Harrison, J., McLaughlin, S., Nipkow, T., Obua, S., Zumkeller, R.: A revision of the proof of the kepler conjecture. Discrete & Computational Geometry 44(1), 1–34 (2010)
2. Hales, T.C.: Introduction to the flyspeck project. In: Coquand, T., Lombardi, H., Roy, M.-F. (eds.) Mathematics, Algorithms, Proofs. Dagstuhl Seminar Proceedings, vol. 05021, Internationales Begegnungs- und Forschungszentrum für Informatik (IBFI), Schloss Dagstuhl, Germany (2006)
3. Peyrl, H., Parrilo, P.A.: Computing sum of squares decompositions with rational coefficients. Theor. Comput. Sci. 409(2), 269–281 (2008)
4. Kaltofen, E.L., Li, B., Yang, Z., Zhi, L.: Exact certification in global polynomial optimization via sums-of-squares of rational functions with rational coefficients. JSC 47(1), 1–15 (2012); In memory of Wenda Wu (1929–2009)
5. Monniaux, D., Corbineau, P.: On the generation of Positivstellensatz witnesses in degenerate cases. In: van Eekelen, M., Geuvers, H., Schmaltz, J., Wiedijk, F. (eds.) ITP 2011. LNCS, vol. 6898, pp. 249–264. Springer, Heidelberg (2011)

6. Besson, F.: Fast reflexive arithmetic tactics the linear case and beyond. In: Altenkirch, T., McBride, C. (eds.) TYPES 2006. LNCS, vol. 4502, pp. 48–62. Springer, Heidelberg (2007)

7. Harrison, J.: Verifying nonlinear real formulas via sums of squares. In: Schneider, K., Brandt, J. (eds.) TPHOLs 2007. LNCS, vol. 4732, pp. 102–118. Springer, Heidelberg (2007)

8. Akbarpour, B., Paulson, L.C.: Metitarski: An automatic theorem prover for real-valued special functions. J. Autom. Reason. 44(3), 175–205 (2010)

9. Solovyev, A., Hales, T.C.: Formal verification of nonlinear inequalities with taylor interval approximations. CoRR, abs/1301.1702 (2013)

10. Gao, S., Avigad, J., Clarke, E.M.: Delta-complete decision procedures for satisfiability over the reals. CoRR, abs/1204.3513 (2012)

11. Lasserre, J.B.: Global optimization with polynomials and the problem of moments. SIAM Journal on Optimization 11(3), 796–817 (2001)

12. Parrilo, P.A., Sturmfels, B.: Minimizing polynomial functions. DIMACS Ser. Discrete Math. Theoret. Comput. Sci, vol. 60, pp. 83–99. Amer. Math. Soc., Providence (2003)

13. Waki, H., Kim, S., Kojima, M., Muramatsu, M.: Sums of squares and semidefinite programming relaxations for polynomial optimization problems with structured sparsity. SIAM Journal on Optimization 17, 218–242 (2006)

14. Zumkeller, R.: Rigorous Global Optimization. PhD thesis, École Polytechnique (2008)

15. Hansen, E.R.: Sharpening interval computations. Reliable Computing 12(1), 21–34 (2006)

16. Cartis, C., Gould, N.I.M., Toint, P.L.: Adaptive cubic regularisation methods for unconstrained optimization. part i: motivation, convergence and numerical results. Math. Program. 127(2), 245–295 (2011)

17. Montaz Ali, M., Khompatraporn, C., Zabinsky, Z.B.: A numerical evaluation of several stochastic algorithms on selected continuous global optimization test problems. J. of Global Optimization 31(4), 635–672 (2005)

18. Sankaranarayanan, S., Sipma, H.B., Manna, Z.: Scalable analysis of linear systems using mathematical programming. In: Cousot, R. (ed.) VMCAI 2005. LNCS, vol. 3385, pp. 25–41. Springer, Heidelberg (2005)

19. Adje, A., Gaubert, S., Goubault, E.: Coupling policy iteration with semi-definite relaxation to compute accurate numerical invariants in static analysis. Logical Methods in Computer Science 8(1), 1–32 (2012)

20. Fleming, W.H., McEneaney, W.M.: A max-plus-based algorithm for a Hamilton-Jacobi-Bellman equation of nonlinear filtering. SIAM J. Control Optim. 38(3), 683–710 (2000)

21. Akian, M., Gaubert, S., Lakhoua, A.: The max-plus finite element method for solving deterministic optimal control problems: basic properties and convergence analysis. SIAM J. Control Optim. 47(2), 817–848 (2008)

22. McEneaney, W.M., Deshpande, A., Gaubert, S.: Curse-of-complexity attenuation in the curse-of-dimensionality-free method for HJB PDEs. In: Proc. of the 2008 American Control Conference, Seattle, Washington, USA, pp. 4684–4690 (June 2008)

23. McEneaney, W.M.: A curse-of-dimensionality-free numerical method for solution of certain HJB PDEs. SIAM J. Control Optim. 46(4), 1239–1276 (2007)

24. Gaubert, S., McEneaney, W.M., Qu, Z.: Curse of dimensionality reduction in max-plus based approximation methods: Theoretical estimates and improved pruning algorithms. In: CDC-ECC [36], pp. 1054–1061

25. Allamigeon, X., Gaubert, S., Magron, V., Werner, B.: Certification of inequalities involving transcendental functions: combining sdp and max-plus approximation. To appear in the Proceedings of the European Control Conference, ECC 2013, Zurich (2013)
26. Lasserre, J.B., Putinar, M.: Positivity and optimization for semi-algebraic functions. SIAM Journal on Optimization 20(6), 3364–3383 (2010)
27. Yamashita, M., Fujisawa, K., Nakata, K., Nakata, M., Fukuda, M., Kobayashi, K., Goto, K.: A high-performance software package for semidefinite programs: Sdpa7. Technical report, Dept. of Information Sciences, Tokyo Institute of Technology, Tokyo, Japan (2010)
28. Akian, M., Gaubert, S., Kolokoltsov, V.N.: Set coverings and invertibility of functional galois connections. In: Litvinov, G.L., Maslov, V.P. (eds.) Idempotent Mathematics and Mathematical Physics, Contemporary Mathematics, pp. 19–51. American Mathematical Society (2005)
29. Calafiore, G., Dabbene, F.: Reduced vertex set result for interval semidefinite optimization problems. Journal of Optimization Theory and Applications 139, 17–33 (2008), doi:10.1007/s10957-008-9423-1
30. Riener, C., Theobald, T., Andrén, L.J., Lasserre, J.B.: Exploiting symmetries in sdp-relaxations for polynomial optimization. CoRR, abs/1103.0486 (2011)
31. Waki, H., Kim, S., Kojima, M., Muramatsu, M., Sugimoto, H.: Algorithm 883: Sparsepop—a sparse semidefinite programming relaxation of polynomial optimization problems. ACM Trans. Math. Softw. 35(2) (2008)
32. Montanher, T.M.: Intsolver: An interval based toolbox for global optimization, Version 1.0 (2009), www.mathworks.com
33. Hansen, E.R., Greenberg, R.I.: An interval newton method. Applied Mathematics and Computation 12(2-3), 89–98 (1983)
34. Brisebarre, N., Joldeş, M.: Chebyshev interpolation polynomial-based tools for rigorous computing. In: Proceedings of the 2010 International Symposium on Symbolic and Algebraic Computation, ISSAC 2010, pp. 147–154. ACM, New York (2010)
35. Lasserre, J.B., Thanh, T.P.: Convex underestimators of polynomials. In: CDC-ECE [36], pp. 7194–7199
36. Proceedings of the 50th IEEE Conference on Decision and Control and European Control Conference, CDC-ECC 2011, Orlando, FL, USA, December 12-15. IEEE (2010)

Verifying a Plaftorm for Digital Imaging:
A Multi-tool Strategy*

Jónathan Heras[1], Gadea Mata[2], Ana Romero[2],
Julio Rubio[2], and Rubén Sáenz[2]

[1] School of Computing, University of Dundee, UK
[2] Department of Mathematics and Computer Science, University of La Rioja, Spain
jonathanheras@computing.dundee.ac.uk,
{gadea.mata,ana.romero,julio.rubio,ruben.saenz}@unirioja.es

Abstract. Fiji is a Java platform widely used by biologists and other experimental scientists to process digital images. In our research, made together with a biologists team, we use Fiji in some pre-processing steps before undertaking a homological digital processing of images. In a previous work, we have formalised the correctness of the programs which use homological techniques to analyse digital images. However, the verification of Fiji's pre-processing step was missed. In this paper, we present a *multi-tool* approach (based on the combination of Why/Krakatoa, Coq and ACL2) filling this gap.

1 Introduction

Fiji [27] is a Java platform widely used by biologists and other experimental scientists to process digital images. In our research, made together with a biologists team, we use Fiji in some pre-processing steps before undertaking a homological digital processing of images.

Due to the fact that the reliability of results is instrumental in biomedical research, we are working towards the certification of the programs that we use to analyse biomedical images – here, certification means verification assisted by computers. In a previous work, see [16,17], we have formalised two homological techniques to process biomedical images. However, in both cases, the verification of Fiji's pre-processing step was not undertaken.

Being a software built by means of plug-ins developed by several authors, Fiji is messy, very flexible (program pieces are used in some occasions with a completely different objective from the one they were designed), contains many redundancies and dead code, and so on. In summary, it is a big software system which has not been devised to be formally verified. So, this endeavour is challenging.

* Partially supported by Ministerio de Educación y Ciencia, project MTM2009-13842-C02-01, and by the European Union's 7th Framework Programme under grant agreement nr. 243847 (ForMath).

J. Carette et al. (Eds.): CICM 2013, LNAI 7961, pp. 66–81, 2013.

There are several approaches to verify Java code; for instance, proving the correctness of the associated Java bytecode, see [22]. In this paper, we use Krakatoa [11] to specify and prove the correctness of Fiji/Java programs. This experience allows us to evaluate both the verification of *production* Fiji/Java code, and the Krakatoa tool itself in an unprepared scenario.

Krakatoa uses some automated theorem provers (as Alt-Ergo [5] or CVC3 [3]) to discharge the proof obligations generated by means of the Why tool [11]. When a proof obligation cannot be solved by means of the automated provers, the corresponding statement is generated in Coq [9]. Then, the user can try to prove the missing property by interacting with this proof assistant.

In this picture, we add the ACL2 theorem prover [20]. ACL2 is an automated theorem prover but more powerful than others. In many aspects, working with ACL2 is more similar to interactive provers than to automated ones, see [20]. Instead of integrating ACL2 in the architecture of Why/Krakatoa, we have followed another path leaving untouched the Why/Krakatoa code. Our approach reuses a proposal presented in [2] to translate first-order Isabelle/HOL theories to ACL2 through an XML specification language called XLL [2]. We have enhanced our previous tools to translate Coq theories to the XLL language, and then apply the tools developed in [2] to obtain ACL2 files. In this way, we can use, unmodified, the Why/Krakatoa framework; the Coq statements are then translated (if needed) to ACL2, where an automated proof is tried; if it succeeds, Coq is only an intermediary specification step; otherwise, both ACL2 or Coq can be interactively used to complete the proof.

The organization of the paper is as follows. The used tools together with our general way of working are briefly presented in Section 2. Section 3 deals with a methodology to "tame" production Fiji code in such a way that it is acceptable for Why/Krakatoa – this method is general enough to be applied to any Java code. Section 4 describes an example of the kind of specification we faced. The role of ACL2, and the tools to interoperate between Coq and ACL2, are explained in Section 5. The exposition style along the paper tries to be clear (without much emphasis on formal aspects), driven by real examples extracted from our programming experience in Fiji; in the same vein, Section 6 contains a complete example illustrating the actual role of ACL2 in our setting. The paper ends with a conclusions section and the bibliography.

All the programs and examples presented throughout this paper are available at http://www.computing.dundee.ac.uk/staff/jheras/vpdims/.

2 Context, Tools, Method

2.1 Context

Fiji [27] is a Java program which can be described as a distribution of ImageJ [26]. These two programs help with the research in life sciences and biomedicine since they are used to process and analyse biomedical images. Fiji and ImageJ are open source projects and their functionality can be expanded by means of either a macro scripting language or Java plug-ins. Among the Fiji/ImageJ plug-ins and macros, we can find functionality which allows us to binarise an image via different threshold algorithms, homogenise images through filters such as the "median filter" or obtain the maximum projection of a stack of images.

In the frame of the ForMath European project [1], one of the tasks is devoted to the topological aspects of digital image processing. The objective of that consists in formalising enough mathematics to verify programs in the area of biomedical imaging. In collaboration with the biologists team directed by Miguel Morales, two plug-ins for Fiji have been developed (SynapCountJ [24] and NeuronPersistentJ [23]); these programs are devoted to analyse the effects of some drugs on the neuronal structure. At the end of such analysis, some homological processing is needed (standard homology groups in SynapCountJ and persistent homology in NeuronPersistentJ). As explained in the introduction, we have verified these last steps [16,17]. But all the pre-processing steps, based on already-built Fiji plug-ins and tools, kept unverified. This is the gap we try to fill now, by using the facilities presented in the sequel.

2.2   Tools

Why/Krakatoa: Specifying and Verifying Java Code. The Why/Krakatoa tools [11] are an environment for proving the correctness of Java programs annotated with JML [7] specifications which have been successfully applied in different context, see [4]. The environment involves three distinct components: the Krakatoa tool, which reads the annotated Java files and produces a representation of the semantics of the Java program into Why's input language; the Why tool, which computes proof obligations (POs) for a core imperative language annotated with pre- and post-conditions, and finally several automated theorem provers which are included in the environment and are used to prove the POs. When some PO cannot be solved by means of the automated provers, corresponding statements are automatically generated in Coq [9], so that the user can then try to prove the missing properties in this interactive theorem prover. The POs generation is based on a Weakest Precondition calculus and the validity of all generated POs implies the soundness of the code with respect to the given specification. The Why/Krakatoa tools are available as open source software at http://krakatoa.lri.fr.

Coq and ACL2: Interactive Theorem Proving. Coq [9] is an interactive proof assistant for constructive higher-order logic based on the Calculus of Inductive Construction. This system provides a formal language to write mathematical definitions, executable algorithms and theorems together with an environment for semi-interactive development of machine-checked proofs. Coq has been successfully used in the formalisation of relevant mathematical results; for instance, the recently proven Feit-Thompson Theorem [13].

ACL2 [20] is a programming language, a first order logic and an automated theorem prover. Thus, the system constitutes an environment in which algorithms can be defined and executed, and their properties can be formally specified and proved with the assistance of a mechanical theorem prover. ACL2 has elements of both interactive and automated provers. ACL2 is automatic in the sense that once started on a problem, it proceeds without human assistance. However, non-trivial results are not usually proved in the first attempt, and the user has to lead the prover to a successful proof providing a set of lemmas, inspired by the failed proof generated by ACL2. This system has been used for a variety of important formal methods projects of industrial and commercial interest [15] and for implementing large proofs in mathematics.

2.3   Method

In this section, we present the method that we have applied to verify Fiji code. This process can be split into the following steps.

1. Transforming Fiji code into compilable Krakatoa code.
2. Specifying Java programs.
3. Applying the Why tool.
4. If all the proof obligations are discharged automatically by the provers integrated in Krakatoa, stop; the verification has ended.
5. Otherwise, study the failed attempts, and consider if they are under-specified; if it is the case, go again to step (2).
6. Otherwise, consider the Coq expressions of the still-non-proven statements and transform them to ACL2.
7. If all the statements are automatically proved in ACL2, stop; the verification has ended.
8. Otherwise, by inspecting the failed ACL2 proofs, decide if other specifications are needed (go to item (2)); if it is not the case, decide if the missing proofs should be carried out in Coq or ACL2.

The first step is the most sensitive one, because it is the only point where informal (or, rather, semi-formal) methods are needed. Thus, some unsafe, and manual, code transformation can be required. To minimize this drawback, we apply two strategies:

- First, only well-known transformations are applied; for instance, we eliminate inheritance by "flattening" out the code, but without touching the real behaviour of methods.
- Second, the equivalence between the original code and the transformed one is systematically tested.

Both points together increase the reliability of our approach; a more detailed description of the transformations needed in step (1) are explained in Section 3. Step (2) is quite well-understood, and some remarks about this step are provided in Section 4. Steps (3)-(6) are mechanized in Krakatoa. The role of ACL2 (steps (6)-(8)) is explained in Section 5 and, by means of an example, in Section 6.

3   Transforming Fiji-Java to Krakatoa-Java

In its current state, the Why/Krakatoa system does not support the complete Java programming language and has some limitations. In order to make a Fiji Java program compilable by Krakatoa we have to take several steps.

1. Delete annotations. Krakatoa JML annotations will be placed between *@ and @*\. Therefore, we need to remove other Java Annotations preceded by @.
2. Move the classes that are referenced in the file that we want to compile into the directory *whyInstallationDir/java_api/*. For example, the class *RankFilters* uses the class *java.awt.Rectangle*; therefore, we need to create the folder *awt* inside the *java* directory that already exists, and put the file *Rectangle.java* into it. Moreover, we can remove the body of the methods because only the headers and the fields of the classes will be taken into consideration. We must iterate this process over the classes that we add. The files that we add into the *java_api* directory can contain import, extends and implements clauses although the file that we want to compile cannot do it – Krakatoa does not support these mechanisms. This is a tough process: for instance, to make use of the class *Rectangle*, we need to add fifteen classes.
3. Reproduce the behaviour of the class that we want to compile. Considering that we are not able to use extends and implements clauses, we need to move the code from the upper classes into the one that we want to compile in order to have the same behaviour. For instance, the class *BinaryProcessor* extends from *ByteProcessor* and inside its constructor it calls the constructor of *ByteProcessor*; to solve this problem we need to copy the body of the super constructor at the beginning of the constructor of the class *BinaryProcessor*. If we find the use of interfaces, we can ignore them and remove the implements clause because the code will be implemented in the class that makes use of the interface.
4. Remove import clauses. We need to delete them from the file that we want to compile and change the places where the corresponding classes appear with the full path codes. If for example we are trying to use the class *Rectangle* as we have explained in Step 2, we need to replace it by *java.awt.Rectangle*.
5. Owing to package declarations are forbidden, we need to remove them with the purpose of halting *"unknown identifier packageName"* errors.
6. Rebuild native methods. The Java programming language allows the use of *native* methods, which are written in C or C++ and might be specific to a hardware and operating system platform. For example, many of the methods in the class *Math* (which perform basic numeric operations such as the elementary exponential, logarithm, square root, and trigonometric functions) simply call the equivalent method included in a different class named *StrictMath* for their implementation, and then the code in *StrictMath* of these methods is just a *native* call. Since native methods are not written in Java, they cannot be specified and verified in Krakatoa. Therefore, if our Fiji program uses some native methods, it will be necessary to rewrite them

with our own code. See in Section 6 our implementation (and specification) of the native method `sqrt` computing the square root of a number of type double, based on Newton's algorithm.

7. Add a clause in *if-else* structures in order to remove *"Uncaught exception: Invalid_argument("equal: abstract value")"*. We can find an example in the method `filterEdge` of the class *MedianFilter* where we have to replace the last *else...* clause by *else if(true)....*

8. Remove debugging useless references. We have mentioned in a previous step that we can only use certain static methods that we have manually added to the Why core code and therefore we can remove some debugging instructions like `System.out.println(...)`. We can find the usage of standard output printing statement in the method `write` of the class *IJ*.

9. Modify the declaration of some variables to avoid syntax errors. There can be some compilation errors with the definition of some floats and double values that match the pattern *<number>f* or *<number>d*. We can see an example in the line 180 of the file *RankFilters.java*; we have to transform the code: `float f = 50f;` into `float f = 50`.

10. Change the way that Maximum and Minimum float numbers are written. Those two special numbers are located in the file *Float.java* and there are widely used to avoid overflow errors, but they generate an error due to the *eP* exponent. To stop having errors with expressions like `0x1.fffffeP+127d` we need to convert it into `3.4028235e+38f`.

4   Specifying Programs for Digital Imaging

As already said in Section 2.2, Fiji and ImageJ are open source projects and many different people from many different teams (some of them not being computer scientists) are involved in the development of the different Fiji Java plug-ins. This implies that the code of these programs is in general not suitable for its formal verification and a deep previous transformation process, following the steps explained in Section 3, is necessary before introducing the Java programs into the Why/Krakatoa system. Even after this initial transformation, Fiji programs usually remain complex and their specification in Krakatoa is not a direct process. In this section we present some examples of Fiji methods that we have specified in JML trying to show the difficulties we have faced.

Once that a Fiji Java program has been adapted, following the ideas of Section 3, and is accepted by the Why/Krakatoa application, the following step in order to certify its correctness consists in specifying its behaviour (that is, its precondition and its postcondition) by writing annotations in the Java Modelling Language (JML) [7] . The precondition of a method must be a proposition introduced by the keyword `requires` which is supposed to hold in the pre-state, that is, when the method is called. The postcondition is introduced by the keyword `ensures`, and must be satisfied in the post-state, that is, when the method returns normally. The notation `\result` denotes the returned value. To differentiate the value of a variable in the pre- and post- states, we can use the keyword `\old` for the pre-state.

Let us begin by showing a simple example. The following Fiji method, included in the class *Rectangle*, translates an object by given horizontal and vertical increments dx and dy.

```
/*@ ensures x == \old(x) + dx && y == \old(y) + dy;
  @*/
public void translate(final double dx, final double dy) {
    this.x += dx; this.y += dy;
}
```

The postcondition expresses that the field x is modified by incrementing it by dx, and the field y is increased by dy. In this case no precondition is given since all values of dx and dy are valid, and the keyword \result does not appear because the returned type is void.

Using this JML specification, Why/Krakatoa generates several lemmas (*Proof Obligations*) which express the correctness of the program. In this simple case, the proof obligations are elementary and they can be easily discharged by the automated theorem provers Alt-Ergo [5] and CVC3 [3], which are included in the environment. The proofs of these lemmas guarantee the correctness of the Fiji method translate with respect to the given specification.

Unfortunately, this is not the general situation because, as already said, Fiji code has not been designed for its formal verification and can be very complicated; so, in most cases, Krakatoa is not able to prove the validity of a program from the given precondition and postcondition. In order to formally verify a Fiji method, it is usually necessary to include annotations in the intermediate points of the program. These annotations, introduced by the keyword assert, must hold at the corresponding program point. For loop constructs (while, for, etc), we must give an *inductive invariant*, introduced by the keyword loop_invariant, which is a proposition which must hold at the loop entry and be preserved by any iteration of the loop body. One can also indicate a loop_variant, which must be an expression of type integer, which remains non-negative and decreases at each loop iteration, assuring in this way the termination of the loop. It is also possible to declare new logical functions, lemmas and predicates, and to define *ghost variables* which allow one to monitor the program execution.

Let us consider the following Fiji method included in the class *RankFilters*. It implements Hoare's find algorithm (also known as *quickselect*) for computing the nth lowest number in part of an unsorted array, generalizing in this way the computation of the median element. This method appears in the implementation of the"median filter", a process very common in digital imaging which is used in order to achieve greater homogeneity in an image and provide continuity, obtaining in this way a good binarization of the image.

```
/*@ requires buf!=null && 1<= bufLength <= buf.length && 0<=n <bufLength;
  @ ensures Permut{Old,Here}(buf,0,bufLength-1)
  @     && (\forall integer k; (0<=k<=n-1 ==> buf[k]<=buf[n])
  @               && (n+1<=k<=bufLength-1 ==> buf[k]>=buf[n]))
  @     && \result==buf[n] ;
```

```
@*/
public final static float findNthLowestNumber
                (float[] buf, int bufLength, int n) {
    int i,j;
    int l=0;
    int m=bufLength-1;
    float med=buf[n];
    float dum ;
    while (l<m) {
        i=l ;
        j=m ;
        do {
            while (buf[i]<med) i++ ;
            while (med<buf[j]) j-- ;
            dum=buf[j];
            buf[j]=buf[i];
            buf[i]=dum;
            i++ ; j-- ;
        } while ((j>=n) && (i<=n)) ;
        if (j<n) l=i ;
        if (n<i) m=j ;
        med=buf[n] ;
    }
    return med ;
}
```

Given an array `buf` and two integers `bufLength` and `n`, the Fiji method `findNthLowestNumber` returns the $(n+1)$-th lowest number in the first `bufLength` components of `buf`. The precondition expresses that `buf` is not null, `bufLength` must be an integer between 1 and the length of `buf`, and `n` is an integer between 0 and `bufLength` -1. The definition of the postcondition includes the use of the predicate `Permut`, a predefined predicate, which expresses that when the method returns the (modified) `bufLength` first components of the array `buf` must be a permutation of the initial ones. The array has been reordered such that the components $0, \ldots, n-1$ are smaller than or equal to the component n, and the elements at positions $n+1, \ldots,$ `bufLength` -1 are greater than or equal to that in n. The returned value must be equal to `buf[n]`, which is therefore the $(n+1)$-th lowest number in the first `bufLength` components of `buf`.

In order to prove the correctness of this program, we have included different JML annotations in the Java code. First of all, loop invariants must be given for all `while` and `do` structures appearing in the code. Difficulties have been found in order to deduce the adequate properties for invariants which must be strong enough to imply the program (and other loops) postconditions; automated techniques like discovery of loop invariants [18] will be used in the future. We show as an example the loop invariant (and variant) for the exterior while, which is given by the following properties:

```
/*@ loop_invariant
  @ 0<=1<=n+1 && n-1<=m<=bufLength-1 && 1<=m+2
  @ && (\forall integer k1 k2; (0<=k1<=n && m+1<=k2<=bufLength-1)
  @          ==> buf[k1]<=buf[k2])
  @ && (\forall integer k1 k2; (0<=k1<=1-1 && n<=k2<=bufLength-1)
  @          ==> buf[k1]<=buf[k2])
  @ && Permut{Pre,Here}(buf,0,buf.length-1) && med==buf[n]
  @ && ((1<m)==> ((1<=n)&&(m>=n)));
  @ loop_variant m - 1+2;
  @*/
```

To help the automated provers to verify the program and prove the generated proof obligations it is also necessary to introduce several assertions in some intermediate points of the program and to use ghost variables which allow the system to deduce that the loop variant decreases.

Our final specification of this method includes 78 lines of JML annotations (for only 24 Java code lines). Krakatoa/Why produces 175 proof obligations expressing the validity of the program. The automated theorem prover Alt-Ergo is able to demonstrate all of them, although in some cases more than a minute (in an ordinary computer) is needed; another prover included in Krakatoa, CVC3, is, on the contrary, only capable of proving 171. The proofs of the lemmas obtained by means of Alt-Ergo certify the correctness of the method with respect to the given specification.

In this particular example, the automated theorem provers integrated in Krakatoa are enough to discharge all the proof obligations. In other cases, some properties are not proven, and then one should try to prove them using *interactive* theorem provers, as Coq. In this architecture, we also introduce the ACL2 theorem prover, as explained in the next section.

5   The Role of ACL2

In this section, we present the role played by ACL2 in our infrastructure to verify the correctness of Java programs. The Why platform relies on automated provers, such as Alt-Ergo or CVC3, and interactive provers, such as Coq or PVS, to discharge proof obligations; however, it does not consider the ACL2 theorem prover to that aim. We believe that the use of ACL2 can help in the proof verification process. The reason is twofold.

 – The scope of automated provers is smaller than the one of ACL2; therefore, ACL2 can prove some of the proof obligations which cannot be discharged by automated provers.
 – Moreover, interactive provers lack automation; then, ACL2 can automatically discharge proof obligations which would require user interaction in interactive provers.

We have developed *Coq2ACL2*, a Proof General extension, which integrates ACL2 in our infrastructure to verify Java programs; in particular, we work with

ACL2(r) a variant of ACL2 which supports the real numbers [12] – the formalisation of real analysis in theorem provers is an outstanding topic, see [6]. Coq2ACL2 features three main functions:

F1. it transforms Coq statements generated by Why to ACL2;
F2. it automatically sends the ACL2 statements to ACL2; and
F3. it displays the proof attempt generated by ACL2.

If all the statements are proved in ACL2; then, the verification process is ended. Otherwise, the statements must be manually proved either in Coq or ACL2.

The major challenge in the development of Coq2ACL2 was the transformation of Coq statements to ACL2. There is a considerable number of proposals documented in the literature related to the area of theorem proving interoperability. We have not enough space here to do a thorough review, but we can classify the translations between proof assistants in two groups: *deep* [8, 14, 19] and *shallow* [10, 21, 25].

In our work, we took advantage of a previous shallow development presented in [2], where a framework called *I2EA* to import Isabelle/HOL theories into ACL2 was introduced. The approach followed in [2] can be summarized as follows. Due to the different nature of Isabelle/HOL and ACL2, it is not feasible to replay proofs that have been recorded in Isabelle/HOL within ACL2. Nevertheless, Isabelle/HOL statements dealing with first order expressions can be transformed to ACL2; and then, they can be used as a schema to guide the proof in ACL2.

A key component in the framework presented in [2] was an XML-based specification language called *XLL* (that stands for Xmall Logical Language). XLL was developed to act as an intermediate language to port Isabelle/HOL theories to both ACL2 and an Ecore model (given by UML class definitions and OCL restrictions) – the translation to Ecore serves as a general purpose formal specification of the theory carried out. The transformations among the different languages are done by means of XSLT and some Java programs. We have integrated the Coq system into the I2EA framework as can be seen in Figure 1; in this way, we can reuse both the XLL language and some of the XSLT files developed in [2] to transform (first-order like) Coq statements to ACL2.

In particular, functionality **F1** of Coq2ACL2 can be split into two steps:

1. given a Coq statement, Coq2ACL2 transforms it to an XLL file using a Common Lisp translator program; then,
2. the XLL file is transformed to ACL2 using an XSLT file previously developed in [2].

In this way, ACL2 has been integrated into our environment to verify Java programs. As we will see in the following section, this has meant an improvement to automatically discharge proof obligations.

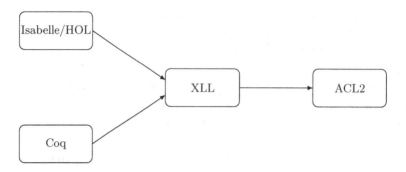

Fig. 1. (Reduced) Architecture of the I2EA framework integrating Coq

6   The Method in Action: A Complete Example

In our work, we deal with images acquired by microscopy techniques from biological samples. These samples have volume and the object of interest is not always in the same plane. For this reason, it is necessary to obtain different planes from the same sample to get more information. This means that several images are acquired in the same XY plane at different levels of Z. To work with this stack of images, it is often necessary to make their *maximum projection*. To this aim, Fiji provides several methods such as maximum intensity or standard deviation to obtain the maximum projection of a set of images.

In this section, we consider the Fiji code for computing the maximum projection of a set of images based on the standard deviation, which uses in particular the method `calculateStdDev` located in the class *ImageStatistics*.

```
double calculateStdDev(double n, double sum, double sum2) {
    double stdDev = 0.0;
    if (n>0.0) {
        stdDev = (n*sum2-sum*sum)/n;
        if (stdDev>0.0)
            stdDev = Math.sqrt(stdDev/(n-1.0));
        else
            stdDev = 0.0;
    } else
    stdDev = 0.0;
}
```

The inputs are **n** (the number of data to be considered), `sum` (the sum of all considered values; in our case, these values will obtained from the pixels in an image) and `sum2` (the sum of the squares of the data values). The method `calculateStdDev` computes the standard deviation from these inputs and assigns it to the field `stdDev`. The specification of this method is given by the following JML annotation.

```
/*@ requires ((n==1.0)==> sum2==sum*sum) && ((n<=0.0) || (n>=1.0)) ;
  @ behaviour negative_n :
  @    assumes   n<=0.0 || (n>0.0 && (n*sum2-sum*sum)/n <=0.0);
  @    ensures stdDev == 0.0;
  @ behaviour normal_behaviour :
  @    assumes n>=1.0 && ((n*sum2-sum*sum)/n  > 0.0);
  @    ensures is_sqrt(stdDev,(double)((n*sum2-sum*sum)/n/(n-1.0)));
@*/
```

The precondition, introduced by the keyword **requires**, expresses that in the case $n = 1$ (that is, there is only one element in the data) the inputs **sum** and **sum2** must satisfy $sum2 = sum * sum$. Moreover we must require that n is less than or equal to 0 or greater than or equal to 1 to avoid the possible values in the interval $(0, 1)$; for **n** in this interval one has $n - 1 < 0$ and then it is not possible to apply the square root function to the given argument $stdDev/(n - 1.0)$. This fact has not been taken into account by the author of the Fiji program because in all real applications the method will be called with n being a natural number; however, to formalise the method we must specify this particular situation in the precondition. For the postcondition we distinguish two different behaviours: if **n** is non-positive or **sum** and **sum2** are such that $n * sum2 - sum * sum < 0$, the field **stdDev** is assigned to 0; otherwise, the standard deviation formula is applied and the result is assigned to the field **stdDev**. The predicate **is_sqrt** is previously defined.

For the proof of correctness of the method **calculateStdDev** in Krakatoa, it is necessary to specify (and verify) the method **sqrt**. The problem here, as already explained in Section 3, is that the method **sqrt** of the class *Math* simply calls the equivalent method in the class *StrictMath*, and the code in *StrictMath* of the method **sqrt** is just a native call and might be implemented differently on different Java platforms. In order to give a JML specification of the method **sqrt** is necessary then to rewrite it with our own code. The documentation of *StrictMath* states *"To help ensure portability of Java programs, the definitions of some of the numeric functions in this package require that they produce the same results as certain published algorithms. These algorithms are available from the well-known network library netlib as the package "Freely Distributable Math Library", fdlibm"*. In the case of the square root, one of these *recommended* algorithms is Newton's method; based on it, we have implemented and specified in JML the computation of the square root of a given (non-negative) input of type double.

```
/*@ requires c>=0 && epsi > 0 ;
  @ ensures \result >=0 &&  (\result*\result>=c)
  @    &&   \result*\result - c < epsi ;
  @*/
public double sqrt(double c, double epsi){
    double t;
    if (c>1) t= c;
        else t=1.1;
    /*@ loop_invariant
```

```
    @ (t >= 0) && (t*t> c) ;
    @*/
    while (t* t - c  >= epsi) {
        t = (c/t + t) / 2.0;
    }
    return t;
}

/*@ requires c>=0   ;
  @ ensures (\result >=0) && (\result*\result>=c)
  @ && (\result*\result - c < 1.2E-7);
  @*/
public double sqrt(double c){
    double eps=1.2E-7;
    return sqrt(c,eps);
}
```

The first method computes the square root of a double x with a given precision epsi; the second one calls the previous method with a precision less than $1.2E-7$. Using JUnit, we have run one million tests between $1E9$ and $1E-9$ to show that the results of our method sqrt have similar precision to those obtained by the *original* method Math.sqrt. Here, we applied the "first test, then verify" approach – intensive testing can be really useful to find bugs (and can save us time) before starting the verification process.

From the given JML specification for the Fiji method calculateStdDev and our sqrt method, Why/Krakatoa produces 52 proof obligations, 9 of them corresponding to lemmas that we have introduced and which are used in order to prove the correctness of the programs. Alt-Ergo is able to prove 50 of these proof obligations, but two of the lemmas that we have defined remain unsolved. CVC3 on the contrary only proves 44 proof obligations.

The two lemmas that Alt-Ergo (and CVC3) are not able to prove are the following ones:

```
/*@ lemma double_div_pos :
  @ \forall double x y; x>0 && y > 0 ==> x / y > 0;
  @*/
/*@ lemma double_div_zero :
  @ \forall double x y; x==0.0 && y > 0 ==> x / y == 0.0;
  @*/
```

In order to discharge these two proof obligations, we can manually prove their associated Coq expressions.

```
Lemma double_div_zero : (forall (x_0_0:R), (forall (y_0:R),
    ((eq x_0_0 (0)%R) /\ (Rgt y_0 (0)%R) -> (eq (Rdiv x_0_0 y_0) (0)%R)))).

Lemma double_div_pos : (forall (x_13:R), (forall (y:R),
    ((Rgt x_13 (0)%R) /\ (Rgt y (0)%R) -> (Rgt (Rdiv x_13 y) (0)%R)))).
```

Both lemmas can be proven in Coq in less than 4 lines, but, of course, it is necessary some experience working with Coq. Therefore, it makes sense to delegate those proofs to ACL2. Coq2ACL2 translates the Coq lemmas to the following ACL2 ones. ACL2 can prove both lemmas without any user interaction (a screenshot of the proof of one of this lemmas in ACL2 is shown in Figure 2).

```
(defthm double_div_zero
  (implies (and (realp x_0_0) (realp y_0) (and (equal x_0_0 0) (> y_0 0)))
           (equal (/ x_0_0 y_0) 0)))


(defthm double_div_pos
  (implies (and (realp x_13) (realp y) (and (> x_13 0) (> y 0)))
           (> (/ x_13 y) 0)))
```

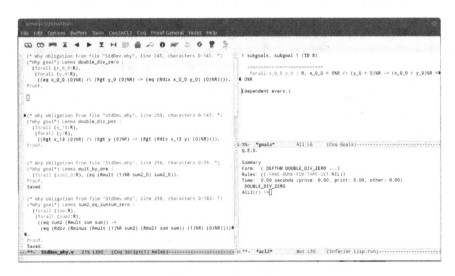

Fig. 2. Proof General with Coq2ACL2 extension. The Coq2ACL2 extension consists of the Coq2ACL2 menu and the right-most button of the toolbar. Left: the Coq file generated by the Why tool. Top Right: current state of the Coq proof. Bottom Right: ACL2 proof of the lemma.

7   Conclusions and Further Work

This paper reports an experience to verify actual Java code, as generated by different-skilled programmers, in a multi-programmer tool called Fiji. As one could suspect, the task is challenging and, in some sense, the objectives are impossible to accomplish, at least in their full extent – after our experiments, we have found that the Fiji system is *unsound*, but the errors are minor (e.g. a variable declared as a real number but which should be declared as an integer) and can be easily corrected.

Nevertheless, we defend the interest of this kind of experimental work. It is useful to evaluate the degree of maturity of the verification tools (Krakatoa, in our case). In addition, by a careful examination of the code really needed for a concrete application, it is possible to isolate the relevant parts of the code, and then it is possible to achieve a complete formalisation. Several examples in our text showed this feature, see Section 4.

In addition to Krakatoa, several theorem provers (Coq and ACL2) have been used to discharge some proof obligations that were not automatically proved by Krakatoa. To this aim, it has been necessary the integration of several tools, and our approach can be considered as semi-formal: we keep transformations as simple as possible, and substantiate the process by systematic testing.

As a further interest of our work, we have reused a previous interoperability-proposal [2], between Isabelle and ACL2, to get an integration of ACL2 (through a partial mapping from Coq to ACL2), without touching the Krakatoa kernel.

Future work includes several improvements in our method. Starting from the beginning, the transformation from real Java code to Krakatoa one could be automated (Section 3 can be understood as a list of requirements to this aim). Then, a formal study of this transformation could be undertaken to increase the reliability of our method. In addition, we can try to automatically reconstruct ACL2 proofs in Coq.

As for applications, more verification is needed to obtain a certified version of, for instance, the SynapCountJ plug-in [24]. The preliminary results presented in this paper allow us to be reasonably optimistic with respect to the feasibility of this objective.

References

1. ForMath: Formalisation of Mathematics, European Project,
   http://wiki.portal.chalmers.se/cse/pmwiki.php/ForMath/ForMath
2. Aransay, J., et al.: A report on an experiment in porting formal theories from Isabelle/HOL to Ecore and ACL2. Technical report (2012),
   http://wiki.portal.chalmers.se/cse/uploads/
   ForMath/isabelle_acl2_report
3. Barrett, C., Tinelli, C.: CVC3. In: Damm, W., Hermanns, H. (eds.) CAV 2007. LNCS, vol. 4590, pp. 298–302. Springer, Heidelberg (2007)
4. Barthe, G., Pointcheval, D., Zanella-Béguelin, S.: Verified Security of Redundancy-Free Encryption from Rabin and RSA. In: Proceedings 19th ACM Conference on Computer and Communications Security (CCS 2012), pp. 724–735 (2012)
5. Bobot, F., Conchon, S., Contejean, E., Iguernelala, M., Lescuyer, S., Mebsout, A.: The Alt-Ergo automated theorem prover (2008), http://alt-ergo.lri.fr/
6. Boldo, S., Lelay, C., Melquiond, G.: Formalization of Real Analysis: A Survey of Proof Assistants and Libraries. Technical report (2013),
   http://hal.inria.fr/hal-00806920
7. Burdy, L., et al.: An overview of JML tools and applications. International Journal on Software Tools for Technology Transf. 7(3), 212–232 (2005)
8. Codescu, M., Horozal, F., Kohlhase, M., Mossakowski, T., Rabe, F., Sojakova, K.: Towards Logical Frameworks in the Heterogeneous Tool Set Hets. In: Mossakowski, T., Kreowski, H.-J. (eds.) WADT 2010. LNCS, vol. 7137, pp. 139–159. Springer, Heidelberg (2012)

9. CoQ development team. The CoQ Proof Assistant, version 8.4. Technical report (2012), http://coq.inria.fr/

10. Denney, E.: A Prototype Proof Translator from HOL to Coq. In: Aagaard, M.D., Harrison, J. (eds.) TPHOLs 2000. LNCS, vol. 1869, pp. 108–125. Springer, Heidelberg (2000)

11. Filliâtre, J.-C., Marché, C.: The Why/Krakatoa/Caduceus Platform for Deductive Program verification. In: Damm, W., Hermanns, H. (eds.) CAV 2007. LNCS, vol. 4590, pp. 173–177. Springer, Heidelberg (2007)

12. Gamboa, R., Kaufmann, M.: Non-Standard Analysis in ACL2. Journal of Automated Reasoning 27(4), 323–351 (2001)

13. Gonthier, G., et al.: A Machine-Checked Proof of the Odd Order Theorem. In: Proceedings 4th Conference on Interactive Theorem Proving (ITP 2013). LNCS (2013)

14. Gordon, M.J.C., Kaufmann, M., Ray, S.: The Right Tools for the Job: Correctness of Cone of Influence Reduction Proved Using ACL2 and HOL4. Journal of Automated Reasoning 47(1), 1–16 (2011)

15. Hardin, D. (ed.): Design and Verification of Microprocessor Systems for High-Assurance Applications. Springer (2010)

16. Heras, J., Coquand, T., Mörtberg, A., Siles, V.: Computing Persistent Homology within Coq/SSReflect. To appear in ACM Transactions on Computational Logic (2013)

17. Heras, J., Poza, M., Rubio, J.: Verifying an Algorithm Computing Discrete Vector Fields for Digital Imaging. In: Jeuring, J., Campbell, J.A., Carette, J., Dos Reis, G., Sojka, P., Wenzel, M., Sorge, V. (eds.) CICM 2012. LNCS, vol. 7362, pp. 216–230. Springer, Heidelberg (2012)

18. Ireland, A., Stark, J.: On the automatic discovery of loop invariants (1997)

19. Jacquel, M., Berkani, K., Delahaye, D., Dubois, C.: Verifying B Proof Rules Using Deep Embedding and Automated Theorem Proving. In: Barthe, G., Pardo, A., Schneider, G. (eds.) SEFM 2011. LNCS, vol. 7041, pp. 253–268. Springer, Heidelberg (2011)

20. Kaufmann, M., Moore, J.S.: ACL2 version 6.0 (2012), http://www.cs.utexas.edu/users/moore/acl2/

21. Keller, C., Werner, B.: Importing HOL Light into Coq. In: Kaufmann, M., Paulson, L.C. (eds.) ITP 2010. LNCS, vol. 6172, pp. 307–322. Springer, Heidelberg (2010)

22. Liu, H., Moore, J.S.: Java Program Verification via a JVM Deep Embedding in ACL2. In: Slind, K., Bunker, A., Gopalakrishnan, G.C. (eds.) TPHOLs 2004. LNCS, vol. 3223, pp. 184–200. Springer, Heidelberg (2004)

23. Mata, G.: NeuronPersistentJ, http://imagejdocu.tudor.lu/doku.php?id=plugin:utilities:neuronpersistentj:start

24. Mata, G.: SynapCountJ, http://imagejdocu.tudor.lu/doku.php?id=plugin:utilities:synapsescountj:start

25. Obua, S., Skalberg, S.: Importing HOL into isabelle/HOL. In: Furbach, U., Shankar, N. (eds.) IJCAR 2006. LNCS (LNAI), vol. 4130, pp. 298–302. Springer, Heidelberg (2006)

26. Rasband, W.S.: ImageJ: Image Processing and Analysis in Java. Technical report, U. S. National Institutes of Health, Bethesda, Maryland, USA (1997-2012)

27. Schindelin, J., et al.: Fiji: an open-source platform for biological-image analysis. Nature Methods 9(7), 676–682 (2012)

A Universal Machine for Biform Theory Graphs

Michael Kohlhase, Felix Mance, and Florian Rabe

Computer Science, Jacobs University Bremen
`initial.lastname@jacobs-university.de`

Abstract. Broadly speaking, there are two kinds of semantics-aware assistant systems for mathematics: proof assistants express the semantic in logic and emphasize deduction, and computer algebra systems express the semantics in programming languages and emphasize computation. Combining the complementary strengths of both approaches while mending their complementary weaknesses has been an important goal of the mechanized mathematics community for some time.

We pick up on the idea of biform theories and interpret it in the MMT/ OMDOC framework which introduced the foundations-as-theories approach, and can thus represent both logics and programming languages as theories. This yields a formal, modular framework of biform theory graphs which mixes specifications and implementations sharing the module system and typing information.

We present automated knowledge management work flows that interface to existing specification/programming tools and enable an OPENMATH Machine, that operationalizes biform theories, evaluating expressions by exhaustively applying the implementations of the respective operators. We evaluate the new biform framework by adding implementations to the OPENMATH standard content dictionaries.

1 Introduction

It is well-known that mathematical practices – conjecturing, formalization, proving, etc. – combine (among others) axiomatic reasoning with computation. Nevertheless, assistant systems for the semantics-aware automation of mathematics can be roughly divided into two groups: those that use *logical languages* to express the semantics and focus on *deduction* (commonly called **proof assistants**), and those that use *programming languages* to express the semantics and focus on *computation* (commonly called **computer algebra systems**). Combining their strengths is an important objective in mechanized mathematics.

Our work is motivated by two central observations. Firstly, combination approaches often take a deduction or computation system and try to embed the respective other mode into its operations, e.g., [HT98,DM05] and [HPRR10], respectively. Secondly, most of these systems are usually based on the *homogeneous* method, which fixes one foundation (computational or deductive) with all primitive notions (e.g., types, axioms, or programming primitives) and uses only conservative extensions (e.g., definitions, theorems, or procedures) to model domain objects.

J. Carette et al. (Eds.): CICM 2013, LNAI 7961, pp. 82–97, 2013.

In this paper, we want to employ the *heterogeneous* method, which focuses on encapsulating primitive notions in theories and considers truth relative to a theory. It optimizes reusability by stating every result in the weakest possible theory and using *theory morphisms* to move results between theories in a truth-preserving way. This is often called the *little theories* approach [FGT92]. In computational systems this is mirrored by using programming languages that relegate much of the functionality to an extensible library infrastructure.

In homogeneous approaches, we usually fix a specification language SL and a programming language PL and one implementation for each. In program synthesis, a specification Spec is extended (hooked arrows) to a refined specification Ref, from which a program can be extracted (snaked arrow). Both can be visualized by the diagram on the right where dotted arrows denote the written-in relation. In both cases, the proofs are carried out in a theory of SL, and a non-trivial generation step crosses the

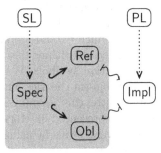

border between the SL-based deduction system (the gray area) and the PL-based computation system, e.g., [HN10] generates programs from Isabelle/HOL proofs.

Dually, we find approaches that emphasize PL over SL. SML-style module systems and object-orientation can be seen as languages that transform parts of SL (namely the type system but not the entailment system) into PL. An example is the transformation of SL=UML diagrams into PL=Java stubs, which are then refined to a Java program. Advanced approaches can transform the whole specification into PL by enriching the programming language as in [KST97] or the programming environment as in [ABB+05].

A third approach is to develop a language SPL that is strong enough to combine the features of specification and programming. Several implementations of λ calculi have been extended with features of programming languages, e.g., Coq [The11] and Isabelle/HOL [NPW02]. The FoCaLiZe language [H+12] systematically combines a functional and object-oriented programming language with a logic, and a compilation process separates the two aspects by producing OCaml and Coq files, respectively. The source files may also contain raw Coq snippets that are not verified by FoCaLiZe but passed on to Coq. In Dedukti [BCH12], rewriting rules are used to enhance a specification language with computational behavior, and the computational aspect can be compiled into a Lua program.

We want to create a heterogeneous framework in which we can represent such homogeneous approaches. We use the MMT language [RK13], which extends the heterogeneous method with language-independence inspired by logical frameworks. The key advantage is that this permit flexibly combining arbitrary specification and programming languages. In MMT, we represent both PL and SL as MMT theories $\overline{\text{SL}}$ and $\overline{\text{PL}}$ (see diagram below), which declare the primitive concepts of the respective language. The dotted lines are represented explicitly using the *meta-theory* relation, and relatively simple mappings (dashed snaked lines) transform between specifications Spec and implementations Impl and the corresponding MMT theories $\overline{\text{Spec}}$ and $\overline{\text{Impl}}$.

Typically SL and PL share some language features, e.g., the type system, which SL enriches with deductive primitives and PL with computational primitives. MMT can represent this by giving a (possibly partial) morphism bifound : $\overline{\text{SL}} \rightarrow \overline{\text{PL}}$ that embeds SL features into PL. Via bifound, $\overline{\text{Impl}}$ can access both $\overline{\text{SL}}$ and $\overline{\text{PL}}$ features, and the fact that $\overline{\text{Impl}}$ implements $\overline{\text{Spec}}$ is represented as an MMT theory morphism (dashed line).

Our framework is inspired by the *biform theories* of [Far07], which extend axiomatic theories with *transformers*: named algorithms that implement axiomatically specified function symbols. We follow the intuition of heterogeneous biform theories but interpret them in MMT. Most importantly, this permits SL and PL to be arbitrary languages represented in MMT.

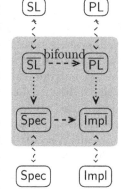

We leverage this by using types and examples in $\overline{\text{Spec}}$ to generate method stubs and test cases in $\overline{\text{Impl}}$. Our interest is not (yet) the corresponding treatment of axioms, which would add formal deduction about the correctness of programs. In particular, we do not provide a formal definition of the meaning of the computational knowledge other than linking symbols to algorithms via theory morphisms.

As a computational backend, we develop what we call the *universal machine*. It extends MMT with a component that collects the individual implementation snippets occurring in a biform MMT theory graph and integrates them into a rule-based rewriting engine. The universal machine keeps track of these and performs computations by applying the available rules.

In the past, a major practical limitation of frameworks like ours has been the development of large libraries of biform theories. Here a central contribution of our work is that the MMT API [Rab13b] (providing, e.g., notations, module system, and build processes) makes it easy to write biform theories in practice. Moreover, the API is designed to make integration with other applications easy so that the universal machine can be easily reused by other systems.

We evaluate this infrastructure in an extensive case study: We represent a collection of OPENMATH content dictionaries in MMT (i.e., SL = OpenMath) and provide implementations for the symbols declared in them using the programming language Scala (i.e., PL = Scala). The resulting biform theory graph integrates OPENMATH CDs with the Scala code snippets implementing the symbols.

2   Representing Languages in MMT

In this section we introduce the MMT language and directly apply it to modeling the pieces of our running example by representing OPENMATH and Scala in MMT.

2.1   The MMT Language and System

MMT [RK13] is a knowledge representation format focusing on modularity and logic-independence. It is accompanied by the MMT API, which implements the language focusing on scalable knowledge management and system interoperability. For our purposes, the simplified fragment of MMT (which in particular omits named imports and sharing between imports) given in Figure 1 suffices.

Module   ::= theory T : ModuleName Statement*
        | view V : ModuleName → ModuleName Statement*
Statement ::= constant c [: Term] [= Term] [#Notation]
        | include ModuleName
Term   ::= c | x | number | OMA(Term$^+$) | OMBIND(Term; x^+; Term)
Notation ::= (number[string...] | string)*

Fig. 1. A Fragment of the MMT Grammar

We will briefly explain the intuitions behind the concepts and then exemplify them in the later sections, where we represent OPENMATH CDs and Scala classes as MMT theories.

An MMT *theory* theory T : M Σ defines a theory T with meta-theory M consisting of the statements in Σ. The *meta-theory* relation between theories is crucial to obtain logic-independence: The meta-theory gives the language, in which the theory is written. For example, the meta-theory of a specification is the specification logic, and the meta-theory of a program is the programming language – and the logic and the programming language are represented as MMT theories themselves (possibly with further meta-theories). Thus, MMT achieves a uniform representation of logics and programming languages as well as their theories and programs.

MMT theories form a category, and an MMT view V : T_1 → T_2 Σ defines a theory morphism V from T_1 to T_2 consisting of the statements in Σ. In such a view, Σ may use the constants declared in T_2 and must declare one definition for every definition-less constant declared in T_1. Views uniformly capture the relations "T_2 interprets/implements/models T_1". For example, if T_1 represents a specification and T_2 a programming language, then views T_1 → T_2 represent implementations of T_1 in terms of T_2 (via the definitions in Σ).

Theories and views are subject to the MMT module system. Here we will restrict attention to the simplest possible case of unnamed inclusions between modules: If a module T contains a statement include S, then all declarations of S are included into T.

Within modules, MMT uses constants to represent atomic named declarations. A constant's optional type and definiens are arbitrary terms. Due to the freedom of using special constants declared in the meta-theory, a type and a definiens are sufficient to uniformly represent diverse statements of formal languages such as function symbols, examples, axioms, inference rules. Moreover, constants have an optional notation which is used by MMT to parse and render objects. We will not go into details and instead explain notations by example, when we use them later on.

MMT terms are essentially the OPENMATH objects [BCC+04] formed from the constants included into the theory under consideration. This is expressive enough to subsume the abstract syntax of a wide variety of formal systems. We will only consider the fragment of MMT terms formed from constants c, variables x, numbers literals, applications $OMA(f, t_1, \ldots, t_n)$ of f to the t_i, and bindings $OMBIND(b; x_1, \ldots, x_n; t)$ where a binder b binds the variables x_i in the scope t.

2.2   Content Dictionaries as MMT Theories

OPENMATH declares symbols in named content dictionaries that have global scope (unlike MMT theories where symbols must be imported explicitly). Consequently, references to symbols must reference the CD and the symbol name within that CD. The official OPENMATH CDs [OMC] are a collection of content dictionaries for basic mathematics. For example, the content dictionary arith1 declares among others the symbols plus, minus, times, and divide for arithmetic in any mathematical structure – e.g., a commutative group or a field – that supports it.

Each symbol has a type using the STS type system [Dav00]. The types describe what kinds of application (rarely: binding) objects can be formed using the symbol. For example, its type licenses the application of plus to any sequence of arguments, which should come from a commutative semigroup. Moreover, each symbol comes with a textual description of the meaning of the thus-constructed application, and sometimes axioms about it, e.g., commutativity in the case of plus.

We represent every OPENMATH CD as an MMT theory, whose meta-theory is a special MMT theory OpenMath. Moreover, every OPENMATH symbol is represented as an MMT constant. All

OPENMATH	MMT
CD	theory
symbol	constant
property F	constant $OMA(FMP, F)$

constants are definition-less, and it remains to describe their types and notations. Mathematical properties that are given as formulas are also represented as MMT constants using a special type.

Meta-Theory and Type System. OpenMath must declare all those symbols that are used to form the types of OPENMATH symbols. This amounts to a formalization of the STS type system [Dav00] employed in the OPENMATH CDs. However, because the details STS are not obvious and not fully specified, we identify the strongest type system that we know how to formalize and of which STS is a weakening. Here strong/weak means that the typing relation holds rarely/often, i.e., every STS typing relation also holds in our weakened version. The types in this system are: *i)* Object *ii)* $OMA(mapsto, Object, \ldots, Object, A, Object)$ where A

is either Object or naryObject *iii)* binder. Here binder is the type of symbols that take a context C and an Object in context C and return an Object. This type system ends up being relatively simple and is essentially an arity-system.[1]

Moreover, we add a special symbol FMP to represent mathematical properties as follows: A property asserting F is represented as a constant with definiens $\mathrm{OMA}(\mathsf{FMP}, F)$.[2] Intuitively, we can think of FMP as a partial function that can only be applied to true formulas. We do not need symbols for the formation of formulas F because they are treated as normal symbols that are introduced in CDs such as logic1.

```
theory OpenMath
    constant mapsto # 1×... → 2
    constant Object
    constant naryObject
    constant binder
    constant FMP
```

Fig. 2. MMT Theory OpenMath

This results in the following MMT theory OpenMath in Figure 2. There, the notation of mapsto means that it takes first a sequence or arguments with separator × followed by the separator → and one more argument.

```
theory arith1 : OpenMath              theory NumbersTest : OpenMath
    plus : naryObject → Object            include arith1
        # 1+...                           include fns1
    minus : Object × Object → Object      include set1
        # 1 − 2                           include relations1
    plus : naryObject → Object            maptest = FMP
        # 1*...                               {0,1,2} map (x ↦ −x∗x+2∗x+3) = {3,4}
    ...
```

Fig. 3. OPENMATH CDs in MMT

Notations. In order to write OPENMATH objects conveniently – in particular, to write the examples mentioned below – we add notations to all OPENMATH symbols. OPENMATH does not explicitly specify notations for the symbols in the official CDs. However, we can gather many implied notations from the stylesheets provide to generate presentation MATHML. Most of these can be mapped to MMT notations in a straightforward fashion. As MMT notations are one-dimensional, we make reasonable adjustments to two-dimensional MATHML notations such as those for matrices and fractions.

Example 1. We will use a small fragment of our case study (see Section 5) as a running example. The left listing in Fig. 3 gives a fragment of the MMT theory representing the CD arith1. Here the notation of plus means that it takes a sequence or arguments with separator +, and the one of minus that it takes two arguments separated by −.

---

[1] In fact, we are skeptical whether any fully formal type system for all of OPENMATH can be more than an arity system.

[2] We do not use a propositions-as-types representation here because it would make it harder to translate OpenMath to other languages.

The right listing uses the module system to import some CDs and then give an example of a true computation as an axiom. It uses the symbols set1?set, fns1?lambda, and relation1?eq and the notations we declare for them.

2.3   Scala Classes as MMT Theories

Scala [OSV07] combines features of object-oriented and functional programming languages. At the module and statement level, it follows the object-oriented paradigm and is similar to Java. At the expression level, it supplements Java-style imperative features with simple function types and inductive types.

A *class* is given by its list of member declarations, and we will only make use of 3 kinds of members: *types*, immutable typed *values*, and *methods*, which are essentially values of functional type.

Values have an optional definiens, and a class is *concrete* if all members have one, otherwise abstract. Scala introduces special concepts that can be used instead of classes without constructor arguments: *trait* in the abstract and *object* in the concrete case. Traits permit *multiple inheritance*, i.e., every class can inherit from multiple traits. Objects are singleton classes, i.e., they are at the same time a class and the only instance of this class. An object and a trait may have the same name, in which case their members correspond to the static and the non-static members, respectively, of a single Java class.

The representation of Scala classes proceeds very similarly to that of OPEN-MATH CDs above (see Figure 4). In particular, we use a special meta-theory Scala that declares the primitive concepts needed for our Scala expressions. Then we represent Scala classes as MMT theories and members as constants. While OPENMATH CDs always have the flavor of specifications, Scala classes can have the flavor of specifications (abstract classes/traits) or implementations (concrete classes/objects).

Scala	MMT
trait T	theory $\overline{T}$
type member	constant of type type
value member	constant
method member	constant of functional type
object O of type T	theory morphism $\overline{T} \to$ Scala
members of O	assignment to the corresponding $\overline{T}$-constant
extension between classes	inclusion between theories

Fig. 4. Scala Classes as MMT Theories

Meta-Theory and Type System. Our meta-theory Scala could declare symbols for every primitive concept used in Scala expressions. However, most of the complexity of Scala expressions stems from the richness of the term language. While the representation of terms would be very useful for verification systems, it does not contribute much to our goals of computation and biform development. Therefore, we focus on the simpler type language. Moreover, we omit many theoretically important technicalities (e.g., singleton and existential types) that have little practical bearing. Indeed, many practically relevant types (e.g., function and collection types) are derived notions defined in the Scala library.

Therefore, we represent only the relevant fragment of Scala in Scala. Adding further features later is easy using the MMT module system. For all inessential (sub-)expressions, we simply make use MMT escaping: MMT expressions can seamlessly escape into arbitrary non-MMT formats.

Thus, we use the MMT theory Scala in Figure 5, which gives mainly the important type operators and their introductory forms. Where applicable, we use MMT notations that mimic Scala's concrete syntax. This has the added benefit that the resulting theory is hardly Scala-specific and thus can be reused easily for other programming languages. It would be straightforward to add typing rules to this theory by using a logical framework as the meta-theory of Scala, but this is not essential here.

```
theory Scala
  constant type
  constant Any
  constant Function # (1,...)=> 2
  constant Lambda # (1,...)=> 2
  constant List # List[1]
  constant list # List(1,...)
  constant BigInt
  constant Double
  constant Boolean
  constant String
```

Fig. 5. The MMT Theory Scala

Representing Classes. It is now straightforward to represent a Scala trait T containing only 1. type members, 2. value members whose types only use symbols from Scala, 3. method members whose argument and return types only use symbols from Scala as an MMT theory $\overline{T}$ with meta-theory Scala.

1. type n yields constant n: type
2. val n: A yields constant n: $\overline{A}$
3. def n(x1:A_1,..,x_r:A_r):A yields constant n: $(\overline{A_1},...,\overline{A_r})$=>$\overline{A}$

Here $\overline{A}$ is the structural translation of the Scala type A into an MMT expression, which replaces every Scala primitive with the corresponding symbol in Scala.

Similarly, we represent every object O defining (exactly) the members of T as an MMT view $\overline{O} : \overline{T} \to$ Scala. The member definitions in O give rise to assignments in $\overline{O}$ as follows:

1. type n = t yields constant n = $\overline{t}$
2. val n: A = a yields constant n = $"a"$
3. def n(x_1:A_1,...,x_r:A_r):A = a yields constant n = $(x_1$:$\overline{A_1}$,...,x_r:$\overline{A_r}$):$\overline{A}$ = $"a"$

Here "E" represents the escaped representation of the literal Scala expression E. Note that we do not escape the λ-abstraction in the implementation of comp. The resulting partially escaped term is partially parsed and analyzed by MMT. This has the advantage that the back-translation from MMT to Scala can reuse the same variable names that the Scala programmer had chosen.

Example 2. A Scala class for monoids (with universe, unit, and composition) and an implementation in terms of the integers are given as the top two code fragments in Figure 6, their MMT representations in the lower two.

| ```
trait Monoid {
  type U
  val unit: U
  def comp(x1: U, x2: U): U
}
``` | ```
object Integers extends Monoid {
  type U = BigInt
  val unit = 0
  def comp(x1: U, x2: U) = x1 + x2
}
``` |
| ```
theory Monoid : Scala
  constant U : type
  constant unit : U
  constant comp : (U,U) => U
``` | ```
view Integers : Monoid -> Scala
  constant U = BigInt
  constant unit = "0"
  constant comp = (x1:U, x2:U) => "x1 + x2"
``` |

Fig. 6. Scala and MMT representations of Monoids and Integers

Representing the Module Systems. The correspondence between MMT theory inclusions and Scala class extensions is not exact due to what we call the import name clash in [RK13]: It arises when modules M_1 and M_2 both declare a symbol c and M imports both M_1 and M_2. OPENMATH and MMT use qualified names for scoped declarations (e.g., $M_1?c$ and $M_2?c$) so that the duplicate use of c is inconsequential. But Scala – typical for programming languages – identifies the two constants if they have the same type.

There are a few ways to work around this problem, and the least awkward of them is to qualify all field names when exporting MMT theories to Scala. Therefore, the first declaration in the trait Monoid is actually type Monoid_U and similar for all other declarations. Vice versa, when importing Scala classes, we assume that all names are qualified in this way.

It remains future work to align larger fragments of the module systems, which would also include named imports and sharing.

3   Biform Theory Development in MMT

We can now combine the representations of OPENMATH and Scala in MMT into a biform theory graph. In fact, we will obtain this combination as an example of a general principle of combining a logic and a programming language.

Bifoundations. Consider a logic represented as an MMT theory L and a programming language represented (possibly partially as in our case with Scala) as an MMT theory P. Moreover, consider an MMT theory morphism $s : L \to P$. Intuitively, s describes the meaning of L-specifications in terms of P.

Definition 1. A *bifoundation* is a triple $(L, P, s : L \to P)$.

Now consider a logical theory T represented as an MMT theory with meta-theory L. This yields the diagram in the category of MMT theories, which is given on the right. Then, inspired by [Rab13a], we introduce the following definition of what it means to implement T in P:

Definition 2. A *realization* of T over a bifoundation (L, P, s) is a morphism $r : T \to P$ such that the resulting triangle commutes.

Note that in MMT, there is a canonical pushout $s(T)$ of T along s. Thus, using the canonical property of the pushout, realizations r are in a canonical bijection with morphisms $r' : s(T) \to P$ that are the identity on P.

A Bifoundation for OPENMATH *CDs and Scala.* We obtain a bifoundation by giving an MMT morphism $s :$ OpenMath $\to$ Scala. This morphism hinges upon the choice for the Scala type that interprets the universal type Object. There are two canonical choices for this type, and the resulting morphisms are given in Figure 7. Firstly, we can choose the universal Scala type Any. This leads to a semantic bifoundation where we interpret every OPENMATH object by its Scala counterpart, i.e., integers as integers, lists as lists, etc. Secondly, we can choose a syntactic bifoundation where every object is interpreted as itself. This requires using a conservative extension ScalaOM of Scala that defines inductive types Term of OPENMATH objects and Context of OPENMATH contexts. Such an extension is readily available because it is part of the MMT API.

view Semantic: OpenMath −> ScalaOM	view Syntactic: OpenMath −> ScalaOM
constant Object = Any	constant Object = Term
constant mapsto = Function	constant mapsto = Function
naryObject = List[Any]	naryObject = List[Term]
binder = (Context, Term) => Any	binder = (Context, Term) => Term
FMP = (x:Any) => "assert(x == true)"	FMP = (x:Term) => "assert(x == OMS(logic1.true))"

Fig. 7. Two Bifoundations From Scala to OPENMATH

In both cases, n-ary arguments are easily interpreted in terms of lists and functions as functions. The case for binders is subtle: In both cases, we must interpret binders as Scala functions that take a syntactic object in context. Therefore, even the semantic foundation requires ScalaOM as the codomain.

Finally, we map mathematical properties to certain Scala function calls, e.g., assertions. In the semantic case, we assert the formula to be true. In the syntactic case, we assert it to be equal to the symbol true from the OPENMATH CD logic1. Here, OMS is part of the MMT API.

Of course, in practice, only the simplest of FMPs actually hold in the sense that a simple Scala computation could prove them. However, our interpretation of FMPs is still valuable: It naturally translates examples given in the OPEN-MATH CDs to Scala test cases that can be run systematically and automatically. Moreover, in the syntactic case, we have the additional option to collect the asserted formulas and to maintain them as input for verification tools.

4 Mechanizing Biform Theory Graphs

We are particularly interested in the syntactic bifoundation given above. It corresponds to the well-understood notion of a syntactic model of a logic. Thus, it has the advantage of completeness in the sense that the algorithms given in T-realizations can be used to describe deductive statements about T. In this section, we make this more precise and generalize it to arbitrary logics.

Abstract Rewrite Rules. First we introduce an abstract definition of rule that serves as the interface between the computational and the deductive realm. We need one auxiliary definition:

Definition 3. An **arity** is an element of $\{n, n* : n \in N\} \cup \{\mathsf{binder}\}$.

We use n ($n*$) for symbols that can be applied to n arguments (and a sequence argument), and we use binder for symbols that form binding object. For example, 2 is the arity of binary symbols and $0*$ the arity of symbols with an arbitrary sequence of arguments. This is a simplification of the arities we give in [KR12] and use in MMT, which permit sequences anywhere in the argument list and gives binders different arities as well.

Now let us fix an arbitrary set of MMT theories and write $\mathcal{C}$ for the set of constants declared in them. We write $\mathcal{T}$ for the set of closed MMT terms using only constants from $\mathcal{C}$, and $\mathcal{T}(x_1, \ldots, x_n)$ for the set of terms that may additionally use the variables $x_1, \ldots, x_n$. Then we define:

Definition 4. A **rule** r for a constant c with arity $n \in \mathbb{N}$ is a mapping $\mathcal{T}^n \to \mathcal{T}$. Such a rule is **applicable** to any $t \in \mathcal{T}$ of the form $\mathtt{OMA}(c, t_1, \ldots, t_n)$. In that case, its intended meaning is the formula $t = r(t_1, \ldots, t_n)$.

A **rule** for a constant c with arity $n*$ is a mapping $\mathcal{T}^n \times (\bigcup_{i=0}^{\infty} \mathcal{T}^i) \to \mathcal{T}$. Such a rule is **applicable** to any $t \in \mathcal{T}$ of the form $\mathtt{OMA}(c, t_1, \ldots, t_k)$ for $k \geq n$. In that case, its intended meaning is the formula $t = r(t_1, \ldots, t_k)$.

A **rule** for a constant c with arity binder is a mapping $\{(G, t) | G = x_1, \ldots, x_n \wedge t \in \mathcal{T}(G)\} \to \mathcal{T}$. Such a rule is **applicable** to any $t \in \mathcal{T}$ of the form $\mathtt{OMBIND}(c; G; t')$. In that case, its intended meaning is the formula $t = r(G, t')$.

A **rule base** R is a set of rules for some constants in $\mathcal{C}$. We write $R(c, a)$ for the set of rules in R for the constant c with arity a.

Our rules are different from typical rewrite rules [BN99] of the form $t_1 \rightsquigarrow t_2$ in two ways. Firstly, the left hand side is more limited: A rule for c is applicable exactly to the terms t_1 whose head is c. This corresponds to the intuition of a rule implementing the constant c. It also makes it easy to find the applicable rules within a large rule base. Secondly, the right hand side is not limited at all: Instead of a term t_2, we use an arbitrary function that returns t_2. This corresponds to our open-world assumption: Constants are implemented by arbitrary programs (written in any programming language) provided by arbitrary sources.

In the special case without binding, our rules are essentially the same as those used in [Far07], where the word *transformer* is used for the function $r(-)$.

It is now routine to obtain a rewrite system from a rule base:

Definition 5. Given a rule base R, R-rewriting is the reflexive-transitive closure of the relation $\rightsquigarrow \subseteq \mathcal{T} \times \mathcal{T}$ given by:

$$\frac{r \in R(c,0)}{c \rightsquigarrow r()} \qquad \frac{t_i \rightsquigarrow t_i' \text{ for } i=0,\ldots,n}{\mathtt{OMA}(t_0,\ldots,t_n) \rightsquigarrow \mathtt{OMA}(t_0',\ldots,t_n')} \qquad \frac{r \in R(c,n) \text{ or } r \in R(c,i*) \text{ for } i \leq n}{\mathtt{OMA}(c,t_1,\ldots,t_n) \rightsquigarrow r(t_1,\ldots,t_n)}$$

$$\frac{t_i \rightsquigarrow t_i' \text{ for } i=1,2}{\mathtt{OMBIND}(t_1;G;t_2) \rightsquigarrow \mathtt{OMBIND}(t_1';G;t_2')} \qquad \frac{r \in R(c,\text{binder})}{\mathtt{OMBIND}(c;G;t) \rightsquigarrow r(G,t)}$$

R-rewriting is not guaranteed to be confluent or terminating. This is unavoidable due to our abstract definition of rules where not only the set of constants and rules are unrestricted but even the choice of programming language. However, this is usually no problem in practice if each rule has evaluative flavor, i.e., if it transforms a more complex term into a simpler one.

Realizations as Rewriting Rules. Consider a realization r of T over the bifoundation (OpenMath, ScalaOM, Syntactic), and let ρ be the corresponding Scala object. Then for every constant c with type $\mathtt{OMA}(\mathtt{mapsto}, \mathtt{Object}, \ldots, \mathtt{Object})$ declared in T, we obtain a rule r_c by putting $r_c(t_1,\ldots,t_n)$ to be the result of evaluating the Scala expression $\rho.c(t_1,\ldots,t_n)^3$. We obtain rules for constants with other types accordingly. More generally, we define:

Definition 6. Given a theory T, an **arity assignment** maps every T-constant to an arity.

Given an arity assignment, a realization $T \to$ ScalaOM is called **syntactic** if the type of every T-constant with arity a is mapped to the following Scala type: `(Term,...,Term) => Term` if $a = n$; `(Term,...,Term, List[Term]) => Term` if $a = n*$; and `(Context,Term) => Term` if $a = $ binder.

A syntactic realization $r : T \to$ ScalaOM induces for every constant c of T a rule r_c in a straightforward way. If c has arity n, the rule r_c maps $(t_1,\ldots,t_n)$ to the result of evaluating the Scala expression $r(c)(t_1,\ldots,t_n)$, where $r(c)$ is the Scala function that r assigns to c. Technically, r_c is only a partial function because evaluation might fail or not terminate; in that case, we put $r_c(t_1,\ldots,t_n) = \mathtt{OMA}(c,t_1,\ldots,t_n)$. For other arities, r_c is defined accordingly.

Definition 7. *We write* Rules(r) *for the rule base containing for each constant c declared in T the rule r_c.*

A general way of obtaining arity assignments for all theories T with a fixed meta-theory L is to give an MMT morphism $e : L \to$ OpenMath. e can be understood as a *type-erasure translation* that forgets all type information and merges all types into one universal type. Then the arities of the T-constants are determined by the OPENMATH types in the pushout $e(T)$. Therefore, we can often give bifoundations for which all realizations are guaranteed to be syntactic, the bifoundation (OpenMath, ScalaOM, Syntactic) being the trivial example.

---

[3] Technically, in practice, we need to catch exceptions and set a time-out to make r_c a total function, but that is straightforward.

Def. 7 applies only to realizations in terms of Scala. However, it is straightforward to extend it to arbitrary programming languages. Of course, MMT – being written in Scala – can directly execute Scala-based realizations whereas for any other codomain it needs a plugin that supplies an interpreter.

The Universal Machine. We use the name *universal machine* for the new MMT component that maintains the rule base arising as the union of all sets $Rules(r)$ for all syntactic realizations r with domain ScalaOM in MMT's knowledge base. Here "universal" refers to the open-world perspective that permits the extension with new logics and theories as well as programming languages and implementations.

The universal machine implements the rewrite system from Def. 5 by exhaustively applying rules (which are assumed to be confluent) and exposes it as a single API function, called *simplification*. The MMT system does not perform simplification at any specific point.

Instead, it is left to other components like plugins and applications to decide if and when simplification should be performed. In the MMT API, any term may carry metadata, and this is used to mark each subterm that has already been simplified. Thus, different components may call simplification independently without causing multiple traversals of the same subterm.

Additionally, the API function is exposed in two ways. Firstly, MMT accepts simplification requests via HTTP post, where input and output are given as strings using MMT notations or as OPENMATH XML elements. Secondly, simplification is integrated with the Scala interactive interpreter, where users can type objects using MMT notations and simplification is performed automatically. It is straightforward to connect further frontends.

5   Building a Biform Library

We evaluate the new MMT concepts by building a biform MMT theory graph based on the bifoundation (OpenMath, Scala, Syntactic), which represents > 30 of the official OpenMath CDs in MMT and provides Scala implementations and test cases for > 80 symbols. This development is available as an MMT project and described in more detail at https://tntbase.mathweb.org/repos/oaff/openmath.

MMT projects [HIJ+11] already support different dimensions of knowledge, such as source, content, and presentation, as well as build processes that transform developments between dimensions. We add one new dimension for generated programs and workflows for generating it.

Firstly, we write MMT theories representing the OPENMATH CDs such as the one given on the left of Fig. 3. Specifically, we represent the arith, complex, fns, integer, interval, linalg, list, logic, minmax, nums, relation, rounding, set, setname, and units CDs along with appropriate notations.

Secondly, we write views from these CDs to ScalaOM. Then a new MMT build process generates all corresponding Scala classes. Typically, users write view stubs in MMT and then fill out the generated Scala stubs using an IDE of their choice. Afterwards MMT imports the Scala stubs and merges the user's changes into the MMT views. This is exemplified in Fig. 8. Here the left side gives a fragment of an MMT view out of arith1, which implements arithmetic on numbers. (We also give other views out of arith1, e.g., for operations on matrices.) The implementation for plus is still missing whereas the one for minus is present. The right side shows the generated Scala code with the editable parts marked by comments.

```
view NumberArith :                   object NumberArith extends arith1 {
  arith1 −> ScalaOM =                  def arith1_plus(args: List[Term]) : Term = {
    plus = (args: List[Term]) "            // start NumberArith?plus
      "                                    // end NumberArith?plus
                                         }
                                         def arith1_minus(a: Term, b: Term) : Term = {
    minus = (a: Term, b: Term) "           // start NumberArith?minus
      (a,b) match {                         (a,b) match {
        case (OMI(x), OMI(y)) =>               case (OMI(x), OMI(y)) => OMI(x − y)
          OMI(x − y)                         }
      }                                    // end NumberArith?minus
      "                                  }
                                       }
```

Fig. 8. Partial Realization in MMT and Generated Scala Code

Finally, we write MMT theories for extensions of the OPENMATH CDs with examples as on the right in Fig. 3. We also give realizations for them, which import the realizations of the extended CDs. Here MMT generates assertions for each FMP.

To apply these workflows to large libraries, we have added three build processes to MMT that can be integrated easily with make files or MMT IDEs. extract walks over an MMT project and translates realizations into Scala source files containing the corresponding objects. This permits editing realizations using Scala IDEs. integrate walks over the Scala source files and merges all changes made to the realizations back into the MMT files. load walks over the Scala source files, compiles them, loads the class files, and registers the rule bases $Rules(r)$ with the universal machine. Optionally, it runs all test cases and generates a report.

6   Conclusion

We described a formal framework and a practical infrastructure for biform theory development, i.e., the integration of deductive theories and computational definitions of the functions specified in them. The integration is generic and permits arbitrary logics and programming languages; moreover, the same module system is used for specifications and implementations.

We have instantiated our design with a biform development of the OPENMATH content dictionaries in Scala. Future work will focus on the development of larger biform libraries and the use of further logics and programming languages. In particular, we want to explore how to treat richer type systems and to preserve their information in the generated Scala code.

Regarding the integration of deduction and computation we focused only on "soft verification", i.e., linking function symbols with unverified implementations. We only extracted the computational content of examples (which results in test cases) and omitted the more difficult problem of axioms. We believe that future work can extend our approach to generate computation rules by spotting axioms of certain shapes such as those in inductive definitions or rewrite rules. Moreover, given a verifier for the used programming language, it will be possible to generate the verification obligations along with the generated programs.

Acknowledgements. The work reported here was prompted by discussions with William Farmer and Jacques Carette. An initial version of the universal machine was developed in collaboration with Vladimir Zamdzhiev.

References

ABB+05. Ahrendt, W., Baar, T., Beckert, B., Bubel, R., Giese, M., Hähnle, R., Menzel, W., Mostowski, W., Roth, A., Schlager, S., Schmitt, P.: The KeY Tool. Software and System Modeling 4, 32–54 (2005)

BCC+04. Buswell, S., Caprotti, O., Carlisle, D., Dewar, M., Gaetano, M., Kohlhase, M.: The Open Math Standard, Version 2.0. Technical report, The Open Math Society (2004), http://www.openmath.org/standard/om20

BCH12. Boespflug, M., Carbonneaux, Q., Hermant, O.: The $\lambda\Pi$-calculus modulo as a universal proof language. In: Pichardie, D., Weber, T. (eds.) Proceedings of PxTP 2012: Proof Exchange for Theorem Proving, pp. 28–43 (2012)

BN99. Baader, F., Nipkow, T.: Term Rewriting and All That. Cambridge University Press (1999)

Dav00. Davenport, J.: A small OpenMath type system. Bulletin of the ACM Special Interest Group on Symbolic and Automated Mathematics (SIGSAM) 34(2), 16–21 (2000)

DM05. Delahaye, D., Mayero, M.: Dealing with Algebraic Expressions over a Field in Coq using Maple. Journal of Symbolic Computation 39(5), 569–592 (2005)

Far07. Farmer, W.M.: Biform theories in chiron. In: Kauers, M., Kerber, M., Miner, R., Windsteiger, W. (eds.) MKM/CALCULEMUS 2007. LNCS (LNAI), vol. 4573, pp. 66–79. Springer, Heidelberg (2007)

FGT92. Farmer, W., Guttman, J., Thayer, F.: Little Theories. In: Kapur, D. (ed.) Conference on Automated Deduction, pp. 467–581 (1992)

H+12. Hardin, T., et al.: The focalize essential (2012), http://focalize.inria.fr/

HIJ+11. Horozal, F., Iacob, A., Jucovschi, C., Kohlhase, M., Rabe, F.: Combining Source, Content, Presentation, Narration, and Relational Representation. In: Davenport, J.H., Farmer, W.M., Urban, J., Rabe, F. (eds.) Calculemus/MKM 2011. LNCS, vol. 6824, pp. 212–227. Springer, Heidelberg (2011)

HN10.   Haftmann, F., Nipkow, T.: Code Generation via Higher-Order Rewrite Systems. In: Blume, M., Kobayashi, N., Vidal, G. (eds.) FLOPS 2010. LNCS, vol. 6009, pp. 103–117. Springer, Heidelberg (2010)

HPRR10.   Heras, J., Pascual, V., Romero, A., Rubio, J.: Integrating Multiple Sources to Answer Questions in Algebraic Topology. In: Autexier, S., Calmet, J., Delahaye, D., Ion, P.D.F., Rideau, L., Rioboo, R., Sexton, A.P. (eds.) AISC 2010. LNCS, vol. 6167, pp. 331–335. Springer, Heidelberg (2010)

HT98.   Harrison, J., Théry, L.: A Skeptic's Approach to Combining HOL and Maple. Journal of Automated Reasoning 21, 279–294 (1998)

KR12.   Kohlhase, M., Rabe, F.: Semantics of OpenMath and MathML3. Mathematics in Computer Science 6(3), 235–260 (2012)

KST97.   Kahrs, S., Sannella, D., Tarlecki, A.: The definition of extended ML: A gentle introduction. Theoretical Computer Science 173(2), 445–484 (1997)

NPW02.   Nipkow, T., Paulson, L.C., Wenzel, M.T. (eds.): Isabelle/HOL. LNCS, vol. 2283. Springer, Heidelberg (2002)

OMC.   OpenMath Content Dictionaries, `http://www.openmath.org/cd/`

OSV07.   Odersky, M., Spoon, L., Venners, B.: Programming in Scala. artima (2007)

Rab13a.   Rabe, F.: A Logical Framework Combining Model and Proof Theory. Mathematical Structures in Computer Science (to appear, 2013), `http://kwarc.info/frabe/Research/rabe_combining_10.pdf`

Rab13b.   Rabe, F.: The MMT API: A Generic MKM System. In: Carette, J., Aspinall, D., Lange, C., Sojka, P., Windsteiger, W. (eds.) CICM 2013. LNCS (LNAI), vol. 7961, pp. 339–343. Springer, Heidelberg (2013)

RK13.   Rabe, F., Kohlhase, M.: A Scalable Module System. Information and Computation (2013), conditionally accepted `http://arxiv.org/abs/1105.0548`

The11.   The Coq Development Team. The coq proof assistant: Reference manual. Technical report, INRIA (2011)

Mathematical Practice, Crowdsourcing, and Social Machines

Ursula Martin and Alison Pease

Queen Mary University of London
Ursula.Martin@qmul.ac.uk, Alison.Pease@eecs.qmul.ac.uk

Abstract. The highest level of mathematics has traditionally been seen as a solitary endeavour, to produce a proof for review and acceptance by research peers. Mathematics is now at a remarkable inflexion point, with new technology radically extending the power and limits of individuals. Crowdsourcing pulls together diverse experts to solve problems; symbolic computation tackles huge routine calculations; and computers check proofs too long and complicated for humans to comprehend.

The *Study of Mathematical Practice* is an emerging interdisciplinary field which draws on philosophy and social science to understand how mathematics is produced. Online mathematical activity provides a novel and rich source of data for empirical investigation of mathematical practice - for example the community question-answering system *mathoverflow* contains around 40,000 mathematical conversations, and *polymath* collaborations provide transcripts of the process of discovering proofs. Our preliminary investigations have demonstrated the importance of "soft" aspects such as analogy and creativity, alongside deduction and proof, in the production of mathematics, and have given us new ways to think about the roles of people and machines in creating new mathematical knowledge. We discuss further investigation of these resources and what it might reveal.

Crowdsourced mathematical activity is an example of a "social machine", a new paradigm, identified by Berners-Lee, for viewing a combination of people and computers as a single problem-solving entity, and the subject of major international research endeavours. We outline a future research agenda for mathematics social machines, a combination of people, computers, and mathematical archives to create and apply mathematics, with the potential to change the way people do mathematics, and to transform the reach, pace, and impact of mathematics research.

1 Introduction

For centuries, the highest level of mathematical research has been seen as an isolated creative activity, whose goal is to identify mathematical truths, and justify them by rigorous logical arguments which are presented for review and acceptance by research peers.

Yet mathematical discovery also involves soft aspects such as creativity, informal argument, error and analogy. For example, in an interview in 2000 [1]

J. Carette et al. (Eds.): CICM 2013, LNAI 7961, pp. 98–119, 2013.
© Springer-Verlag Berlin Heidelberg 2013

Andrew Wiles describes his 1989 proof of Fermat's theorem in almost mystical terms "... and sometimes I realized that nothing that had ever been done before was any use at all. Then I just had to find something completely new; it's a mystery where that comes from." Michael Atiyah remarked at a workshop in Edinburgh in 2012 [2] "I make mistakes all the time" and "I published a theorem in topology. I didn't know why the proof worked, I didn't understand why the theorem was true. This worried me. Years later we generalised it—we looked at not just finite groups, but Lie groups. By the time we'd built up a framework, the theorem was obvious. The original theorem was a special case of this. We got a beautiful theorem and proof."

Computer assisted proof formed some of the earliest experiments in artificial intelligence: in 1955 Newell, Shaw and Simon's Logic Theorist searched forward from axioms to look for proofs of results taken from Russell and Whitehead's 1911 Principia Mathematica. Simon reported in a 1994 interview [74] that he had written to Russell (who died in 1970, aged 97), who "wrote back that if we'd told him this earlier, he and Whitehead could have saved ten years of their lives. He seemed amused and, I think, pleased." By the mid-1980s a variety of approaches and software tools, such as the theorem provers HOL, NuPrl and Nqthm, had started to be developed for practical reasoning about programs: [42] is a thorough account of the early history. This laid the foundation for a flourishing academic and industry community, and currently verification to ensure error-free systems is a major endeavour in companies like Intel and Microsoft [37], as well as supporting specialist small companies. At the same time theorem provers are now being used by an influential community of mathematicians. Tom Hales and his team have almost completed a ten-year formalisation of their proof of the Kepler conjecture, using several theorem provers to confirm his major 1998 paper [36]. In September 2012 Georges Gonthier announced that after a six year effort his team had completed a formalisation, in the Coq theorem prover, of one of the most important and longest proofs of 20th century algebra, the 255 page odd-order theorem [32]. He summarised the endeavour as:

Number of lines ˜ 170 000
Number of definitions ˜ 15 000
Number of theorems ˜ 4 200
Fun ˜ enormous!

The growth in the use of computers in mathematics, and in particular of computer proof, has provoked debate, reflecting the contrast between the "logical" and "human" aspects of creating mathematics: see [56] for a survey. For example in an influential paper in 1979, De Millo, Lipton and Perlis [28], argued that "Mathematical proofs increase our confidence in the truth of mathematical statements only after they have been subjected to the social mechanisms of the mathematical community", and expressed concern over "symbol chauvinism". Similar concerns were raised in the mathematical community over the use of a computer by Appel and Haken [10] to settle the long standing four colour conjecture. Indeed, Hume, in his 1739 Treatise on Human Nature [40] p231, identified the importance of the social context of proof:

There is no Algebraist nor Mathematician so expert in his science, as to place entire confidence in any truth immediately upon his discovery of it, or regard it as any thing, but a mere probability. Every time he runs over his proofs, his confidence encreases; but still more by the approbation of his friends; and is rais'd to its utmost perfection by the universal assent and applauses of the learned world. [*sic*]

The sociology of science addresses such paradoxes in the understanding of the scientific process, and a comprehensive account is given by sociologist Donald MacKenzie in his 2001 book "Mechanizing Proof" [52]. He concludes that used to extend human capacity the computer is benign, but that "trust in the computer cannot entirely replace trust in the human collectivity". In recent years "the study of mathematical practice" has emerged from the work of Pólya and Lakatos as a subdiscipline drawing upon the work of sociologists, cognitive scientists, philosophers and the narratives of mathematicians themselves, to study exactly what it is that mathematicians do to create mathematics. Section 2 of this paper contains a fuller account.

The mathematical community were "early adopters" of the internet for disseminating papers, sharing data, and blogging, and in recent years have developed systems for "crowdsourcing" (albeit among a highly specialised crowd) the production of mathematics through collaboration and sharing, providing further evidence for the social nature of mathematics. To give just a few examples:

- A number of senior mathematicians produce influential and widely read blogs. In the summer of 2010 a paper was released plausibly claiming to prove one of the major challenges of theoretical computer science, that $P \neq NP$: it was withdrawn after rapid analysis organised by senior scientist-bloggers, and coordinated from Richard Lipton's blog. Fields Medallist Sir Tim Gowers used his blog to lead an international debate about mathematics publishing.
- *polymath* collaborative proofs, a new idea led by Gowers, use a blog and wiki for collaboration among mathematicians from different backgrounds and have led to major advances [35]
- discussion fora allow rapid informal interaction and problem-solving; in three years the community question answering system for research mathematicians *mathoverflow* has 23,000 users and has hosted 40,000 conversations
- the widely used "Online Encyclopaedia of Integer Sequences" (OEIS) invokes subtle pattern matching against over 200,000 user-provided sequences on a few digits of input to propose matching sequences: so for example input of (3 1 4 1) returns π (and other possibilities) [3]
- the *arXiv* holds around 750K preprints in mathematics and related fields. By providing open access ahead of journal submission, it has markedly increased the speed of refereeing, widely identified as a bottleneck to the pace of research [27]
- Innocentive [4], a site hosting open innovation and crowdsourcing challenges, has hosted around 1,500 challenges with a 57% success rate, of which around 10% were tagged as mathematics or ICT.

As well as having a remarkable effect on mathematical productivity, these systems provide substantial and unprecedented evidence for studying mathematical practice, allowing the augmentation of traditional ethnography with a variety of empirical techniques for analysing the texts and network structures of the interactions. In Section 3 we describe two of our own recent preliminary studies, of *mathoverflow* and *polymath*, which provide evidence for the theories of Pólya and Lakatos, and shed new light on mathematical practice, and on the current or future computational tools that might enhance it. Analysing the content of a sample of questions and responses, we find that *mathoverflow* is very effective, with 90% of our sample of questions answered completely or in part. A typical response is an informal dialogue, allowing error and speculation, rather than rigorous mathematical argument: a surprising 37% of our sample discussions acknowledged error. Looking at one of the recent *mini-polymath* problems, we find only 24% of the 174 comments formed the development of the final proof, with the remainder comprising a high proportion of examples (33%) alongside conjectures and social glue. We conclude that extending the power and reach of *mathoverflow* or *polymath* through a combination of people and machines raises new challenges for artificial intelligence and computational mathematics, in particular how to handle error, analogy and informal reasoning.

Of course, mathematics is not the only science in which productive new human collaborations are made possible by machines. Over the past twenty years researchers in e-science have devised systems such as Goble's myExperiment [71] for managing scientific workflow, especially in bioinformatics, so that data, annotations, experiments, and results can be documented and shared across a uniform platform, rather than in a mixture of stand alone software systems and formats. Michael Nielsen, one of the founders of *polymath*, in his 2011 book "Reinventing discovery" [61] discusses a number of examples of crowdsourced and citizen science. Alongside *polymath*, he describes Galaxy Zoo, which allows members of the public to look for features of interest in images of galaxies, and has led to new discoveries, and Foldit, an online game where users solve protein folding problems.

Considered more broadly, such systems are exemplars of "Social machines", a broad new paradigm identified by Berners-Lee in his 1999 book "Weaving the Web" [16], for viewing a combination of people and computers as a single problem-solving entity. Berners-Lee describes a dream of collaborating through shared knowledge:

> Real life is and must be full of all kinds of social constraint — the very processes from which society arises. Computers can help if we use them to create abstract social machines on the Web: processes in which the people do the creative work and the machine does the administration. . . The stage is set for an evolutionary growth of new social engines. The ability to create new forms of social process would be given to the world at large, and development would be rapid.

Current social machines provide platforms for sharing knowledge and leading to innovation, discovery, commercial opportunity or social benefit: the combination

of mobile phones, Twitter and google maps used to create real-time maps of the
effects of natural disasters has been a motivating example. Future more ambitious social machines will combine social involvement and sophisticated automation, and are now the subject of major research, for example in Southampton's
SOCIAM project [5] following an agenda laid out by Hendler and Berners-Lee
[38]. In Section 4 we look at collaborative mathematics systems through the lens
of social machines research, presenting a research agenda that further develops
the results of work on the practice of mathematics.

2   The Study of Mathematical Practice

The study of mathematical practice emerged as a fledgling discipline in the
1940's when mathematician and educator Georg Pólya formulated problem-
solving heuristics designed to aid mathematics students. These heuristics, such as
"rephrase the question", and "draw a diagram" were based on Pólya's intuition
about rules of thumb which he himself followed during his research, and have
been influential in mathematics education (although not without critics, who argue that meta-heuristics are needed to determine when a particular route is likely
to be fruitful [48,62,73]). Pólya's idea, that it is possible to identify heuristics
which describe mathematical research – a logic of discovery – was extended by
Imre Lakatos, fellow countryman and philosopher of mathematics and science.[1]
Lakatos used in-depth analyses of extended historical case studies to formulate
patterns of reasoning which characterised conversations about a mathematical
conjecture and its proof. These patterns focused on interactions between mathematicians and, in particular, on the role that counterexamples play in driving
negotiation and development of concepts, conjectures and proofs.

Lakatos demonstrated his argument by presenting a rational reconstruction
of the development of Euler's conjecture that for any polyhedron, the number of
vertices (V) minus the number of edges (E) plus the number of faces (F) is equal
to two; and Cauchy's proof of the conjecture that the limit of any convergent
series of continuous functions is itself continuous. He outlined six methods for
modifying mathematical ideas and guiding communication: surrender, monster-
barring, exception-barring, monster-adjusting, lemma-incorporation, and proofs
and refutations. These are largely triggered by counterexamples, or problematic
entities, and result in a modified proof, conjecture or concept. For instance,
the methods of *monster-barring* and *monster-adjusting* exploit ambiguity or
vagueness in concept definitions in order to attack or defend a conjecture, by
(re)defining a concept in such a way that a problematic object is either excluded
or included. With *monster-barring*, the ambiguous concept is central to the conjecture and defines the domain of application, such as a "polyhedron" (in Euler's conjecture), a "finite group" (in Lagrange's theorem), or an "even number"
(in Goldbach's conjecture). Here, Lakatos presents the picture-frame, for which

---

[1] Lakatos translated Pólya's [67] and other mathematical works into Hungarian before
developing his own logic of discovery, intended to carry on where Pólya left off [46,
p. 7].

V - E + F = 16 - 32 + 16 = 0 (see figure 1): this is "monster-barred" as being an invalid example of a polyhedron, and the definition of polyhedron tightened to exclude it. With *monster-adjusting*, the ambiguous concept is a sub-concept (appears in the *definition* of the central concept), such as "face", "identity", or "division" (following the polyhedron/finite group/even number examples). (Re)defining this sub-concept can provide an alternative way of viewing a problematic object in such a way that it ceases to be problematic: Lakatos gives the example of Kepler's star-polyhedron, which is a counterexample if V - E + F is 12 - 30 + 12 = -6 (where its faces are seen as star-pentagons), but can be salvaged if we see V - E + F as 32 - 90 +60 = 2 (where its faces are seen as triangles) (see figure 1). The result of both of these methods is a preserved conjecture statement, where the meaning of the terms in it have been revised or clarified.

Fig. 1. Controversial polyhedra: A picture-frame, on the left, for which V - E + F = 16 - 32 + 16 = 0, and Kepler's star-polyhedron, on the right, for which V - E + F can be 12 - 30 + 12 = -6 (if it has star-pentagon faces) or 32 - 90 +60 = 2 (if it has triangular faces)

In Lakatos's exception-barring method, a counterexample is seen as an exception, triggering a refinement to the conjecture, and in his lemma-incorporation and proofs and refutations methods, problematic objects are found and examined to see whether they are counterexamples to a conjecture or a proof step, which are then revised accordingly.

Lakatos held an essentially optimistic view of mathematics, in which the process of mathematics traditionally thought of as impenetrable and inexplicable by rational laws, considered to be lucky guess work or intuition, is seen in a rationalist light, thereby opening up new arenas of rational thought. He challenged Popper's view [70] that philosophers can form theories about how to evaluate conjectures, but not how to generate them, which should be left to psychologists and sociologists. Rather, Lakatos believed that philosophers could theorise about both of these aspects of the scientific and mathematical process. He challenged Popper's view in two ways - arguing that *(i)* there *is* a logic of discovery, the process of generating conjectures and proof ideas *is* subject to rational laws; and *(ii)* the distinction between discovery and justification is misleading as each affects the other; *i.e.*, the way in which we discover a conjecture affects our proof (justification) of it, and proof ideas affect what it is that we are trying to prove (see [49]). This happens to such an extent that the boundaries of each are blurred. These ideas have a direct translation into automated proof research, suggesting that conjecture and concept generation are subject to rationality as well as proof, and therefore systems can (perhaps even should) be developed which integrate these theory-development aspects alongside proof generation.

At the heart of both Pólya and Lakatos's work was the idea that the mechanisms by which research mathematics progresses – as messy, fallible, and speculative as this may be – can usefully be studied via analysis of informal mathematics. This idea has been welcomed and extended by a variety of disciplines; principally philosophy, history sociology, cognitive science and mathematics education [9,21,26,54]. The development of computer support for mathematical reasoning provides further motivation for studying the processes behind informal mathematics, particularly in the light of the criticisms this has sometimes received. Sociologist Goffman [31] provides a useful distinction here, of front and backstage activities, where activities in the front are services designed for public consumption, and those in the back constitute the private preparation of the services. Hersh [39] extends this distinction to mathematics, where textbook or publication-style "finished mathematics" takes frontstage, and the informal workings and conversations about "mathematics in the making" is hidden away backstage. Pólya employed a similar distinction, and Lakatos warned of the dangers of hiding the backstage process, either from students (rendering the subject impenetrable) or from experts (making it more difficult to develop concepts or conjectures which may arise out of earlier versions of a theorem statement). Computer support for mathematics, such as computer algebra or computational mathematics, has typically been for the frontstage. A second, far less developed, approach is to focus on the backstage, including the mistakes, the dead ends and the unfinished, and to try to extract principles which are sufficiently clear as to allow an algorithmic interpretation: the study of mathematical practice provides a starting point for this work.

Implicit or explicit in much work on mathematical practice is the recognition that mathematics takes place in a social context. Education theorist, Paul Ernest [29], sees mathematics as being socially constructed via conversation; a conversation which is as bound by linguistic and social conventions as any other discourse. Thus, if such conventions are violated (by other cultures, or, perhaps, by machines) then shared understanding is lost and – mirroring Kuhnian paradigm shift – new conventions may need to be formed which accommodate the rogue participant. Kitcher [44], a philosopher of mathematics, elaborates what a mathematical practice might mean, suggesting a socio-cultural definition as consisting in a language and four socially negotiated sets: accepted statements, accepted reasonings, questions which are considered to be important and meta-mathematical views such as standards of proof and the role of mathematics in science (agreement over the content of these sets helps to define a mathematical culture). Mackenzie [52] looked at the role of proof, especially computer proof, and his student Barany [12] used ethnographic methods to trace the cycle of development and flow of mathematical ideas from informal thoughts, to seminar, to publication, to dissemination and classroom, and back to informal thoughts. He sees (re)representations in varying media such as notes, blackboard scribbles, physical manifestations or patterns of items on a desk, as necessary, for the knowledge to be decoded and encoded into socially and cognitively acceptable forms. In particular, Barany investigated the relationship between the material (the "pointings, tappings, rubbings, and writings" of mathematics [12, p.9]) and the abstract, arguing that each constrains the other. Other developments in the study of mathematical practice include work on visualisation, such as diagrammatic reasoning in mathematics [30,53]; analogies, such as between mathematical theories and axiom sets [13,72]; and mathematical concept development, such as ways to determine potential fruitfulness of rival definitions [75,76]. At the heart of many of these analyses lies the question of what proof is for, and the recognition that it plays multiple roles; explaining, convincing, evaluating, aiding memory, and so on, complementing or replacing traditional notions of proof as a guarantee of truth). This in turn gives an alternative picture of machines as members of a mathematical community.

3   Mathematical Practice and Crowdsourced Mathematics

In this section we outline preliminary results from our own ongoing programme of work which uses collaborative online systems as an evidence base for further understanding of mathematical practice. We studied a sample of *mathoverflow* questions and the ensuing discussions [57], and the third *mini-polymath* problem [65], looking at the kinds of activities taking place, the relative importance of each, and evidence for theories of mathematical practice described in the previous section, especially the work of Pólya [67] and Lakatos [46].

   mathoverflow and *polymath* are similar in that they are examples of the backstage of collaborative mathematics. They provide records of mathematicians collaborating through nothing more than conversation, underpinned by varying levels of shared expertise and context. While participants may invoke results from computational engines, such as GAP or Maple, or cite the literature, neither system contains any formal links to software or databases. The usual presentation of mathematics in research papers is the frontstage, in a standardised precise and rigorous style: for example, the response to a conjecture is either a counterexample, or a proof of a corresponding theorem, structured by means of intermediate definitions, theorems and proofs. By contrast these systems present the backstage of mathematics: facts or short chains of inference that are relevant to the question, but may not answer it directly, justified by reference to mathematical knowledge that the responder expects the other participants to have.

3.1   Mathoverflow

Discussion fora for research mathematics have evolved from the early newsnet newsgroups to modern systems based on the *stackexchange* architecture, which allow rapid informal interaction and problem-solving. In three years *mathover-flow.net* has accumulated 23,000 users and hosted 40,000 conversations. Figure 2 shows part of a *mathoverflow* conversation [8], in answer to a question about the existence of certain kinds of chains of subgroups. The highly technical nature of research mathematics means that, in contrast to activities like GalaxyZoo, this is not currently an endeavour accessible to the public at large: a separate site *math.stackexchange.com* is a broader question and answer site "for people studying math at any level and professionals in related fields". Within *mathover-flow*, house rules give detailed guidance, and stress clarity, precision, and asking questions with a clear answer. Moderation is fairly tight, and some complain it constrains discussion.

The design of such systems has been subject to considerable analysis (see, for instance, [15]), and *meta.mathoverflow* contains many reflective discussions. A key element is user ratings of questions and responses, which combine to form reputation ratings for users. These have been studied by psychologists Tausczik and Pennebaker [77,78], who concluded that *mathoverflow* reputations offline (assessed by numbers of papers published) and in *mathoverflow* were consistently and independently related to the *mathoverflow* ratings of authors' submissions, and that while more experienced contributors were more likely to be motivated by a desire to help others, all were motivated by building their *mathoverflow* reputation.

I think you can get arbitrarily long chains of this type in the simple groups $PSL(2, p)$ for p prime.

We make use of the maximal dihedral subgroups of order $p - 1$. For given N, choose p such that $(p - 1)/2 = q^N r$ with q an odd prime and $r > 2$. Then there is a chain of subgroups

$$H_0 < H_1 < \cdots H_N < G$$

where H_i is dihedral of order $2q^i r$.

link | flag | cite

answered **Apr 21 2011 at 19:14**
Derek Holt
6,713 ● 1 ● 12 ● 25

Thanks, that's really neat. The "r" makes sure anything containing H0 lands in HN. Why does q have to be odd? If we require H0 to be order not dividing 120 (so make r larger and avoid the A4,A5,S4 subgroups), can we take q=2 anyways? – Jack Schmidt Apr 21 2011 at 19:55

Yes you are right, q need not be odd. – Derek Holt Apr 21 2011 at 21:47

Wow, that's pretty cool. I wouldn't have noticed these long chains without your nice use of maximal subgroups. Probably these aren't examples that are easily verified in GAP, but I'll take your word for it for now, and then convince myself that it works. Thank you!! ..and thank you to everyone else who gave helpful suggestions. – William DeMeo Apr 21 2011 at 23:39

Fig. 2. A typical *mathoverflow* conversation

Within *mathoverflow* we identified the predominant kinds of questions as: **Conjecture (36%)**, which ask whether or under what circumstances a statement is true; **What is this (28%)**, which describe an object or phenomenon and ask what is known about it; and **Example (14%)** which ask for examples of a phenomenon or an object with particular properties. Other smaller categories ask for an explicit formula or computation technique, for alternatives to a known proof , for literature references, for help in understanding a difficulty or apparent contradiction,[2] or for motivation.

Analysing the answers in our sample shed further light on how the system was being used. *mathoverflow* is very effective, with 90% of our sample successful, in that they received responses that the questioner flagged as an "answer", of which 78% were reasonable answers to the original question, and a further 12% were partial or helpful responses that moved knowledge forward in some way. The high success rate suggests that, of the infinity of possible mathematical questions, questioners are becoming adept at choosing those for *mathoverflow* that are amenable to its approach.

The presentation is often speculative and informal, a style which would have no place in a research paper, reinforced by conversational devices that are accepting of error and invite challenge, such as "I may be wrong but...", "This isn't quite right, but roughly speaking...". Where errors are spotted, either by the person who made them or by others, the style is to politely accept and correct them: corrected errors of this kind were found in 37% of our sample.[3]

In 34% of the responses explicit examples were given, as evidence for, or counterexamples to, conjectures: thus playing exactly the role envisaged by Lakatos. We return to this below. In 56% of the responses we found citations to the literature. This includes both finding papers that questioners were unaware of, and extracting results that are not explicit in the paper, but are straightforward (at least to experts) consequences of the material it contains.

It is perhaps worth commenting on things that we did not see. As we shall see in the next section, in developing "new" mathematics considerable effort is put into the formation of new concepts and definitions: we saw little of this in *mathoverflow*, where questions by and large concern extending or refining existing knowledge and theories. We see little serious disagreement in our *mathoverflow* sample: perhaps partly because of the effect of the "house rules", but also because of the style of discussion, which is based on evidence from the shared research background and knowledge of the participants: there is more discussion and debate in *meta.mathoverflow*, which has a broader range of non-technical questions about the development of the discipline and so on.

3.2 Polymath

In 2009 the mathematician Timothy Gowers asked "Is massively collaborative mathematics possible?" [34], and with Terence Tao initiated experiments which

---

[2] Several questions concerned why Wikipedia and a published paper seemed to contradict each other.

[3] This excludes "conjecture" questions where the responses refutes the conjecture. We looked at *discussions* of error: we have no idea how many actual errors there are!

invited contributions on a blog to solving open, difficult conjectures. Participants were asked to follow guidelines [7], which had emerged from an online collaborative discussion, and were intended to encourage widespread participation and a high degree of interaction, with results arising from the rapid exchange of informal ideas, rather than parallelisation of sub-tasks. These included "It's OK for a mathematical thought to be tentative, incomplete, or even incorrect" and "An ideal polymath research comment should represent a 'quantum of progress' ". While *mathoverflow* is about asking questions, where typically the questioner believes others in the community may have the answer, *polymath* is about collaborating to solve open conjectures.

There have now been seven Polymath discussions, with some still ongoing, leading to significant advances and published papers, under the byline of "D J H Polymath" [69]. Analysis by Gowers [35], and by HCI researchers Cranshaw and Kittur [43], indicates that *polymath* has enabled a level of collaboration which, before the internet, would probably have been impossible to achieve; the open invitation has widened the mathematical community; and the focus on short informal comments has resulted in a readily available and public record of mathematical progress. As noted by Gowers, this provided "for possibly the first time ever (though I may well be wrong about this) the first fully documented account of how a serious research problem was solved, complete with false starts, dead ends etc." [33]. Four annual *mini-polymath* projects (so far) have selected problems from the current International Mathematical Olympiad: thus in contrast to the open-ended research context of *polymath*, participants trust the question to be solvable without advanced mathematical knowledge.

We investigated *mini-polymath* 3, which used the following problem.

Let S be a finite set of at least two points in the plane. Assume that no three points of S are collinear. A *windmill* is a process that starts with a line l going through a single point $P \in S$. The line rotates clockwise about the pivot P until the first time that the line meets some other point Q belonging to S. This point Q takes over as the new pivot, and the line now rotates clockwise about Q, until it next meets a point of S. This process continues indefinitely.

Show that we can choose a point P in S and a line l going through P such that the resulting windmill uses each point of S as a pivot infinitely many times.

It was solved over a period of 74 minutes by 27 participants through 174 comments on 27 comment threads. People mostly followed the rules, which were largely self regulating due to the speed of responses: a long answer in response to an older thread was likely to be ignored as the main discussion had moved on. Some sample comments included:

1 If the points form a convex polygon, it is easy.

2 Can someone give me *any* other example where the windmill cycles without visiting all the points? The only one I can come up with is: loop over the convex hull of S

3 One can start with any point (since every point of S should be pivot infinitely often), the direction of line that one starts with however matters!

4 Perhaps even the line does not matter! Is it possible to prove that any point and line will do?

5 The first point and line P_0, l_0 cannot be chosen so that P_0 is on the boundary of the convex hull of S and l_0 picks out an adjacent point on the convex hull. Maybe the strategy should be to take out the convex hull of S from consideration; follow it up by induction on removing successive convex hulls.

6 Since the points are in general position, you could define "the wheel of p", w(p) to be radial sequence of all the other points p!=p around p. Then, every transition from a point p to q will "set the windmill in a particular spot" in q. This device tries to clarify that the new point in a windmill sequence depends (only) on the two previous points of the sequence.

Within *mini-polymath* 3, we classified the main activity of each of the 174 comments as either:

Example 33% (1, 2 above). Examples and counterexamples played a key role: in understanding and exploring the problem, in clarifying explanations, and in exploring concepts and conjectures about the problem. In the early stages of understanding the problem, a number of participants were misled by the use of the term "windmill" to think of the rotating line as a half-line, a misunderstanding that led to counterexamples to the result they were asked to prove.[4]

Conjecture 20% (3, 4 above). This category included exploration of the limits of the initial question and various sub-conjectures. We identified conjectures made by analogy; conjectures that generalised the original problem; sub-conjectures towards a proof; and conjectured properties of the main windmill concept.

Proof 14% (5 above) Proof strategies found included induction, generalisation, and analogy.

Concept 10% (6 above) As well as standard concepts from Euclidean geometry and the like, even in such a relatively simple proof, new concepts arise by analogy; in formulating conjectures; or from considering examples and counterexamples. For example, analogies involving "windmills" led to the misapprehension referred to above.

Other 23% These typically concerned cross referencing to other comments; clarification; and social interjections, both mathematically interesting and purely social, including smiley faces and the like. All help to create a friendly, collaborative, informal and polite environment.

3.3   What do We Learn about Mathematical Practice?

Both *mathoverflow* and *mini-polymath* provide living examples of the backstage of mathematics.

While the utility of Pólya's ideas in an educational setting has been contested, *mini-polymath* shows many examples of his problem-solving heuristics operating in a collaborative, as opposed to individual, setting: for example we see participants rephrasing the question, using case splits and trying to generalise the problem. This is hardly surprising, as the questions themselves may have been designed to be solved by these techniques.

Both *mathoverflow* and *mini-polymath* afford precisely the sort of openness that Lakatos advocated in the teaching and presentation of mathematics (described above). We have seen the striking number of examples used in both *mathoverflow* and *mini-polymath* : this accords with the emphasis which Lakatos placed on examples. He emphasised fallibility and ambiguity in mathematical development, addressing semantic change in mathematics as the subject develops, the role that counterexamples play in concept, conjecture and proof development, and the social component of mathematics via a dialectic of ideas. Although his theory was highly social, it was not necessarily collaborative. For reasons of space we single out here Lakatos's notion of "monster-adjusting" examples: others are considered in [65].

Monster-adjusting occurs when an object is seen as a supporting example of a conjecture by one person and as a counterexample by another; thus exposing two rival interpretations of a concept definition. The object then becomes a trigger for concept development and clarification. Thus in our *mathoverflow* example this occurs, relative to the larger conversation not displayed, in the comment and adjustment of Figure 2 around "Why does q have to be odd?" In our *mini-polymath* study the monster-adjusting occurs in clarifying the rotating line of the question as a full line not a half-line: the problematic object is an equilateral triangle with one point in the centre; this exposes different interpretations of the concept of the rotating line.

While with sufficient ingenuity most of the examples we found in both systems could be assigned to one or more of Lakatos's categories, the process is quite subtle, and dependent on context in a way not always taken into account in Laktos's work: the *mathoverflow* example taken alone could also be seen a variation of Lakatos's exception-barring, where the conjecture is strengthened by lifting unnecessary conditions.

While Lakatos identifies the role that hidden assumptions play, and suggests ways of diagnosing and repairing flawed assumptions, he does not suggest how they might arise. Here we can go beyond Lakatos and hypothesise as to what might be the underlying reason for mistaken assumptions or rival interpretations. Lakoff and colleagues [47] and Barton [14] have explored the close connection between language and thought, and shown that images and metaphors used in ordinary language shape mathematical (and all other types of) thinking. We hypothesise that the misconception of a line as a half-line may be due to the naming of the concept; which triggered images of windmills with sails which pivoted around a central tower and extended in one direction only.[5]

We expect the use and development of online discussion to provide researchers into mathematical practice with large new bodies of data of informal reasoning in the wild. While it is an open question whether online mathematics is representative of other mathematical activity, it is certainly the case that this is one type of activity. This is validated by peer reviewed collective publications arising out of online discussions and by the user-base of 23,000 people on MathOverflow (a small but significant proportion of the world's research mathematicians).[6] It is also an open question as to whether it is *desirable* for online mathematical collaboration to model offline work, given the new potential of the online world. As a form of mathematical practice, it will inform (evolving) theories of (evolving) mathematical practices and – crucially – provides a much-needed way of empirically evaluating them.

The interdisciplinary study of mathematical practice is still very young, particularly when considered relative to its older, more respectable sibling, the philosophy of mathematics (~70 years versus ~2,300 years).[7] Different disciplines will focus on different aspects of the sites: philosophers will concern themselves with their fundamental question of how mathematics progresses; sociologists on the dynamics of the discussion and the socio-cultural-technical context in which it takes place; linguists may analyse the language used, and compare it to other forms of communication; mathematicians might reflect on whether there is a significant difference between massively collaborative maths and ordinary mathematics research; cognitive scientists will look for evidence of hypothesised cognitive behaviours, and so on. However, these questions are deeply interrelated. We predict that multi-disciplinary collaboration in constructing theories of mathematical practice will increase, and that online discussion sites will play an important role in uncovering processes and mechanisms behind informal mathematical collaboration. There is a an exciting potentially symbiotic relationship-in-the-making between the study of mathematical practice and that of computer support for mathematics.

---

[5] The IMO presents tremendous opportunity for cultural and linguistic analysis, as each problem is translated into at least five different languages, and candidate problems are evaluated partially for the ease with which they can be translated, and the process of translating a problem is taken extremely seriously.

[6] Estimates vary from ~80,000 (an estimate by Jean-Pierre Bourguignon based on the number of people who are in a profession which attaches importance to mathematics research and hold a Mathematics PhD or equivalent [17]), to ~140,000 (the number of people in the Mathematics Genealogy Project who got their PhD between 1960-2012), to ~350,000 (the number of people estimated still living, on the Math Reviews authors database): see http://mathoverflow.net/questions/5485/how-many-mathematicians-are-there

[7] We calculated the 2325 year age gap based on Polya's [68] in 1945 marking the beginning of MP and Plato's [66] in 380 BC on the theory of forms and the status of mathematical objects, marking the beginning of PoM.

4   Mathematics as a Social Machine: The Next Steps

The goal of social machines research is to understand the underlying computational and social principles, and devise a framework for building and deploying them.

While *polymath* and *mathoverflow* are fairly recent, the widely used "Online Encyclopaedia of Integer Sequences" (www.oeis.org) is a more long-standing example of a social machines for mathematics. Given a few digits of input, it proposes sequences which match it, through invoking subtle pattern matching against over 220,000 user-provided sequences: so, for example, user input of (3 1 4 1) returns π, and 556 other possibilities, each supported by links to the mathematical literature. Viewed as a social machine, it involves users with queries or proposed new entries; a wiki for discussions; volunteers curating the system; governance and funding mechanisms through a trust; alongside traditional computer support for a database, matching engine and web interface, with links to other mathematical data sources, such as research papers. While anyone can use the system, proposing a new sequence requires registration and a short CV, which is public, serving as a reputation system.

One can imagine many kinds of mathematics social machines: the kinds of parameters to be considered in thinking about them in a uniform way include, for example:

– precise versus loose queries and knowledge
– human versus machine creativity
– specialist or niche users versus general users
– logical precision versus cognitive appeal for output
– formal versus natural language for interaction
– checking versus generating conjectures or proofs
– formal versus informal proof
– "evolution" versus "revolution" for developing new systems
– governance, funding and longevity

Current social and not-so-social machines occupy many different points in this design space. Each dimension raises broad and enduring challenges, whether in traditional logic and semantics, human computer interaction, cognitive science, software engineering or information management.

4.1   Mathematical Elements

Likely mathematical elements of a mathematics social machine would include the following, all currently major research activities in their own right.

"Traditional" machine resources available, include software for symbolic and numeric mathematics such as GAP or Maple, theorem provers such as Coq or HOL, and bodies of data and proofs arising from such systems. Our work highlights the importance of including databases of examples, perhaps incorporating user tagging, and also of being able to mine libraries for data and deductions beyond the immediate facts they record: see in particular the work of Urban [79] on machine learning from such libraries. The emerging field of mathematical knowledge management [20] addresses ontologies and tools for sharing and mining such resources, for example providing "deep" search or executable papers. Such approaches should in future be able to provide access to the mathematical literature, especially in the light of ambitious digitisation plans currently being developed by the American Mathematical Society and the Sloan Foundation [6].

The presentation in *mathoverflow* and *polymath* is linear and text based. Machine rendering of mathematical text has been a huge advance in enabling mathematicians to efficiently represent their workings *in silico*, which in turn has enabled online rapid-fire exchange of ideas, but technology for going beyond the linear structure to capture the more complex structure of a proof attempt, or to represent diagrams, is less developed. At the end of the first *polymath* discussion there were 800 comments, and disentangling these for newcomers to the discussion or to write up the proof for publication can be problematic. Representing the workflow in realtime using argumentation visualization software, which provides a graphical representation, would help prospective participants to more easily understand the discussion and to more quickly identify areas to which they can contribute: initial experiments using the Online Visualization of Argument software [8] developed by Chris Reed and his group at the University of Dundee, are promising.

Turning to the less formal side of mathematics, current challenges raised by the mathematical community, for example see [2], include the importance of collaborative systems that "think like a mathematician", can handle unstructured approaches such as the use of "sloppy" natural language, support the exchange of informal knowledge and intuition not recorded in papers, and engage diverse researchers in creative problem-solving. This mirrors the results of research into mathematical practice: the importance of human factors, and of handling informal reasoning, error, and uncertainty. Turning messy human knowledge into a usable information space, and reasoning across widely differing user contexts and knowledge bases is only beginning to emerge as a challenge in artificial intelligence applied to mathematics, for example in the work of Bundy [18] on "soft" aspects such as creativity, analogy and concept formation and the handling of error by ontology repair [58], or work in cognitive science which studies the role of metaphor in the evolution and understanding of mathematical concepts [47].

Automated theory formation systems which automatically invent concepts and conjectures are receiving increasing attention. Examples include Lenat's AM [51], which was designed to both construct new concepts and conjecture relationships between them, and Colton's HR system [24,25]. HR uses production rules to form new concepts from old ones; measures of interestingness to drive a heuristic search; empirical pattern-based conjecture making techniques to find relationships between concepts, and third party logic systems to prove conjectures or find counterexamples. Other examples include the IsaScheme system by Montano Rivas [60], which employs a scheme-based approach to mathematical theory exploration; the IsaCosy system by Johansson *et al.* [41] which performs

---

[8] http://ova.computing.dundee.ac.uk/

inductive theory formation by synthesising conjectures from the available constants and free variables; and the MATHsAiD system by McCasland [55], which applies inference rules to user-provided axioms, and classifies the resulting proved statements as facts (results of no intrinsic mathematical interest), lemmas (statements which might be useful in the proof of subsequent theorems), or theorems (either routine or significant results). A survey of next generation automated theory formation is given in [64], including Pease's philosophically-inspired system HRL [63], which provides a computational representation of Lakatos's theory [46], and Charnley's cognitively-inspired system [22] based on Baar's theory of the Global Workspace [11].

Social expectations in *mathoverflow*, and generally in research mathematics, are of a culture of open discussion, and knowledge is freely shared provided it is attributed: for example, it is common practice in mathematics to make papers available before journal submission. As with mathematics as a whole, information accountability in principle in a mathematics social machine comes from a shared understanding that the arguments presented, while informal, are capable of refinement to a rigorous proof. In *mathoverflow*, as described in [77], social expectation and information accountability are strengthened through the power of off-line reputation: users are encouraged to use real names, and are likely to interact through professional relationships beyond *mathoverflow*. A further challenge for social computation will be scaling these factors up to larger more disparate communities who have less opportunity for real-world interaction; dealing in a principled way with credit and attribution as the contributions that social computation systems make become routinely significant; and incorporating models where contributions are traded rather than freely given.

4.2   Social Machines: The Broader Context

The research agenda laid out by social machines pioneers like Hendler, Berners Lee and Shadbolt is ambitious [38], with a goal of devising overarching principles to understand, design, build and deploy social machines. Viewing mathematics social machines in this way has the potential to provide a unifying framework for disparate ideas and activities.

Designing social computations. Social machine models view users as "entities" (cf agents or peers) and allow consideration of social interaction, enactment across the network, engagement and incentivisation, and methods of software composition that take into account evolving social aggregation. For mathematics this has far reaching implications — handling known patterns of practice, and enabling others as yet unimagined, as well as handling issues such as error and uncertainty, and variations in user beliefs.

Accessing data and information. Semantic web technology enables databases supporting provenance, annotation, citation and sophisticated search. Mathematics data includes papers, records of mathematical objects from systems such as Maple, and scripts from theorem provers. There has been considerable research in mathematical knowledge management [45], but current experiments in social machines for mathematics have little such support. Yet effective search, mining and data re-use would transform both mathematics research and related areas of software verification. Research questions are both technical, for example tracking provenance or ensuring annotation remains timely and correct [23], and social, for example many *mathoverflow* responses cite published work, raising the question of why users prefer asking *mathoverflow* to using a search engine.

Accountability, provenance and trust. Participants in social machines need to be able to trust the processes and data they engage with and share. Key concepts are *provenance*, knowing how data and results have been obtained, which contributes to *accountability*, ensuring that the source of any breakdown in trust can be identified and mitigated [80]. There is a long tradition of openness in mathematical research which has made endeavours like *polymath* or the *arXiv* possible and effective — for example posting drafts on the *arXiv* ahead of journal submission is reported as speeding up refereeing and reducing priority disputes [2]. Trusting mathematical results requires considering provenance of the proof, a major issue in assessing the balance between formal and informal proofs, and the basis for research into proof certificates [59]. Privacy and trust are significant for commercial or government work, where revealing even broad interests may already be a security concern.

Interactions among people, machines and data. Interactions among people, machines and data are core to social machines, which have potential to support novel forms of interaction and workflow which go beyond current practice, a focus of current social machine research [38]. Social mathematics shows a variety of communities, interactions and purposes, looking for information, solving problems, clarifying information and so on, displaying much broader interactions than those supported by typical mathematical software. In particular such workflows need to take account of informality and mistakes [50].

In conclusion, social machines both provide new ways of doing mathematics and the means for evaluating theories of mathematical practices. Improved knowledge of human interactions and reasoning in mathematics will suggest new ways in which artificial intelligence and computational mathematics can intersect with mathematics. We envisage that the challenges raised will include developing better computational support for mathematicians and modelling soft aspects of mathematical thinking such as errors, concept development and value judgements. There is much to be done, and a substantial body of research lies ahead of us, but the outcomes could transform the nature and production of mathematics.

Acknowledgements. Both authors acknowledge EPSRC support from EP/H500162 and EP/F02309X, and Pease in addition from EP/J004049. We both thank the School of Informatics at the University of Edinburgh for kind hospitality, where this work was done while the first author was on sabbatical, and thank, in particular, Alan Bundy and members of the DReaM group, and Robin Williams and members of the Social Informatics group, for many helpful discussions and insights.

References

1. A Wiles, interview on PBS (2000),
   http://www.pbs.org/wgbh/nova/physics/andrew-wiles-fermat.html
2. http://events.inf.ed.ac.uk/sicsa-mcp/
3. OEIS Foundation Inc., The On-Line Encyclopedia of Integer Sequences (2011),
   http://oeis.org
4. www.innocentive.com (retrieved October 2012)
5. SOCIAM: the theory and practice of social machines,
   http://tinyurl.com/cmmwbcf
6. http://ada00.math.uni-bielefeld.de/mediawiki-1.18.1
   (retrieved October 2012)
7. General polymath rules,
   http://polymathprojects.org/general-polymath-rules/
8. Which finite nonabelian groups have long chains of subgroups as intervals in their
   subgroup lattice? (April 2011), http://mathoverflow.net/questions/62495
9. Aberdein, A.: The uses of argument in mathematics. Argumentation 19, 287–301
   (2005)
10. Appel, K., Haken, W.: The solution of the four-color-map problem. Sci.
    Amer. 237(4), 108–121, 152 (1977)
11. Baars, B.: A cognitive theory of consciousness. Cambridge University Press (1988)
12. Barany, M.J., MacKenzie, D.: Chalk: Materials and concepts in mathematics re-
    search. New Representation in Scientific Practice (forthcoming)
13. Bartha, P.: By Parallel Reasoning. Oxford University Press, New York (2010)
14. Barton, B.: The Language of Mathematics: Telling Mathematical Tales. Mathe-
    matics Education Library, vol. 46. Springer (2009)
15. Begel, A., Bosch, J., Storey, M.-A.: Social Networking Meets Software Develop-
    ment: Perspectives from GitHub, MSDN, Stack Exchange, and TopCoder. IEEE
    Software 30(1), 52–66 (2013)
16. Berners-Lee, T., Fischetti, M.: Weaving the web - the original design and ultimate
    destiny of the World Wide Web by its inventor. HarperBusiness (2000)
17. Bourguignon, J.-P.: Mathematicians in France and in the world. L'Explosion des
    Mathematiques, pp. 92–97. SMF et SMAI (July 2002)
18. Bundy, A.: Automated theorem provers: a practical tool for the working mathe-
    matician? Ann. Math. Artif. Intell. 61(1), 3–14 (2011)
19. Carette, J., Dixon, L., Coen, C.S., Watt, S.M. (eds.): Calculemus/MKM 2009.
    LNCS, vol. 5625. Springer, Heidelberg (2009)
20. Carette, J., Farmer, W.M.: A Review of Mathematical Knowledge Management.
    In: [19], pp. 233–246
21. Cellucci, C., Gillies, D.: Mathematical reasoning and heuristics. King's College
    Publications, London (2005)
22. Charnley, J.: A Global Workspace Framework for Combined Reasoning. PhD the-
    sis, Imperial College, London (2010)
23. Cheney, J., Ahmed, A., Acar, U.A.: Provenance as dependency analysis. Mathe-
    matical Structures in Computer Science 21(6), 1301–1337 (2011)
24. Colton, S.: Automated Theory Formation in Pure Mathematics. Springer (2002)
25. Colton, S., Muggleton, S.: Mathematical applications of Inductive Logic Program-
    ming. Machine Learning 64, 25–64 (2006)
26. Corfield, D.: Assaying Lakatos's philosophy of mathematics. Studies in History and
    Philosophy of Science 28(1), 99–121 (1997)

27. Crowley, J., Hezlet, S., Kirby, R., McClure, D.: Mathematics journals: what is valued and what may change. Notices Amer. Math. Soc. 58(8), 1127–1130 (2011); Report on the workshop held at MSRI, Berkeley, CA (February 14-16, 2011)
28. DeMillo, R.A., Lipton, R.J., Perlis, A.J.: Social processes and proofs of theorems and programs. Commun. ACM 22(5), 271–280 (1979)
29. Ernest, P.: Social constructivism as a philosophy of mathematics. State University of New York Press, Albany (1997)
30. Giaquinto, M.: Visual Thinking in Mathematics. Clarendon Press, Oxford (2007)
31. Goffman, E.: The presentation of self in everyday life. Doubleday Anchor Books, New York (1959)
32. Gonthier, G.: Engineering mathematics: the odd order theorem proof. In: Giacobazzi, R., Cousot, R. (eds.) POPL, pp. 1–2. ACM (2013)
33. Gowers, T.: Polymath1 and open collaborative mathematics, http://gowers.wordpress.com/2009/03/10/polymath1-and-open-collaborative-mathematics/
34. Gowers, T.: Is massively collaborative mathematics possible? (March 2009), http://gowers.wordpress.com/2009/01/27/is-massively-collaborative-mathematics-possible/
35. Gowers, T., Nielsen, M.: Massively collaborative mathematics. Nature 461(7266), 879–881 (2009)
36. Hales, T.C., Harrison, J., McLaughlin, S., Nipkow, T., Obua, S., Zumkeller, R.: A revision of the proof of the kepler conjecture. Discrete & Computational Geometry 44(1), 1–34 (2010)
37. Harrison, J.: Handbook of Practical Logic and Automated Reasoning. Cambridge University Press (2009)
38. Hendler, J., Berners-Lee, T.: From the semantic web to social machines: A research challenge for ai on the world wide web. Artif. Intell. 174(2), 156–161 (2010)
39. Hersh, R.: Mathematics has a front and a back. Synthese 88(2), 127–133 (1991)
40. Hume, D.: A treatise of human nature. Penguin Books (1969)
41. Johansson, M., Dixon, L., Bundy, A.: Conjecture synthesis for inductive theories. Journal of Automated Reasoning (2010) (forthcoming)
42. Jones, C.B.: The early search for tractable ways of reasoning about programs. IEEE Annals of the History of Computing 25(2), 26–49 (2003)
43. Justin Cranshaw, J., Kittur, A.: The polymath project: Lessons from a successful online collaboration in mathematics. In: CHI, Vancouver, BC, Canada, May 7-12 (2011)
44. Kitcher, P.: The Nature of Mathematical Knowledge. Oxford University Press, Oxford (1983)
45. Kohlhase, M., Rabe, F.: Semantics of OpenMath and MathML3. Mathematics in Computer Science 6(3), 235–260 (2012)
46. Lakatos, I.: Proofs and Refutations. Cambridge University Press, Cambridge (1976)
47. Lakoff, G., Núñez, R.: Where Mathematics Comes From: How the Embodied Mind Brings Mathematics into Being. Basic Books, New York (2000)
48. Larsen, S., Zandieh, M.: Proofs and refutations in the undergraduate mathematics classroom. Educational Studies in Mathematics 67(3), 205–216 (2008)
49. Larvor, B.: Lakatos: An Introduction. Routledge, London (1998)
50. Lehmann, J., Varzinczak, I.J., Bundy, A.: Reasoning with context in the semantic web. J. Web Sem. 12, 1–2 (2012)
51. Lenat, D.B.: Automated theory formation in mathematics. In: Proceedings of the 5th International Joint Conference on Artificial Intelligence, pp. 833–842. Morgan Kaufmann, Cambridge (1977)

52. MacKenzie, D.: Mechanizing proof. MIT Press, Cambridge (2001)
53. Mancosu, P., Jørgensen, K.F., Pedersen, S.A. (eds.): Visualization, Explanation and Reasoning Styles in Mathematics. Springer, Dordrecht (2005)
54. Mancosu, P. (ed.): The Philosophy of Mathematical Practice. Oxford University Press, USA (2008)
55. McCasland, R., Bundy, A.: MATHsAiD: a Mathematical Theorem Discovery Tool. In: Proceedings of SYNASC, pp. 17–22. IEEE (2006)
56. Martin, U.: Computers, reasoning and mathematical practice. In: Computational logic (Marktoberdorf, 1997). NATO Adv. Sci. Ser. F Comput. Systems Sci., vol. 165, pp. 301–346. Springer, Berlin (1999)
57. Martin, U., Pease, A.: The mathematics social machine will be social! In: SOHU-MAN 2013 (to appear, 2013)
58. McNeill, F., Bundy, A.: Dynamic, automatic, first-order ontology repair by diagnosis of failed plan execution. Int. J. Semantic Web Inf. Syst. 3(3), 1–35 (2007)
59. Nigam, V., Miller, D.: A Framework for Proof Systems. J. Autom. Reasoning 45(2), 157–188 (2010)
60. Montano-Rivas, O., McCasland, R., Dixon, L., Bundy, A.: Scheme-based theorem discovery and concept invention. Expert Systems with Applications 39(2), 1637–1646 (2011)
61. Nielsen, M.: Reinventing discovery: the new era of networked science. Princeton University Press (2011)
62. Nunokawa, K.: Applying Lakatos' theory to the theory of mathematical problem solving. Educational Studies in Mathematics 31(3), 269–293 (1996)
63. Pease, A.: A Computational Model of Lakatos-style Reasoning. PhD thesis, School of Informatics, University of Edinburgh (2007), http://hdl.handle.net/1842/2113
64. Pease, A., Colton, S., Charnley, J.: Automated theory formation: The next generation. IFCoLog Lectures in Computational Logic (forthcoming, 2013)
65. Pease, A., Martin, U.: Seventy four minutes of mathematics: An analysis of the third mini-polymath project. In: Proc. AISB Symp. on Mathematical Practice and Cognition II, pp. 19–29 (2012)
66. Plato: The Republic. OUP, Oxford (1993)
67. Pólya, G.: How to solve it. Princeton University Press (1945)
68. Pólya, G.: Mathematical Discovery. John Wiley and Sons, New York (1962)
69. Polymath, D.H.J.: A new proof of the density Hales-Jewett theorem. Ann. of Math. (2) 175(3), 1283–1327 (2012)
70. Popper, K.R.: Objective Knowledge. OUP, Ely House (1972)
71. De Roure, D., Goble, C.A.: Software design for empowering scientists. IEEE Software 26(1), 88–95 (2009)
72. Schlimm, D.: Two ways of analogy: Extending the study of analogies to mathematical domains. Philosophy of Science 75, 178–200 (2008)
73. Schoenfeld, A.H.: Pólya, problem solving, and education. Mathematics Magazine 60(5), 283–291 (1987)
74. Simon, H.: Machine discovery. Foundations of Science 2, 171–200 (1997)
75. Tappenden, J.: Mathematical concepts and definitions. In: Mancosu, P. (ed.) The Philosophy of Mathematical Practice, pp. 256–275. Oxford University Press, Oxford (2008)
76. Tappenden, J.: Mathematical concepts: Fruitfulness and naturalness. In: Mancosu, P. (ed.) The Philosophy of Mathematical Practice, pp. 276–301. Oxford University Press, Oxford (2008)

77. Tausczik, Y.R., Pennebaker, J.W.: Predicting the perceived quality of online mathematics contributions from users' reputations. In: Tan, D.S., Amershi, S., Begole, B., Kellogg, W.A., Tungare, M. (eds.) CHI, pp. 1885–1888. ACM (2011)

78. Tausczik, Y.R., Pennebaker, J.W.: Participation in an online mathematics community: differentiating motivations to add. In: Poltrock, S.E., Simone, C., Grudin, J., Mark, G., Riedl, J. (eds.) CSCW, pp. 207–216. ACM (2012)

79. Urban, J., Vyskočil, J.: Theorem proving in large formal mathematics as an emerging AI field. In: Bonacina, M.P., Stickel, M.E. (eds.) Automated Reasoning and Mathematics. LNCS, vol. 7788, pp. 240–257. Springer, Heidelberg (2013)

80. Weitzner, D.J., Abelson, H., Berners-Lee, T., Feigenbaum, J., Hendler, J., Sussman, G.J.: Information accountability. Commun. ACM 51(6), 82–87 (2008)

Automated Reasoning Service for **HOL Light**

Cezary Kaliszyk[1] and Josef Urban[2]

[1] University of Innsbruck, Austria
[2] Radboud University, Nijmegen

Abstract. HOL(y)Hammer is an AI/ATP service for formal (computer-understandable) mathematics encoded in the HOL Light system, in particular for the users of the large Flyspeck library. The service uses several automated reasoning systems combined with several premise selection methods trained on previous Flyspeck proofs, to attack a new conjecture that uses the concepts defined in the Flyspeck library. The public online incarnation of the service runs on a 48-CPU server, currently employing in parallel for each task 25 AI/ATP combinations and 4 decision procedures that contribute to its overall performance. The system is also available for local installation by interested users, who can customize it for their own proof development. An Emacs interface allowing parallel asynchronous queries to the service is also provided. The overall structure of the service is outlined, problems that arise are discussed, and an initial account of using the system is given.

1 Introduction and Motivation

HOL Light [10] is one of the best-known interactive theorem proving (ITP) systems. It has been used to prove a number of well-known mathematical theorems[1] and to formalize the proof of the Kepler conjecture targeted by the Flyspeck project [9]. The whole Flyspeck development, together with the required parts of the HOL Light library consists of about 14.000 theorems and 1800 concepts. Motivated by the development of large-theory automated theorem proving [12,18,26,31] and its growing use for ITPs like Isabelle [19] and Mizar [29,30], we have recently implemented translations from HOL Light to ATP (automated theorem proving) formats, developed a number of premise-selection techniques for HOL Light, and experimented with the strongest and most orthogonal combinations of the premise-selection methods and various ATPs. This work, described in [15], has shown that 39% of the 14185 Flyspeck theorems could be proved in a push-button mode (without any high-level advice and user interaction) in 30 seconds of real time on a fourteen-CPU workstation.

The experiments that we did emulated the Flyspeck development (when user always knows all the previous proofs[2] at a given point, and wants to prove the next theorem), however they were all done in an offline mode which is suitable

---

[1] http://www.cs.ru.nl/~freek/100/
[2] The Flyspeck processing order is used to define precisely what "previous" means. See [15] for details.

J. Carette et al. (Eds.): CICM 2013, LNAI 7961, pp. 120–135, 2013.

for such experimentally-driven research. The ATP problems were created in large batches using different premise-selection techniques and different ATP encodings (untyped first-order [23], polymorphic typed first-order [4], and typed higher-order [8]), and then attempted with different ATPs (17 in total) and different numbers of the most relevant premises. Analysis of the results interleaved with further improvements of the methods and data have gradually led to the current strongest combination of the AI/ATP methods.

This strongest combination now gives to a HOL Light/Flyspeck user a 39% chance (when using 14 CPUs, each for 30s) that he will not have to search the library for suitable lemmas and figure out the proof of the next toplevel theorem by himself. For smaller (proof-local) lemmas such likelihood should be correspondingly higher. To really provide this strong automated advice to the users, the functions that have been implemented for the experiments need to be combined into a suitable AI/ATP tool. Our eventual goal (from which we are of course still very far) should be an easy-to-use service, which in its online form offers to formal mathematics (done here in HOL Light, over the Flyspeck-defined concepts) what services like Wolfram Alpha offer for informal/symbolic mathematics. Some expectations (linked to the recent success of the IBM Watson system) are today even higher[3]. Indeed, we believe that developing stronger and stronger AI/ATP tools similar to the one presented here is a necessary prerequisite (providing the crucial semantic understanding/reasoning layer) for building larger Watson-like systems for mathematics that will (eventually) understand (nearly-)natural language and (perhaps reasonably semanticized versions/alternatives of) LATEX. The more user-friendly and smarter such AI/ATP systems become, the higher also the chance that mathematicians (and exact scientists) will get some nontrivial benefits (apart from the obvious verification/correctness argument, which however so far convinced only a few) from encoding mathematics (and exact science) directly in a computer-understandable form.

This paper describes the first instance of such a HOL Light/Flyspeck-based AI/ATP service. The service – HOL(y)Hammer[4] (HH) – is now available in its strongest form as a public online system, running on a 48-CPU server spawning for each query 25 different AI/ATP combinations and four decision procedures. This functionality is described in Section 2, together with short examples of interaction (Emacs, command-line queries). The service can be also installed locally, and trained on user's private developments. This is described in Section 3. The advantages of the two approaches are briefly compared in Section 4, and Section 5 concludes and discusses future work.

2   The Online Service Description

The overall architecture of the system is shown in Figure 1. The service receives a query (a formula to prove, possibly with local assumptions) generated by one

---

[3] See for example Jonathan Borwein's article: `http://theconversation.edu.au/if-i-had-a-blank-cheque-id-turn-ibms-watson-into-a-maths-genius-1213`

[4] See [33] for an example of future where AIs turn into deities.

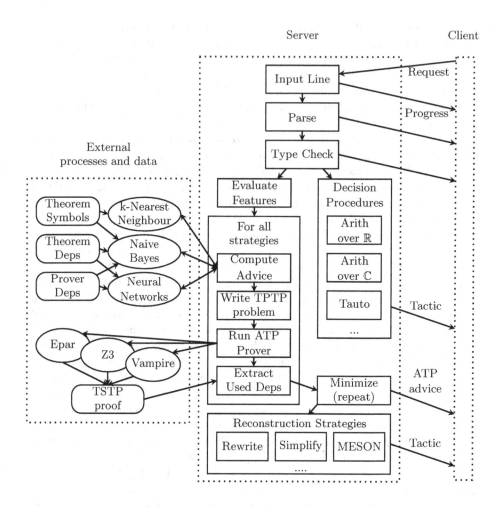

Fig. 1. Online service architecture overview

of the clients/frontends (Emacs, web interface, HOL session, etc.). If the query produces a parsing (or type-checking) error, an exception is raised, and an error message is sent as a reply. Otherwise the parsed query is processed in parallel by the (time-limited) AI/ATP combinations and the native HOL Light decision procedures (each managed by its forked HOL Light process, and terminated/killed by the master process if not finished within its global time limit). Each of the AI/ATP processes computes a specific feature representation of the query, and sends such features to a specific instance of a premise advisor trained (using the particular feature representation) on previous proofs. Each of the advisors replies with a specific number of premises, which are then translated to a suitable ATP format, and written to a temporary file on which a specific ATP is run. The successful ATP result is then (pseudo-)minimized, and handed over to the combination of proof-reconstruction procedures. These procedures again run in parallel, and if any of them is successful, the result is sent as a particular tactic application to the frontend. In case a native HOL Light decision procedure finds a proof, the result (again a particular tactic application) can be immediately sent to the frontend. The following subsections explain this process in more detail.

2.1 Feature Extraction and Premise Selection

Given a (formal) mathematical conjecture, the selection of suitable premises from a large formal library is an interesting AI problem, for which a number of methods have been tried recently [16, 26]. The strongest methods use machine learning on previous problems, combined in various ways with heuristics like SInE [12]. To use the machine learning systems, the previous problems have to be described as training examples in a suitable format, typically as a set of (input) features characterizing a given theorem, and a set of labels (output features) characterizing the proof of the theorem. Devising good feature/label characterizations for this task is again an interesting AI problem (see, e.g. [30]), however already the most obvious characterizations like the conjecture symbols and the names of the theorems used in the conjecture's proof are useful. This basic scheme can be extended in various ways; see [15] for the feature-extraction functions (basically adding various subterm and type-based characteristics) and label-improving methods (e.g., using minimized ATP proofs instead of the original Flyspeck proofs whenever possible) that we have so far used for HOL Light.

On average, for each feature-extraction method there are in total about 30.000 possible conjecture-characterizing features extracted from the theorems in the Flyspeck development. The output features (labels) are in the simplest setting just the names of the 14185 Flyspeck theorems[5] extracted from the proofs with a modified (proof recording [13]) HOL Light kernel. These features and labels are (for each extraction method) serially numbered in a stable way (using hashtables), producing from all Flyspeck proofs the training examples on which the premise selectors are trained. The learning-based premise selection methods currently used are those available in the SNoW [5] sparse learning toolkit (most prominently sparse naive Bayes) together with a custom implementation of the k-nearest neighbor (k-NN) learner. Training a particular learning method on all (14185) characterizations extracted from the Flyspeck proofs takes from 1 second for k-NN (a lazy learner that essentially just loads all the 14185 proof characterizations) and 6 seconds for naive Bayes using labels from minimized ATP proofs, to 25 seconds for naive Bayes using the labels from the original Flyspeck proofs.[6] The trained premise selectors are then run as daemons (using

---

[5] In practice, the Flyspeck theorems are further preprocessed to provide better learning precision, for example by splitting conjunctions and detecting which of the conjuncts are relevant in which proof. Again, see [15] for the details. The most recent number of labels used is thus 16082.

[6] The original Flyspeck proofs are often using theorems that are in some sense redundant, resulting in longer proof characterizations (and thus longer learning). This is typically a consequence of using larger building blocks (e.g., decision procedures, drawing in many dependencies) when constructing the ITP proofs.

their server modes) that accept queries in the language of the numerical features over which they have been trained, producing for each query their ranking of all the labels (corresponding to the available Flyspeck theorems).

Given a new conjecture, the first step of each of the forked HOL Light AI/ATP managing process is thus to compute the features of the conjecture according to a particular feature extraction method, compute (using the corresponding hashtable) the numerical representation of the features, and send these numeric features as a query to the corresponding premise-selection daemon. The daemon then replies (again, the speed depending on the learning method and the feature/label size) within a fraction of a second with its ranking, which is translated back (using the corresponding table) to the ranking of the HOL Light theorems. Each of the AI/ATP combinations then uses its particular number (optimized so that the methods in the end complement each other as much as possible) of the best-ranked theorems, passing them together with the conjecture to the function that translates such set of HOL Light formulas to a suitable ATP format.

2.2   Translation to ATP Formats and Running ATPs

As mentioned in Section 1, several ATP formalisms are used today by ATP and SMT systems. However the (jointly) most useful proof-producing systems in our experiments turned out to be E [22] version 1.6 (run under the Epar [28] strategy scheduler), Vampire [21] 2.6, and Z3 [6] 4.0. All these systems accept the TPTP untyped first-order format (FOF). Even when the input formalism (the HOL logic [20] - polymorphic version of Church's simple type theory) and the output formalism (TPTP FOF) are fixed, there are in general many methods [3] how to translate from the former to the latter, each method providing different tradeoffs between soundness, completeness, ATP efficiency, and the overall (i.e., including HOL proof reconstruction) efficiency. The particular method chosen by us in [15] and used currently also for the service is the polymorphic tagged encoding [3]. To summarize, the higher-order features (such as lambda abstraction, application) of the HOL formulas are first encoded (in a potentially incomplete way) in first-order logic (still using polymorphic types), and then type tags are added in a way that usually guarantees type safety during the first-order proof search.

This translation method is in general not stable on the level of single formulas, i.e., it is not possible to just keep in a global hashtable the translated FOF version for each original HOL formula, as done for example for the MizAR ATP service. This is because a particular optimization (by Meng and Paulson [17]) is used for translating higher-order constants, creating for each such constant c a first-order function that has the minimum arity with which c is used in the particular set of HOL formulas that is used to create the ATP (FOF) problem. So once the particular AI/ATP managing process advises its N most-relevant HOL Light theorems for the conjecture, this set of theorems and the conjecture are as a whole passed to the translation function, which for each AI/ATP instance may produce slightly different FOF encoding on the formula level. The encoding function is still reasonably fast (fractions of a second when using hundreds of formulas), and still has the property that both the FOF formula names and the FOF formulas (also those inferred during the ATP proof search) can (typically) be decoded back into the original HOL names and formulas (allowing later HOL proof reconstruction).

Each AI/ATP instance thus produces its specific temporary file (the FOF ATP problem) and runs its specific ATP system on it with its time limit. The time limit is currently set globally to 30 seconds for each instance, however (as usual in strategy scheduling setups) this could be made instance-specific too, based on further analysis of the time performance of the particular instances. Vampire and Epar already do such scheduling internally: the current version of Epar runs a fixed schedule of 14 strategies, while Vampire runs a problem-dependent schedule of several to dozen of strategies. Assuming one strategy for Z3 and on average eight strategies for Vampire, this means (counting the combinations in Table 1) that for each HOL query there are now 249 different proof-data/feature-extraction/learning/premise-slicing/ATP-strategy instantiations tried by the online service within the 30 seconds of the real time allowed for the query. Provided sufficient complementarity of such instantiations, this significantly raises the overall power of the service.

2.3   The AI/ATP Combinations Used

The 25 currently used combinations of the machine learner, proof data, number of top premises used, the feature extraction method, and the ATP system are shown in Table 1. The proof data are either just the data from the (minimized) ATP proofs (ATP0, ..., ATP3) created by a particular (MaLARea-style [31], i.e., re-using the proofs found in previous iteration for further learning) iteration of the experimenting, possibly preferring either the Vampire or Epar proofs (V_pref, E_pref), or a combination of such data from the ATP proofs with the original HOL proofs, obtained by slightly different versions of the HOL proof recording. Such combination typically uses the HOL proof only when the ATP proof is not available, see [15] for details. The standard feature extraction method combines the formula's symbols, standard-normalized subterms and normalized types into its feature vector. The standard normalization here means that each variable name is in each formula replaced by its normalized HOL type. The all-vars-same and all-vars-diff methods respectively just rename all formula variables into one common variable, or keep them all different. This obviously influences the concept of similarity used by the machine learners (see [15] for more discussion). The 40-NN and 160-NN learners are k-nearest-neighbors, run with $k = 40$ and $k = 160$. The reason for running these 25 particular combinations is that they together (computed in a greedy fashion) currently provide the greatest coverage of the solvable Flyspeck problems. This obviously changes quite often, whenever some of the many components of this AI architecture gets strengthened.

Table 1. The 25 AI/ATP combinations used by the online service

Learner	Proofs	Premises	Features	ATP
Bayes	ATP2	0092	standard	Vampire
Bayes	ATP2	0128	standard	Epar
Bayes	ATP2	0154	standard	Epar
Bayes	ATP2	1024	standard	Epar
Bayes	HOL0+ATP0	0512	all-vars-same	Epar
Bayes	HOL0+ATP0	0128	all-vars-diff	Vampire
Bayes	ATP1	0032	standard	Z3
Bayes	ATP1_V_pref	0128	all-vars-diff	Epar
Bayes	ATP1_V_pref	0128	standard	Z3
Bayes	HOL0+ATP0	0032	standard	Z3
Bayes	HOL0+ATP0	0154	all-vars-same	Epar
Bayes	HOL0+ATP0	0128	standard	Epar
Bayes	HOL0+ATP0	0128	standard	Vampire
Bayes	ATP1_E_pref	0128	standard	Z3
Bayes	ATP0_V_pref	0154	standard	Vampire
40-NN	ATP1	0032	standard	Epar
160-NN	ATP1	0512	standard	Z3
Bayes	HOL3+ATP3	0092	standard	Vampire
Bayes	HOL3+ATP3	0128	standard	Epar
Bayes	HOL3+ATP3	0154	standard	Epar
Bayes	HOL3+ATP3	1024	standard	Epar
Bayes	ATP3	0092	standard	Vampire
Bayes	ATP3	0128	standard	Epar
Bayes	ATP3	0154	standard	Epar
Bayes	ATP3	1024	standard	Epar

2.4   Use of Decision Procedures

Some goals are hard for ATPs, but are easy for the existing decision procedures already implemented in HOL Light. To make the service more powerful, we also try to directly use some of these HOL Light decision procedures on the given conjecture. A similar effect could be achieved also by mapping some of the HOL Light symbols (typically those encoding arithmetics) to the symbols that are reserved and treated specially by SMT solvers and ATP systems. This is now done for example in Isabelle/Sledgehammer [18], with the additional benefit of the combined methods employed by SMTs and ATPs over various well-known theories. Our approach is so far much simpler, which also means that we do not have to ensure that the semantics of such special theories remains the same (e.g., $1/0 = 0$ in HOL Light). The HOL Light decision procedures might often not be powerful enough to prove whole theorems, however for example the REAL_ARITH[7] tactic is called on 2678 unique (sub)goals in Flyspeck, making such tools a useful addition to the service.

---

[7] http://www.cl.cam.ac.uk/~jrh13/hol-light/HTML/REAL_ARITH.html

Each decision procedure is spawned in a separate instance of HOL Light using our parallel infrastructure, and if any returns within the timeout, it is reported to the user. The decision procedures that we found most useful for solving goals are:[8]

- TAUT[9] — Propositional tautologies.
  `(A ==> B ==> C) ==> (A ==> B) ==> (A ==> C)`
- INT_ARITH[10] — Algebra and linear arithmetic over $\mathbb{Z}$ (including $\mathbb{R}$).
  `&2 * &1 = &2 + &0`
- COMPLEX_FIELD — Field tactic over $\mathbb{C}$ (including multivariate $\mathbb{R}^{11}$).
  `(Cx (&1) + Cx(&1)) = Cx(&2)`

Additionally the decision procedure infrastructure can be used to try common tactics that could solve the goal. One that we found especially useful is simplification with arithmetic (`SIMP_TAC[ARITH]`), which solves a number of simple numerical goals that the service users ask the server.

2.5   Proof Minimization and Reconstruction

When an ATP finds (and reports in its proof) a subset of the advised premises that prove the goal, it is often the case that this set is not minimal. By re-running the prover and other provers with only this set of proof-relevant premises, it is often possible to obtain a proof that uses less premises. A common example are redundant equalities that may be used by the ATP for early (but unnecessary) rewriting in the presence of many premises, and avoided when the number of premises is significantly lower (and different ordering is then used, or a completely different strategy or ATP might find a very different proof). This (pseudo/cross-minimization) procedure is run recursively, until the number of premises needed for the proof no longer decreases. Minimizing the number of premises improves the chances of the HOL proof reconstruction, and the speed of (re-)processing large libraries that contain many such reconstruction tactics.[12]

Given the minimized list of advised premises, we try to reconstruct the proof. As mentioned in Section 2.1, the advice system may internally use a number of theorem names (now mostly produced by splitting conjunctions) not present in standard HOL Light developments. It is possible to call the reconstruction tactics with the names used internally in the advice system; however this would create

---

[8] The reader might wonder why the above mentioned REAL_ARITH is not among the tactics used. The reason is that even though REAL_ARITH is used a lot in HOL Light formalizations, INT_ARITH is simply more powerful. It solves 60% more Flyspeck goals automatically without losing any of those solved by REAL_ARITH. As with the AI/ATP instances, the usage of decision procedures is optimized to jointly cover as many problems as possible.

[9] http://www.cl.cam.ac.uk/~jrh13/hol-light/HTML/TAUT.html

[10] http://www.cl.cam.ac.uk/~jrh13/hol-light/HTML/INT_ARITH.html

[11] http://www.cl.cam.ac.uk/~jrh13/hol-light/HTML/REAL_FIELD.html

[12] Premise minimization has been for long time used to improve the quality and refactoring speed of the Mizar articles. It is now also a standard part of Sledgehammer.

proof scripts that are not compatible with the original developments. We could directly address the theorem sub-conjuncts (using, e.g., "nth (CONJUNCTS thm) n") however such proof scripts look quite unnatural (even if they are indeed faster to process by HOL Light). Instead, we now prefer to use the whole original theorems (including all conjuncts) in the reconstruction.

Three basic strategies are now tried to reconstruct the proof: REWRITE[13] (rewriting), SIMP[14] (conditional rewriting) and MESON [11] (internal first-order ATP). These three strategies are started in parallel, each with the list of HOL theorems that correspond to the minimized list of ATP premises as explained above. The strongest of these tactics – MESON – can in one second reconstruct 79.3% of the minimized ATP proofs. While this is certainly useful, the performance of MESON reconstruction drops below 40% as soon as the ATP proof uses at least seven premises. Since the service is getting stronger and stronger, the ratio of MESON-reconstructable proofs is likely to get lower and lower. That is why we have developed also a fine-grained reconstruction method – HH_RECON [14], which uses the quite detailed TPTP proofs produced by Vampire and E. This method however still needs an additional mechanism that maintains the TPTP proof as part of the user development: either dedicated storage, or on-demand ATP-recreation, or translation to a corresponding fine-grained HOL Light proof script. That is why HH_RECON is not yet included by default in the service.

2.6   Description of the Parallelization Infrastructure

An important aspect of the online service is its parallelization capability. This is needed to efficiently process multiple requests coming in from the clients, and to execute the large number of AI/ATP instances in parallel within a short overall wall-clock time limit. HOL Light uses a number of imperative features of OCaml, such as static lists of constants and axioms, and a number of references (mutable variables). Also a number of procedures that are needed use shared references internally. For example the MESON procedure uses list references for variables. This makes HOL Light not thread safe. Instead of spending lots of time on a thread-safe re-implementation, the service just (in a pragmatic and simple way, similar to the Mizar parallelization [27]) uses separate processes (Unix fork), which is sufficient for our purposes. Given a list of HOL Light tasks that should be performed in parallel and a timeout, the managing process spawns a child process for each of the tasks. It also creates a pipe for communicating with each child process. Progress, failures or completion information are sent over the pipe using OCaml marshalling. This means that it is enough to have running just one managing instance of HOL Light loaded with Flyspeck and with the advising infrastructure. This process forks itself for each client query, and the child then spawns as many AI/ATP, minimization, reconstruction, and decision procedure instances as needed. The service currently runs on a 48-core server with AMD Opteron 6174 2.2 GHz CPUs, 320 GB RAM, and 0.5 MB L2 cache per CPU.

---

[13] http://www.cl.cam.ac.uk/~jrh13/hol-light/HTML/REWRITE_TAC.html
[14] http://www.cl.cam.ac.uk/~jrh13/hol-light/HTML/SIMP_TAC.html

2.7   Use of Caching

Even though the service can asynchronously process a number of parallel requests, it is not immune to overloading by a large number of requests coming in simultaneously. In such cases, each response gets less CPU time and the requests are less likely to succeed within the 30 seconds of wall-clock time. Such overloading is especially common for requests generated automatically. For example the Wiki service that is being built for Flyspeck [24] may ask many queries practically simultaneously when an article in the wiki is re-factored, but many of such queries will in practice overlap with previously asked queries. Caching is therefore employed by the service to efficiently serve such repeated requests.

Since the parallel architecture uses different processes to serve different requests, a file-system based cache is used (using file-level locking). For any incoming request the first job done by the forked process handling the request is to check whether an identical request has already been served, and if so, the process just re-sends the previously computed answer. If the request is not found in the cache, a new entry (file) for it is created, and any information sent to the client (apart from the progress information) is also written to the cache entry. This means that all kinds of answers that have been sent to the client can be cached, including information about terms that failed to parse or typecheck, terms solved by ATP only, minimization results and replaying results, including decision procedures. The cache stored in the filesystem has the additional advantage of persistence, and in case of updating the service the cache can be easily invalidated by simply removing the cache entries.

2.8   Modes of Interaction with the Service

Figure 2 shows an Emacs session with several HOL Light goals.[15] The online advisor has been asynchronously called on the goals, and just returned the answer for the fifth goal and inserted the corresponding tactic call at an appropriate place in the buffer. The relevant Emacs code (customized for the HOL Light mode distributed with Flyspeck) is available online[16] and also distributed with the local HOL(y)Hammer install. It is a modification of the similar code used for communicating with the MizAR service from Emacs.

An experimental web editor interacting both with HOL Light and with the online advisor is described in [24]. The simplest option (useful as a basis for more sophisticated interfaces) is to interact with the service in command line, for example using netcat, as shown for two following two queries. The first query is solved easily by INT_ARITH, while the other requires nontrivial premise and proof search. Table 2 gives an overview of the service use so far (the queries came from 67 unique IP addresses).

---

[15] A longer video of the interaction is at http://mws.cs.ru.nl/~urban/ha1.mp4
[16] https://raw.github.com/JUrban/hol-advisor/master/hol-advice.el

```
g `CARD {2, 3} = 2`;; (* ATP Proof: *)
e(REWRITE_TAC[TRUTH;NOT_CLAUSES_WEAK;Geomdetail.CARD_SET2;ARITH_EQ]);;

g `0 = 1`;; (* No ATP proof found *)

g `!n s. n simplex s ==> FINITE {f | f face_of s}`;; (* ATP proof: *)
e(MESON_TAC[SIMPLEX_IMP_POLYTOPE;FINITE_POLYTOPE_FACES]);;

g `ODD 0 ==> ODD (S (S 0))`;; (* ATP proof: *)
e(SIMP_TAC[ARITH]);;

g `!f s t. f face_of s /\ convex t ==> f INTER t face_of s INTER t`;;
(* ATP proof: *)
e(REWRITE_TAC[FACE_OF_SLICE]);;

g `f continuous_on s /\ closed s /\ f continuous_on s1 /\ closed s1
    ==> measurable_on f (s UNION s1)`;; (* ATP proof: *)
e(SIMP_TAC[CONTINUOUS_ON_UNION;CONTINUOUS_IMP_MEASURABLE_ON_CLOSED_SUB
;SET;CLOSED_UNION]);;

g `interior(closure s) = {} /\ interior(closure t) = {}
    ==> interior(closure(s UNION t)) = {}`;; (* ATP proof: *)
-:**- tst.hl      Top (1,22)   Git:master  (HOL Light +l Abbrev)
* Result (51.69s): FACE_OF_SLICE
* Replaying: SUCCESS (0.75s):REWRITE_TAC[FACE_OF_SLICE]
```

Fig. 2. Parallel asynchronous calls of the online advisor from Emacs

```
$ echo 'max a b = &1 / &2 * ((a + b) + abs(a - b))'
  | nc colo12-c703.uibk.ac.at 8080
......
* Replaying: SUCCESS (0.25s): INT_ARITH_TAC
* Loadavg: 48.13 48.76 48.49 52/1151 46604

$ echo '!A B (C:A->bool).((A DIFF B) INTER C=EMPTY) <=> ((A INTER C) SUBSET B)'
  | nc colo12-c703.uibk.ac.at 8080
* Read OK
.............
* Theorem! Time: 14.74s Prover: Z Hints: 32 Str:
  allt_notrivsyms_m10u_all_atponly
* Minimizing, current no: 9
.* Minimizing, current no: 6
* Result: EMPTY_SUBSET IN_DIFF IN_INTER MEMBER_NOT_EMPTY SUBSET SUBSET_ANTISYM
```

Table 2. Statistics of the queries to the online service (Jan 24 - Mar 11 2013)

Total (Unique)	Parsed	Typechecked	Solved	ATP-solved	Reconstructed	Dec. Proc. solved
482	445	382	228	108	86	142

3   The Local Service Description

The service can be also downloaded,[17] installed and used locally, for example when a user is working on a private formalization that cannot be included in the public online service.[18]

Installing the advisor locally now requires two passes through the user's repository. In the first pass, a special module of the advisor stores the names of all the theorems available in the user's repository, together with their features (symbols, terms, types, etc., as explained in Section 2.1). In the second pass, the dependencies between the named theorems are computed, again using the modified proof recording HOL Light kernel that records all the processing steps. Given the exported features and dependencies, local advice system(s) (premise selectors) are trained outside HOL Light. Using the fast sparse learning methods described in Section 2.1, this again takes seconds, depending on the user hardware and the size of the development. The advisors are then run locally (as independent servers) to serve the requests coming from HOL Light. While the first pass is just a fast additional function that can be run by the user at any time on top of his loaded repository, the second pass now still requires full additional processing of the repository. This could be improved in the future by running the proof-recording kernel as a default, as it is done for example in Isabelle.

The user is provided with a tactic (HH_ADVICE_TAC) which runs all the mechanisms described in the Section 2 on the current goal locally. This means that the functions relying on external premise selection and ATPs are tried in parallel, together with a number of decision procedures. The ATPs are expected to be installed on the user's machine and (as in the online service) they are run on the goal translated to the TPTP format, together with a limited number of premises optimized separately for each prover. By default Vampire, Eprover and Z3 are now run, using three-fold parallelization.

The local installation in its simple configuration is now only trained using the naive Bayes algorithm on the training data coming from the HOL Light proof dependencies and the features extracted with the standard method. As shown in [15], the machine learning advice can be strengthened using ATP dependencies, which can be also optionally plugged into the local mode. Further strengthening can be done with combinations of various methods. This is easy to adjust; for example a user with a 24-CPU workstation can re-use/optimize the parallel combinations from Table 1 used by the online service.

4   Comparison of the Online and Local Service

The two related existing services are MizAR and Sledgehammer. MizAR has so far been an online service (accessible via Emacs or web interface), while Sledgehammer has so far required a local install (even though it already calls some ATPs over a

---

[17] http://cl-informatik.uibk.ac.at/users/cek/hh/

[18] The online service could eventually also accommodate private clones, using for example the techniques proposed for the Mizar Wiki in [2].

network). HOL(y)Hammer started as an online service, and the local version has been added recently to answer the demand by some (power)users.

As described in Section 2, the online service now runs 25 different AI/ATP instances and 4 decision procedures for each query. When counting the individual ATP strategies (which may indeed be very orthogonal in systems like Vampire and E), this translates to 249 different AI/ATP attempts for each query. If the demands grows, we can already now distribute the load from the current 48-CPU server to 112 CPUs by installing the service on another 64-CPU server. The old resolution-ATP wisdom is that systems rarely prove a result in higher time limits, since the search space grows very fast. A more recent wisdom (most prominently demonstrated by Vampire) however is that using (sufficiently orthogonal) strategy scheduling makes higher time limits much more useful.[19] And even more recent wisdom is that learning in various ways from related successes and failures further improves the systems' chances when given more resources.[20] All this makes a strong case for developing powerful online computing services that can in short bursts focus its great power on the user queries, which are typically related to many previous problems. Also in some sense, the currently used AI/ATP methods are only scratching the surface. For example, further predictive power is obtained in MaLARea [31] by computing thousands of interesting finite models, and using evaluation in them as additional semantic features of the formulas. ATP prototypes like MaLeCoP [32] can already benefit from accumulated fine-grained learned AI guidance at every inference step that they make. The service can try to make the best (re-)use of all smaller lemmas that have been proved so far (as in [25]). And as usual in machine learning, the more data are centrally accumulated for such methods, the stronger the methods become. Finally, it is hard to overlook the recent trend of light-weight devices for which the hard computational tasks are computed by large server farms (cloud computing).

The arguments for installing the service locally are mainly the option to use the service offline, and so far also the fact that the online service does not yet accept and learn on (possibly private) user developments. The latter is just a matter of additional implementation work. For example the MizAℝ service already now keeps a number of (incompatible) MML versions over which the query can be formulated, and techniques have been recently developed for the Mizar wiki that provide very fast and space-efficient cloning of large libraries and private additions over them managed by the server. As usual, the local install will also require the tools involved to work on all kinds of architectures, which is often an issue, particularly with software that is mostly developed in academia.

5   Conclusion and Future Work

HOL(y)Hammer is one of the strongest AI/ATP services currently available. It uses a toolchain of evolving methods that have been continuously improved as

---

[19] In [15], the relative performance of Vampire in 30 and 900 seconds is very different.

[20] See, e.g., the performance graph for the MaLARea 0.4 system in the recent Mizar@Turing12 competition: http://www.tptp.org/CASC/J6/TuringWWWFiles/ResultsPlots.html#MRTProblems

more and more experiments and computations have been done over the Flyspeck corpus in the past six months. The combinations that jointly provide the greatest theorem-proving coverage are employed to answer the queries with parallelization of practically all of the components. The parallelization factor is probably the highest of all existing ATP services, helping to focus the power of many different AI/ATP methods to answer the queries as quickly as possible.

At this moment, there seems to be no end to better premise selection, better translation methods for ATPs (and SMTs, and more advanced combined systems like MetiTarski [1]), better ATP methods (and their AI-based guidance), and better reconstruction methods. Useful work can be also done by making the online service accept private user developments and clones that currently have to rely only on the local installation. An interesting future direction is the use of the service with its large knowledge base and growing reasoning power as a semantic understanding (connecting) layer for experiments with tools that attempt to extract logical meaning from informal mathematical texts. Mathematics, with its explicit semantics, could in fact pioneer the technology of very deep parsing of scientific natural language writings, and their utilization in making stronger and stronger automated reasoning tools about all kinds of scientific domains.

References

1. Akbarpour, B., Paulson, L.C.: MetiTarski: An automatic theorem prover for real-valued special functions. J. Autom. Reasoning 44(3), 175–205 (2010)
2. Alama, J., Brink, K., Mamane, L., Urban, J.: Large formal wikis: Issues and solutions. In: Davenport, J.H., Farmer, W.M., Urban, J., Rabe, F. (eds.) Calculemus/MKM 2011. LNCS, vol. 6824, pp. 133–148. Springer, Heidelberg (2011)
3. Blanchette, J.C., Böhme, S., Popescu, A., Smallbone, N.: Encoding Monomorphic and Polymorphic Types. In: Piterman, N., Smolka, S.A. (eds.) TACAS 2013. LNCS, vol. 7795, pp. 493–507. Springer, Heidelberg (2013), http://www21.in.tum.de/~blanchet/enc_types_paper.pdf
4. Blanchette, J.C., Paskevich, A.: TFF1: The TPTP typed first-order form with rank-1 polymorphism, http://www21.in.tum.de/~blanchet/tff1spec.pdf
5. Carlson, A., Cumby, C., Rosen, J., Roth, D.: The SNoW Learning Architecture. Technical Report UIUCDCS-R-99-2101, UIUC Computer Science Department, 5 (1999)
6. de Moura, L., Bjørner, N.: Z3: An efficient SMT solver. In: Ramakrishnan, C.R., Rehof, J. (eds.) TACAS 2008. LNCS, vol. 4963, pp. 337–340. Springer, Heidelberg (2008)
7. Furbach, U., Shankar, N. (eds.): IJCAR 2006. LNCS (LNAI), vol. 4130. Springer, Heidelberg (2006)
8. Van Gelder, A., Sutcliffe, G.: Extending the TPTP language to higher-order logic with automated parser generation. In: Furbach, Shankar (eds.) [7], pp. 156–161
9. Hales, T.C.: Introduction to the Flyspeck project. In: Coquand, T., Lombardi, H., Roy, M.-F. (eds.) Mathematics, Algorithms, Proofs. Dagstuhl Seminar Proceedings, vol. 05021. Internationales Begegnungs- und Forschungszentrum für Informatik (IBFI), Schloss Dagstuhl, Germany (2005)
10. Harrison, J.: HOL Light: A tutorial introduction. In: Srivas, M., Camilleri, A. (eds.) FMCAD 1996. LNCS, vol. 1166, pp. 265–269. Springer, Heidelberg (1996)

11. Harrison, J.: Optimizing Proof Search in Model Elimination. In: McRobbie, M.A., Slaney, J.K. (eds.) CADE 1996. LNCS, vol. 1104, pp. 313–327. Springer, Heidelberg (1996)

12. Hoder, K., Voronkov, A.: Sine qua non for large theory reasoning. In: Bjørner, N., Sofronie-Stokkermans, V. (eds.) CADE 2011. LNCS, vol. 6803, pp. 299–314. Springer, Heidelberg (2011)

13. Kaliszyk, C., Krauss, A.: Scalable LCF-style proof translation. Accepted to ITP 2013, http://cl-informatik.uibk.ac.at/~cek/import.pdf

14. Kaliszyk, C., Urban, J.: PRocH: Proof reconstruction for HOL Light. In: Bonacina, M.P. (ed.) CADE 2013. LNCS, vol. 7898, pp. 267–274. Springer, Heidelberg (2013), http://mws.cs.ru.nl/~urban/proofs.pdf

15. Kaliszyk, C., Urban, J.: Learning-assisted automated reasoning with Flyspeck. CoRR, abs/1211.7012 (2012)

16. Kühlwein, D., van Laarhoven, T., Tsivtsivadze, E., Urban, J., Heskes, T.: Overview and evaluation of premise selection techniques for large theory mathematics. In: Gramlich, B., Miller, D., Sattler, U. (eds.) IJCAR 2012. LNCS, vol. 7364, pp. 378–392. Springer, Heidelberg (2012)

17. Meng, J., Paulson, L.C.: Translating higher-order clauses to first-order clauses. J. Autom. Reasoning 40(1), 35–60 (2008)

18. Paulson, L.C., Blanchette, J.: Three years of experience with Sledgehammer, a practical link between automated and interactive theorem provers. In: 8th IWIL (2010) (Invited talk)

19. Paulson, L.C., Susanto, K.W.: Source-level proof reconstruction for interactive theorem proving. In: Schneider, K., Brandt, J. (eds.) TPHOLs 2007. LNCS, vol. 4732, pp. 232–245. Springer, Heidelberg (2007)

20. Pitts, A.: The HOL logic. In: Gordon, M.J.C., Melham, T.F. (eds.) Introduction to HOL: A Theorem Proving Environment for Higher Order Logic. Cambridge University Press (1993)

21. Riazanov, A., Voronkov, A.: The design and implementation of VAMPIRE. AI Commun. 15(2-3), 91–110 (2002)

22. Schulz, S.: E - A Brainiac Theorem Prover. AI Commun. 15(2-3), 111–126 (2002)

23. Sutcliffe, G., Schulz, S., Claessen, K., Van Gelder, A.: Using the TPTP language for writing derivations and finite interpretations. In: Furbach, Shankar (eds.) [7], pp. 67–81

24. Tankink, C., Kaliszyk, C., Urban, J., Geuvers, H.: Formal mathematics on display: A wiki for Flyspeck. In: Carette, J., Aspinall, D., Lange, C., Sojka, P., Windsteiger, W. (eds.) CICM 2013. LNCS (LNAI), vol. 7961, pp. 152–167. Springer, Heidelberg (2013)

25. Urban, J.: MoMM - fast interreduction and retrieval in large libraries of formalized mathematics. International Journal on Artificial Intelligence Tools 15(1), 109–130 (2006)

26. Urban, J.: An Overview of Methods for Large-Theory Automated Theorem Proving (Invited Paper). In: Höfner, P., McIver, A., Struth, G. (eds.) ATE Workshop. CEUR Workshop Proceedings, vol. 760, pp. 3–8. CEUR-WS.org (2011)

27. Urban, J.: Parallelizing Mizar. CoRR, abs/1206.0141 (2012)

28. Urban, J.: BliStr: The Blind Strategymaker. CoRR, abs/1301.2683 (2013)

29. Urban, J., Rudnicki, P., Sutcliffe, G.: ATP and presentation service for Mizar formalizations. J. Autom. Reasoning 50, 229–241 (2013)

30. Urban, J., Sutcliffe, G.: Automated reasoning and presentation support for formalizing mathematics in Mizar. In: Autexier, S., Calmet, J., Delahaye, D., Ion, P.D.F., Rideau, L., Rioboo, R., Sexton, A.P. (eds.) AISC/Calculemus/MKM 2010. LNCS, vol. 6167, pp. 132–146. Springer, Heidelberg (2010)

31. Urban, J., Sutcliffe, G., Pudlák, P., Vyskočil, J.: MaLARea SG1 - Machine Learner for Automated Reasoning with Semantic Guidance. In: Armando, A., Baumgartner, P., Dowek, G. (eds.) IJCAR 2008. LNCS (LNAI), vol. 5195, pp. 441–456. Springer, Heidelberg (2008)

32. Urban, J., Vyskočil, J., Štěpánek, P.: MaLeCoP Machine learning connection prover. In: Brünnler, K., Metcalfe, G. (eds.) TABLEAUX 2011. LNCS, vol. 6793, pp. 263–277. Springer, Heidelberg (2011)

33. Vinge, V.: A Fire Upon the Deep. Tor Books (1992)

Understanding Branch Cuts of Expressions

Matthew England, Russell Bradford, James H. Davenport, and David Wilson

University of Bath, Bath, BA2 7AY, U.K.
{M.England,R.J.Bradford,J.H.Davenport,D.J.Wilson}@bath.ac.uk
http://people.bath.ac.uk/masjhd/Triangular/

Abstract. We assume some standard choices for the branch cuts of a group of functions and consider the problem of then calculating the branch cuts of expressions involving those functions. Typical examples include the addition formulae for inverse trigonometric functions. Understanding these cuts is essential for working with the single-valued counterparts, the common approach to encoding multi-valued functions in computer algebra systems. While the defining choices are usually simple (typically portions of either the real or imaginary axes) the cuts induced by the expression may be surprisingly complicated. We have made explicit and implemented techniques for calculating the cuts in the computer algebra programme MAPLE. We discuss the issues raised, classifying the different cuts produced. The techniques have been gathered in the **BranchCuts** package, along with tools for visualising the cuts. The package is included in MAPLE 17 as part of the **FunctionAdvisor** tool.

Keywords: branch cuts, simplification, symbolic computation.

1 Introduction

We consider the problem of calculating the branch cuts of expressions in a single complex variable. When defining multi-valued functions mathematicians have a choice of where to define the branch cuts. There are standard choices for most well-known functions [1, 18, 21], usually following the work of Abramowitz and Stegun. These choices were justified in [11] and match the choices within the computer algebra programme MAPLE for all elementary functions except arccot (for reasons explained in [11]). Within this paper we assume branch cut definitions matching those of MAPLE (which may be observed using Maple's FunctionAdvisor by giving the function name without an argument). We note that a different choice would not lead to any fewer or less complicated issues.

Handbooks (including online resources such as [21]) and software usually stop at these static definitions. However, our thesis is that this knowledge should be dynamic; processed for the user so it is suitable for their situation. Hence we consider the problems that follow after the initial choice of definition is settled. This will involve symbolic computation but is also an issue of Mathematical Knowledge Management (following the *process view* of MKM in [9]).

We wish to axiomatically understand the branch cuts of *expressions* in multi-valued functions, such as functions applied to a non-trivial argument, function

J. Carette et al. (Eds.): CICM 2013, LNAI 7961, pp. 136–151, 2013.

compositions, and function combinations (sum, product, relations). Many of the well-known formulae for elementary functions, such as addition formulae for inverse trigonometric functions, are such expressions. Care needs to be taken when working with multi-valued functions since there are different, often unstated, viewpoints possible as discussed in [12,13]. Most computer algebra software (and indeed most users) tend to work with multi-valued functions by defining their single-valued counterparts which will have discontinuities over the branch cuts. As a result, relations true for the multi-valued functions may no longer be true for the single valued counterparts and hence understanding the branch cuts of the relations becomes essential for working with them efficiently.

Despite the importance of understanding such branch cuts, the authors are not aware of any (available) software which calculates them beyond the original definitions. It also seems rare for them to get a detailed mathematical study in their own right, beyond their introduction and simple examples, with [17] one notable exception.

We denote multivalued functions evaluating to sets of values using names with upper cases (i.e. Arctan, Sqrt(z), Log) and denote their single valued counterparts by the normal notation (i.e. arctan, $\sqrt{z}$, log). So, for example, Sqrt(4) = $\{-2, 2\}$ while $\sqrt{4} = 2$. (Given our above choice of branch cut definitions, this now means our notation throughout the paper matches the commands in MAPLE.) We note that when dealing with sets of values for multi-valued functions not all combinations of choices of values of will be meaningful and sometimes the choices for sub-expression values are correlated.

A simple example of the problem described above is that while the identity Sqrt(x)Sqrt(y) = Sqrt(xy) is true (in the sense that the set of all possible products of entries from the two sets on the right is the same as the set on the left), the single valued counterpart $\sqrt{x}\sqrt{y} = \sqrt{xy}$ is not universally true (for example when $x = y = -1$). The regions of truth and failure are determined by the branch cuts of the functions involved.

The standard choices for branch cuts of the elementary functions are reasonably simple, always taking portions of either the real or imaginary axes. Indeed, all the branch cut definitions within MAPLE adhere to this rule (including those from outside the class of elementary functions). However the branch cuts invoked by the expressions built from these can be far more complicated.

Consider for example the composite function $\arcsin(2z\sqrt{1 - z^2})$ which is a term from the double angle formula for arcsin. While $\arcsin(z)$ has simple branch cuts (when z takes values along the real axis, to the left of -1 and to the right of $+1$), the branch cuts of the composite function are curves in the complex plane as demonstrated by the plot of the function on the left of Figure 1.

The cuts can be described by the four sets below which are visualised in the image on the right of Figure 1.

$$\{\Im(z) = 0, 1 < \Re(z)\}, \quad \{\Im(z) = \Im(z), \Re(z) = -(1/2)\sqrt{2 + 4\Im(z)^2}\},$$
$$\{\Im(z) = 0, \Re(z) < -1\}, \quad \{\Im(z) = \Im(z), \Re(z) = (1/2)\sqrt{2 + 4\Im(z)^2}\}. \quad (1)$$

We have implemented techniques for calculating the branch cuts inherited by functions acting on non-trivial arguments, and extended this to calculate the

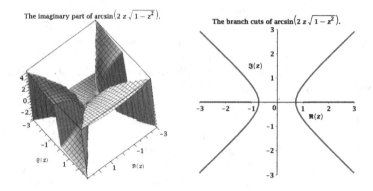

Fig. 1. Plots relating to $\arcsin(2z\sqrt{1-z^2})$

cuts of expressions and relations of such functions. The techniques have been gathered together in a MAPLE package, `BranchCuts` included as part of MAPLE 17 and accessed via the `FunctionAdvisor` tool. Readers with an earlier version can download the code as detailed in Appendix A. Both the sets in (1) and the visualisation on the right of Figure 1 were produced by the package. In fact, all the 2d figures in the paper are produced by the package from the output of the branch cut algorithms, while all the 3d figures are numerical plots of either the real or imaginary parts of the expressions in question.

MAPLE's `FunctionAdvisor` is a handbook for special functions, designed to be both human and machine readable, and interactive, processing the output to fit the query, [10]. It covers topics such as symmetries and series expansions with information for almost all of MAPLE's built in functions. In MAPLE 16 the functionality for branch cut computation was limited. There existed a table with the defining cuts for most functions in terms of a variable z and if a function was given with a different argument it would return the definitions with z replaced by that argument. Presenting branch cuts this way could be unintuitive and in some cases incorrect (for example, when the argument induced its own branch cuts these were not returned). In MAPLE 17 queries to `FunctionAdvisor` on branch cuts use the `BranchCuts` package discussed in this paper, and additionally, a variety of options are now available for visualising the cuts.

The primary motivation for the implementation is a wider project at Bath on *simplification*. The aim is to develop the technology for computer algebra systems to safely apply identities for multi-valued functions on their single valued counterparts. The key idea is to decompose the complex domain using cylindrical algebraic decomposition (CAD) according to the branch cuts of the functions involved, so that the truth value of the proposed identity will be constant in each region of the decomposition and hence may be tested by a sample point. This decomposition approach was introduced in [15] with the method using CAD developed in a series of papers; [2–6, 20]. Many of the results are summarised in [19] with the current state discussed recently in [14]

In this paper we discuss the implementation of the techniques in MAPLE, and the issues raised. We start in Section 2 by giving pseudo-algorithms describing the implementation. These can produce sets of cuts which are a superset of the actual branch cuts, that is, some of the cuts produced may not actually correspond to discontinuities of the functions. This led us to a classification of the different types of output, presented in Section 3. While there has been work on calculating branch cuts before, most notably in [15], our work goes much further with the careful description of the algorithms, their output and how it may be classified. Finally, in Section 4 we consider the use of this work in simplification technology and the effect of the condition that the input to CAD be a semi-algebraic set (list of polynomial equations or inequalities in real variables). Finally, some details on using the actual MAPLE package are provided in Appendix A. Although our implementation is in MAPLE, we note that the ideas presented are relevant for any system to compute branch cuts.

2   Calculating Branch Cuts

2.1   Moving to Real Variables

We first consider representing branch cuts as portions of algebraic curves in two real variables; the real and imaginary parts of a complex variable, z.

Example 1. Consider the function $f(z) = \log(z^2 - 1)$. The function log has branch cuts when its argument lies on the negative real axis hence $f(z)$ has branch cuts when $\Im(z^2 - 1) = 0$ and $\Re(z^2 - 1) < 0$. If we let $x = \Re(z), y = \Im(z)$ then this reduces to $2xy = 0, x^2 - y^2 - 1 < 0$, with solutions $\{y = 0, x \in (-1, 1)\}$ and $\{x = 0, y \text{ free}\}$. Hence the branch cuts are as shown in Figure 2.

This technique is summarised by Algorithm 1. In the implementation steps 1 and 2 are performed by calls to FunctionAdvisor, accessing the table of defining cuts. In step 2 we assume that the defining cuts are portions of either the real or imaginary axis encoded as the choice of which is zero and a range over which

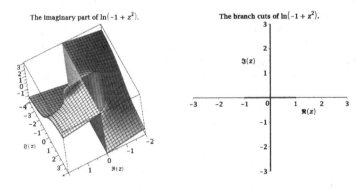

The imaginary part of $\ln(-1 + z^2)$.   The branch cuts of $\ln(-1 + z^2)$.

Fig. 2. Plots relating to $f(z) = \log(z^2 - 1)$ from Examples 1 and 4

the other varies. While not strictly required in theory the assumption is used throughout the implementation. Then in step 6 the semi-algebraic set will consist of one equality and one or two inequalities (depending on whether the range runs to infinity). Each solution in step 7 will consist of an equality defining one of $\{x, y\}$ in terms of the other, and a range for the other variable. Step 7 could be implemented with a variety of techniques. We use MAPLE's standard solving tools and find it most efficient to first solve the equality and then consider each possible solution with the inequalities. In using these tools we are assuming that MAPLE can identify all the solutions, which is not the case for polynomials of high degree. However, we find them sufficient for all practical examples encountered.

Algorithm 1. BC–F–RV1

Input : $f(p(z))$ where p is a polynomial and f has known defining cuts.
Output: The branch cuts of the mathematical function defined $f(p(z))$.

1 **if** f introduces branch cuts **then**
2   Obtain the defining branch cut(s) for f.
3   Set $\Re(z) = x, \Im(z) = y$ to obtain $p(z) = p(x, y)$.
4   Set $\mathcal{R}$ and $\mathcal{I}$ to be respectively the real and imaginary parts of $p(x, y)$.
5   **for** each defining cut C_i **do**
6   Define a semi-algebraic set in (x, y) by substituting $\mathcal{R}$ and $\mathcal{I}$ into C_i.
           Set B_i to be the set of solutions to the semi-algebraic set.
7   **return** The union of the B_i.
8 **else**
9   **return** the empty set.

2.2 Combinations of Functions

We extend Algorithm 1 to study *combinations* of functions (sums, products and relations) by applying the algorithm to each component and then taking the union of the sets of branch cuts in the outputs, as specified in Algorithm 2. In step 3 a suitable algorithm is one beginning BC–F that accepts F_i as input.

Note that the output specification of Algorithm 2 is looser than that of Algorithm 1. One reason for this is that a combination of functions with branch cuts may have their individual branch cuts intersecting, and if the discontinuities introduced are equivalent then these would cancel out as in Example 2. In Section 3 we classify the output of these algorithms, including output relating to these cancellations, (Definition 3).

Example 2. Let $f(z) = \log(z+1) - \log(z-1)$ and use Algorithm 2 to identify the branch cuts. First we use Algorithm 1 to identify the branch cut of the first term as the real axis below -1 and the branch cut of the second to be the real axis below 1. Hence Algorithm 2 returns the union; the real axis below 1 as visualised on the left of Figure 3. However, the function actually only has discontinuities

---

Algorithm 2. BC–C

    Input   : Any combination of functions whose branch cuts can individually be
                studied by an available algorithm.
  Output: A set of cuts, a subset of which are the branch cuts of the
               mathematical function defined by the expression.

1 Set $F_1, \ldots F_n$ to be the functions involved in the expression.
2 for $i = 1 \ldots n$ **do**
3   Set B_i to the output from applying a suitable branch cuts algorithm to F_i.
4 return $\cup_i B_i$

---

on the real axis in the range $(-1, 1)$ as demonstrated by the plot on the right of
Figure 3. Crossing the negative real axis below -1 does induce a discontinuity
in the imaginary part of both terms. However, those discontinuities are equal
and so cancel each other out in the expression for $f(z)$.

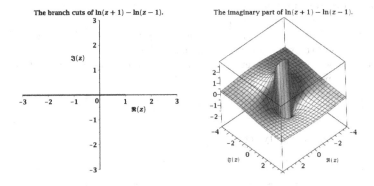

Fig. 3. Plots relating to $f(z) = \log(z+1) - \log(z-1)$ from Example 2

2.3   Allowing Nested Roots

We can extend Algorithm 1 to let p be a rational function by modifying step
7 to multiply up after substituting $\mathcal{R}$ and $\mathcal{I}$ into C_i. The question of zero de-
nominators will only arise if the input p itself has a zero denominator and so we
might assume this issue would have been dealt with previously.

    We can relax the input specification further by allowing nested roots, more
specifically, by letting the argument belong to the class of radical expressions in z
(expressions built up from $+, -, /, *$ and $\sqrt[n]{\ }$ where n is a natural number greater
than 1). This is because such an argument can be modified to give a rational
function from which information on the real and imaginary parts of the original
argument can be inferred, a process known as *de-nesting* the roots. Hence we
can still obtain a semi-algebraic set representing the branch cuts as before.

By de-nesting the roots we may end up with extra solutions which do not define branch cuts of the input function. For example, consider a function with argument $q(z)$ which when squared gives $q^2 = p(z)$, a rational function in z. However, this now represents the solution set $q(z) = \pm p(z)$, i.e. solutions for both branches of the square root, instead of just the desired principal branch. Ideally these erroneous solutions should be identified and removed.

Another issue in relaxing the input specification is that we must now consider the possibility of extra branch cuts arising from the argument itself. Taking these issues into account, we describe Algorithm 3. This is a modification of Algorithm 1 with a relaxed input specification, leading to looser output specification.

Algorithm 3. BC–F–RV2

   Input   : $f(q(z))$ where q is a radical expression and f has known defining cuts.
   Output: A set of cuts, a subset of which are the branch cuts of the
             mathematical function defined by $f(q(z))$.

1 **if** f introduces branch cuts **then**
2 | Obtain the defining branch cuts for f.
3 | Set $z = x + iy$ to obtain $q(z) = q(x, y)$.
4 | De-nest the roots in $q(x, y)$ to obtain $p(x, y)$. Set $\mathcal{R}_p$ and $\mathcal{I}_p$ to be respectively the real and imaginary parts of $p(x, y)$.
5 | Define a semi-algebraic set in (x, y) from $\mathcal{R}_p$ and $\mathcal{I}_p$ using information from the defining cuts.
6 | Set B to be the solutions of the semi-algebraic set.
7 | If possible, remove erroneous solutions arising from the de-nesting.
8 **else**
9 | Set B to be the empty set.
10 Set $A = $BC–C$(q(z))$ (application of Algorithm 2).
11 **return** $A \cup B$.

Various methods for de-nesting roots and removing the erroneous solutions have been studied in [2–4,19]. The **BranchCuts** package currently has only a very limited implementation of the squaring method outlined above, but further work is planned. Note that even this simple implementation can induce the erroneous solutions discussed as outlined by Example 3.

Example 3. Let $f = \log(2\sqrt{z})$ and use Algorithm 3 to identify the branch cuts. First we set $q = 2\sqrt{z} = 2\sqrt{x + iy}$. Then we de-nest by squaring to give $p = q^2 = 4(x + iy)$. In this simple example,

$$\mathcal{R}_p = 4x \qquad \text{and} \qquad \mathcal{I}_p = 4y. \qquad (2)$$

We suppose that $q = \mathcal{R}_q + \mathcal{I}_q i$ and hence

$$p = \mathcal{R}_q^2 - \mathcal{I}_q^2 + 2i(\mathcal{R}_q \mathcal{I}_q) \qquad (3)$$

Since log has defining cuts along the negative real axis we know $\mathcal{R}_q < 0$ and $\mathcal{I}_q = 0$. Upon comparing (2) and (3) we see the second condition implies $y = 0$ and $x = \frac{1}{4}\mathcal{R}_q^2 > 0$. In this example the first condition offers no further information (if the defining cut had not run to ∞ it could have bounded x). Hence we define the semi-algebraic set $\{y = 0, x > 0\}$. We also compute the set $\{y = 0, x < 0\}$, which is the branch cut of $q(z)$ itself, and return the union of the sets which together specify the entire real axis as presented visually on the left of Figure 4.

Unfortunately, the function only actually has discontinuities over the negative real axis, as demonstrated by the plot of the right of Figure 4. The first solution set was erroneous. This is clear since if $x > 0$ and $y = 0$ then $\sqrt{z} > 0$ and so can never lie on the negative real axis. The solution related to the case $q = -\sqrt{p}$ which was not relevant to the problem. (The reason for the factor of 2 in the example is because MAPLE automatically simplifies $\log(\sqrt{z})$ to $\frac{1}{2}\log(z)$ which can be analysed by Algorithm 3 to give exactly the branch cuts of the function.)

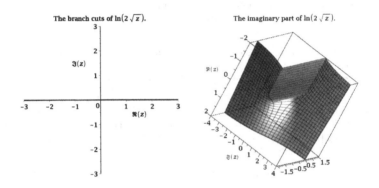

The branch cuts of $\ln(2\sqrt{z})$.   The imaginary part of $\ln(2\sqrt{z})$.

Fig. 4. Plots relating to $f(z) = \log(2\sqrt{z})$ from Example 3

2.4   Using a Complex Parametric Representation

We now consider a second approach to representing branch cuts, first suggested by [15]. Rather than moving to real variables this approach defines cuts using a complex function of a real parameter and a range for that parameter.

Example 4. Let $f(q(z)) = \log(z^2 - 1)$, the function from Example 1. We consider when q takes values on the branch cuts of f by setting $q = a$ where a ranges over the cuts. In this case $z^2 - 1 = a$ can be easily rearranged to give $z(a) = \pm\sqrt{a + 1}$. Hence we can represent the branch cuts by the two roots, each presented with the range $a \in (-\infty, 0)$. By considering the behaviour of the functions for various portions of the parameter range we see that these define the same cuts as presented in Example 1 and visualised on the right of Figure 2.

This technique is summarised by Algorithm 4. In this case the assumption that the defining cuts are portions of either the real or imaginary axis is really required. If $q(z)$ is a radical expression containing nested roots then step 5 will require de-nesting and so the output may be a superset of the actual branch cuts. (For example, the set produced for $\log(2\sqrt{z})$ is equivalent to that produced in Example 3.) Note that Algorithm 4 could have been provided with the input and output specifications of Algorithm 3 (i.e if $q(z)$ were a radical expression then given sufficient computing resources all the branch cuts could be identified as part of a larger set). Instead we have provided the specifications used for the implementation. This does not restrict the possibilities for $q(z)$, instead building in a warning system to ensure the correctness of the output. In particular this allows q to contain any elementary function, returning not the complete (possibly infinite) set of branch cuts but at least those in the principal domain.

Algorithm 4. BC–F–CV

Input : $f(q(z))$ where f has known defining cuts.
Output: A set of cuts which either contain the branch cuts of the
mathematical function defined by $f(q(z))$ as a subset, or are
accompanied with a warning that this is not the case.

1 **if** f introduces branch cuts **then**
2   Obtain the defining branch cuts for f, each a range on an axis.
3   **for** each cut C_i **do**
4   If C_i is on the real axis then set $q(z) = a$, otherwise set $q(z) = ia$.
5   Find the solutions $z(a)$ to this equation. If the complete set of solutions
        cannot be guaranteed then provide a **warning**
6   Set B to be the set of solutions, each given with the range for a from C_i.
7   If possible, remove erroneous solutions arising from any de-nesting.
8 **else**
9   Set B to be the empty set.
10 Set $A =$BC–C$(q(z))$ (application of Algorithm 2).
11 **return** $A \cup B$.

This approach was simple to implement in MAPLE using the `solve` command as a black box for step 5. (As discussed before Algorithm 1, this is making assumptions on the solve tools which would not always be valid, but they are found to be sufficient for all practical examples encountered.) The main advantage of this approach over moving to real variables is that it tends to be much quicker, especially when there are nested roots. The major disadvantage is that the output is usually far more complicated (requires much more space to display), often contains components that map to the same cuts, and is far less intuitive (the curves encoded are not visible algebraically). Example 5 demonstrates some of these features. Despite the often unwieldy output, MAPLE's plotting features allows for the position and nature of the cuts to be quickly made apparent.

For these reasons it is expected that the earlier algorithms are more useful for implementation in other code and use in mathematical study while Algorithm 4 is very useful for getting a quick visual understanding of the branch cuts in an expression and may have much utility in practical applications for this purpose.

Example 5. A classic example within the theory of branch cut calculation and simplification is that of Kahan's teardrop, from [16]. Kahan considers the relation

$$2\text{arccosh}\left(\frac{3+2z}{3}\right) - \text{arccosh}\left(\frac{5z+12}{3(z+4)}\right) = 2\text{arccosh}\left(2(z+3)\sqrt{\frac{z+3}{27(z+4)}}\right)$$

(4)

noting that it is true for all values of z in the complex plane except for a small teardrop shaped region over the negative real axis, as demonstrated by the plot on the left of Figure 5. Both of the approaches to calculating branch cuts outlined above will return a set represented visually by the image on the right of Figure 5. However, the algebraic representations are quite different. When working in real variables the upper half of the teardrop is represented by the set

$$\left\{\Im(z) = \frac{\sqrt{-(2\Re(z)+5)(2\Re(z)+9)}(\Re(z)+3)}{2\Re(z)+5}, \ -\frac{5}{2} < \Re(z), \Re(z) < -\frac{9}{4}\right\}$$

while using the complex parametric approach the same portion of the teardrop is given by

$$z = \frac{-3}{4\left(2a-a^3+2\sqrt{-a^2(-1+a^2)}\right)^{2/3}}\left[\left(2a-a^3+2\sqrt{-a^2(-1+a^2)}\right)^{2/3}\right.$$

$$-3i\sqrt{3}a^2 - 3ia\sqrt{3}\sqrt{-a^2(-1+a^2)} - 3\sqrt[3]{2a-a^3+2\sqrt{-a^2(-1+a^2)}}a$$

$$-3\sqrt[3]{2a-a^3+2\sqrt{-a^2(-1+a^2)}}\sqrt{-a^2(-1+a^2)} - 3a^2$$

$$-3a\sqrt{-a^2(-1+a^2)} + 3i\sqrt[3]{2a-a^3+2\sqrt{-a^2(-1+a^2)}}\sqrt{3}a$$

$$\left.+3i\sqrt[3]{2a-a^3+2\sqrt{-a^2(-1+a^2)}}\sqrt{3}\sqrt{-a^2(-1+a^2)}\right]$$

with a running over the range $(-1, 1)$.

We note that the identity (4) in Example 5 is actually introduced by a fluid mechanics problem and so this example demonstrates how issues relating to branch cuts may be encountered by users of multi-valued functions in other fields. Hence the importance of understanding them fully and the benefit of an accurate and intuitive representation.

3   Classification of Branch Cuts

The work of Section 2 raises several issues and necessitates a classification of the different cuts that can be produced by these methods. It is common to use the generic term **branch cuts** to refer to any curve portions that are defining cuts of functions or output from the algorithms. We classify these, starting with the definition most usually meant by users.

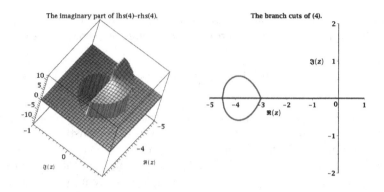

Fig. 5. Plots relating to equation (4) from Example 5

Definition 1. *Let F be an analytic multi-valued function and f its single-valued counterpart. The **branch cuts of the mathematical function** (called **true cuts** for brevity) are the curves or curve segments over which f is discontinuous, corresponding to F moving to another branch of its Riemann surface.*

Hence all the defining branch cuts are true cuts, as are any cuts produced by Algorithm 1. However, as demonstrated by Examples 2 and 3 the other algorithms may give output that does not adhere to this definition.

Definition 2. *Define any branch cuts calculated by the algorithms over which the function is actually continuous as **spurious cuts**.*

(In [15] the authors used the term *removable* instead of spurious, in analogy with removable singularities.) All branch cuts may be classified as either true or spurious. We further classify spurious cuts according to their origin.

Definition 3. *Define those spurious cuts introduced through a de-nesting procedure as **de-nesting cuts**, while those introduced by the intersection of true cuts from different parts of an expression as **formulation cuts**.*

Hence all spurious branch cuts produced by the algorithms in this paper are either de-nesting cuts or formulation cuts. Some spurious cuts may be both (or more accurately there may be two cuts, one of each type, which coincide).

Note that the output of Algorithms 3 and 4 are collections of true cuts and de-nesting cuts, while the output of Algorithm 2 is a collection of true cuts, de-nesting cuts and formulation cuts.

It would be desirable to have algorithms to remove all spurious cuts, or perhaps better, algorithms that do not introduce them in the first place. There has already been work on the removal of certain spurious cuts in [15] and [19, etc.] and this will be the topic of more study. We feel that formulation cuts will be the more difficult to avoid since they are inherent to the formulation of the mathematical function chosen in the expression given to an algorithm. Consider

$$f_\epsilon(z) = \log(z + 1) - \epsilon \log(z - 1).$$

When $\epsilon = 1$ we are in the case of Example 2 and applying Algorithm 2 will result in a branch cut over the real axis below 1, with the portion between -1 and 1 being a true cut and the rest a spurious cut. However, if we let ϵ differ at all from 1 then the spurious cuts will instantly become true. The size of the discontinuities will depend on the magnitude of ϵ but their presence does not. Figure 6 shows the presence of the true cuts occurring when ϵ varies by just one tenth from 1.

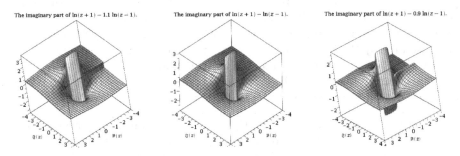

Fig. 6. Plots relating to $f_\epsilon(z) = \log(z+1) - \epsilon \log(z-1)$

4   Semi-algebraic Output for Simplification Technology

As discussed in the introduction, the primary motivation for this work was application in simplification technology, based on decomposing the domain according to the branch cuts of proposed simplifications using CAD. However, most CAD algorithms require the input to be a semi-algebraic set (list of polynomial equations and inequalities), with the polynomials usually defined over the field of rational coefficients. None of the algorithms described so far give such output, however Algorithms 1 and 3 could be easily modified to do so, by terminating early and returning the output of steps 5 and 6 respectively. We denote such an algorithm by BC−F−SA and note that it could be used on combinations via Algorithm 2. For Example 1 BC−F−SA would return $\{2xy = 0, x^2 - y^2 - 1 < 0\}$. However, for more complicated examples, the output may contain far more information than required to describe the cuts.

Example 6. Consider the formula

$$\arctan(z) + \arctan(z^2) = \arctan\left(\frac{z(1+z)}{(1-z^3)}\right). \tag{5}$$

The plot on the left of Figure 7 is a visualisation for the output from either of the approaches in `BranchCuts`, while the true cuts are apparent from the centre plot. Define,

$$f(x,y) = (1-x)\,y^4 - \left(2\,x^3 + 1\right)y^2 - x^5 - x^4 + x^2 + x$$
$$g(x,y) = y^6 - y^5 + 3\,x^2 y^4 - 2\left(x^2 + x\right)y^3$$
$$+ 3\left(2\,x + x^4\right)y^2 - \left(x^4 + 2\,x^3 + 2\,x + 1\right)y + x^6 - 2\,x^3.$$

If we were to instead take the semi-algebraic output, then we would have the following list of semi-algebraic sets,

$$\{x = 0, -y \le -1\}, \{x^2 - y^2 = 0, -2\,xy \le -1\}, \{f(x,y) = 0, g(x,y) \le -1\},$$
$$\{x = 0, y \le -1\}, \{x^2 - y^2 = 0, 2\,xy \le -1\}, \{f(x,y) = 0, g(x,-y) \le -1\}.$$

A full sign-invariant CAD for this problem would ignore the relation signs and just consider the polynomials present. A plot of the polynomials is given on the right of Figure 7 and clearly contains far more information than required to understand the branch cuts. Note that the correctness of the original formula is governed only by the branch cuts and hence the plot on the left: equation (5) is true in the connected region containing the origin and false in the other three full dimensional regions. A CAD allows us to find the regions of truth and falsity axiomatically by testing each cell of the CAD using a sample point.

There are various smarter approaches than calculating a full sign-invariant CAD, such as partial CADs and equational constraints. Work on a CAD based method that can take into account more of the structure of problems of branch cut analysis has recently been reported in [7], and studied further in [8].

5   Summary and Future Work

We have considered the problem of calculating the branch cuts of expressions, presenting two approaches and describing their implementation as part of MAPLE 17. We have classified the output of our algorithms and described how they could be adapted to provide semi-algebraic output for simplification technology. We are currently working on developing such technology based on the new concept of a truth table invariant CAD [7,8]; a decomposition which can be more closely fitted to the semi-algebraic description of branch cuts.

Future work with branch cuts will include the generalisation to many complex variables and the utilisation of better knowledge of branch cuts elsewhere, such as for choosing intelligent plot domains. Most importantly will be the further characterisation of spurious cuts and methods to remove them from the output.

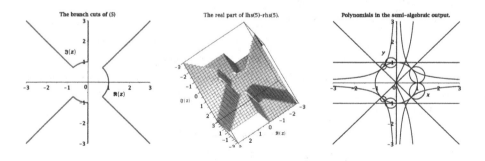

Fig. 7. Plots relating to equation (5) from Example 6

Acknowledgements. This work was supported by the EPSRC grant: EP/J003247/1. The code in MAPLE 17 is a collaboration between the University of Bath and Maplesoft. We would particularly like to thank Edgardo Cheb-Terrab from Maplesoft for his interest in the work and contribution to the code.

References

1. Abramowitz, M., Stegun, I.A.: Handbook of mathematical functions. National Bureau of Standards (1964)
2. Beaumont, J., Bradford, R., Davenport, J.H., Phisanbut, N.: A poly-algorithmic approach to simplifying elementary functions. In: Proc. ISSAC 2004. ACM (2004)
3. Beaumont, J., Bradford, R., Davenport, J.H., Phisanbut, N.: Testing elementary function identities using CAD. Applicable Algebra in Engineering, Communication and Computing 18, 513–543 (2007)
4. Beaumont, J., Phisanbut, N., Bradford, R.: Practical simplification of elementary functions using CAD. In: Dolzmann, A., Seidl, A., Sturm, T. (eds.) Proc. Algorithmic Algebra and Logic, Passau (2005)
5. Bradford, R., Davenport, J.H.: Towards better simplification of elementary functions. In: Proc. ISSAC 2002, pp. 16–22. ACM (2002)
6. Bradford, R., Davenport, J.H.: Better simplification of elementary functions through power series. In: Proc. ISSAC 2003. ACM (2003)
7. Bradford, R., Davenport, J.H., England, M., McCallum, S., Wilson, D.: Cylindrical algebraic decompositions for boolean combinations. In Press: Proc. ISSAC 2013 (2013), Preprint at http://opus.bath.ac.uk/33926/
8. Bradford, R., Davenport, J.H., England, M., Wilson, D.: Optimising problem formulation for cylindrical algebraic decomposition. In: Carette, J., Aspinall, D., Lange, C., Sojka, P., Windsteiger, W. (eds.) CICM 2013. LNCS (LNAI), vol. 7961, pp. 19–34. Springer, Heidelberg (2013)
9. Carette, J., Farmer, W.M.: A review of Mathematical Knowledge Management. In: Carette, J., Dixon, L., Coen, C.S., Watt, S.M. (eds.) Calculemus/MKM 2009. LNCS, vol. 5625, pp. 233–246. Springer, Heidelberg (2009)
10. Cheb-Terrab, E.S.: The function wizard project: A computer algebra handbook of special functions. In: Proceedings of the Maple Summer Workshop. University of Waterloo, Canada (2002)
11. Corless, R.M., Davenport, J.H., Jeffrey, D.J., Watt, S.M.: According to Abramowitz and Stegun. SIGSAM Bulletin 34(3), 58–65 (2000)
12. Davenport, J.H.: What might "Understand a function" mean? In: Kauers, M., Kerber, M., Miner, R., Windsteiger, W. (eds.) MKM/CALCULEMUS 2007. LNCS (LNAI), vol. 4573, pp. 55–65. Springer, Heidelberg (2007)
13. Davenport, J.H.: The challenges of multivalued "functions". In: Autexier, S., Calmet, J., Delahaye, D., Ion, P.D.F., Rideau, L., Rioboo, R., Sexton, A.P. (eds.) AISC 2010. LNCS, vol. 6167, pp. 1–12. Springer, Heidelberg (2010)
14. Davenport, J.H., Bradford, R., England, M., Wilson, D.: Program verification in the presence of complex numbers, functions with branch cuts etc. In: Proc. SYNASC 2012, pp. 83–88. IEEE (2012)
15. Dingle, A., Fateman, R.J.: Branch cuts in computer algebra. In: Proc. ISSAC 1994, pp. 250–257. ACM (1994)

16. Kahan, W.: Branch cuts for complex elementary functions. In: Iserles, A., Powell, M.J.D. (eds.) Proceedings The State of Art in Numerical Analysis, pp. 165–211. Clarendon Press (1987)
17. Markushevich, A.I.: Theory of Functions of a Complex Variable I (translated by Silverman, R.A.). Prentice-Hall (1965)
18. Olver, W.J., Lozier, D.W., Boisvert, R.F., Clark, C.W. (eds.): NIST Handbook of Mathematical Functions. Cambridge University Press (2010), Print companion to [21]
19. Phisanbut, N.: Practical Simplification of Elementary Functions using Cylindrical Algebraic Decomposition. PhD thesis, University of Bath (2011)
20. Phisanbut, N., Bradford, R.J., Davenport, J.H.: Geometry of branch cuts. ACM Communications in Computer Algebra 44(3), 132–135 (2010)
21. National Institute for Standards and Technology. The NIST digital library of mathematical functions, http://dlmf.nist.gov Online companion to [18]

A   The BranchCuts Package in MAPLE 17

The BranchCuts package is part of the MathematicalFunctions package in MAPLE 17, but is usually accessed directly by queries to the FunctionAdvisor. To access the commands individually in Maple 17 use

```
> kernelopts(opaquemodules=false):
> with(MathematicalFunctions:-BranchCuts):
```

Readers with an earlier version of MAPLE can download a file with the code from http://opus.bath.ac.uk/32511/ along with an introductory worksheet demonstrating its installation and use.

Two key commands are available; BCCalc which produces branch cuts using the algorithms of this paper and BCPlot which can make 2d visualisations of the output. There are two mandatory arguments for BCCalc; the expression to be considered and the variable. The key optional argument is the choice of method. Providing method=RealVariables will cause BCCalc to use Algorithms 1 and 3 while providing method=ComplexVariable will use Algorithm 4. The default, chosen for efficiency, uses Algorithm 1 where possible and Algorithm 4 elsewhere. Combinations of functions are dealt with using Algorithm 2.

The specification of the algorithms are checked but not strictly enforced. Instead warnings are provided if the method is not applicable or the output cannot be guaranteed to contain all true cuts. The package can work with any function whose defining cuts (or lack of cuts) is recorded in the FunctionAdvisor table. It covers all elementary functions and many others such as Bessel functions, Jacobi θ-functions and Chebyshev polynomials. These examples are actually multivariate in a computer algebra sense (univariate functions with parameters in a mathematical sense). Their branch cuts can be considered since they only occur with respect to one variable. If the presence of the branch cuts depends on the value of the parameters then the condition is checked. If it cannot be determined true or false (say if the relevant parameter has not been set), then the branch cut is included but a relevant warning is given. For example,

```
> BCCalc( BesselJ(a,sqrt(z^3-1)), z,
          parameters={a}, method=RealVariables );
```

produces the message, `Warning, branch cuts have been calculated which only occur if a::(Not(integer))`, and outputs the six branch cuts

$$\{\Im(z) = 0, \Re(z) < 1\}, \{\Im(z) = 0, 1 < \Re(z)\},$$
$$\{\Re(z) = -\tfrac{1}{3}\sqrt{3}\Im(z), \tfrac{1}{2}\sqrt{3} < \Im(z)\}, \{\Re(z) = -\tfrac{1}{3}\sqrt{3}\Im(z), \Im(z) < \tfrac{1}{2}\sqrt{3}\},$$
$$\{\Re(z) = \tfrac{1}{3}\sqrt{3}\Im(z), -\tfrac{1}{2}\sqrt{3} < \Im(z)\}, \{\Re(z) = \tfrac{1}{3}\sqrt{3}\Im(z), \Im(z) < -\tfrac{1}{2}\sqrt{3}\}.$$

Applying `BCPlot` to this output produces the image on the left of Figure 8. The true cuts for two specific values of a can be observed in the centre and right plots, demonstrating the validity of the warning message.

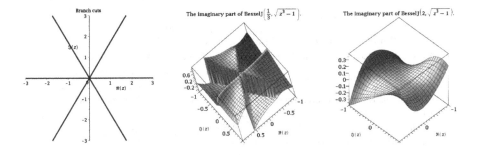

Fig. 8. Plots relating to $\mathrm{BesselJ}(a, \sqrt{z^3 - 1})$

Formal Mathematics on Display:
A Wiki for Flyspeck

Carst Tankink[1], Cezary Kaliszyk[2], Josef Urban[1], and Herman Geuvers[1,3]

[1] ICIS, Radboud Universiteit Nijmegen, Netherlands
[2] Institut für Informatik, Universität Innsbruck, Austria
[3] Technical University Eindhoven, Netherlands

Abstract. The Agora system is a prototype "Wiki for Formal Mathematics", with an aim to support developing and documenting large formalizations of mathematics in a proof assistant. The functions implemented in Agora include in-browser editing, strong AI/ATP proof advice, verification, and HTML rendering. The HTML rendering contains hyperlinks and provides on-demand explanation of the proof state for each proof step. In the present paper we show the prototype Flyspeck Wiki as an instance of Agora for HOL Light formalizations. The wiki can be used for formalizations of mathematics and for writing informal wiki pages about mathematics. Such informal pages may contain islands of formal text, which is used here for providing an initial cross-linking between Hales's informal Flyspeck book, and the formal Flyspeck development.

The Agora platform intends to address distributed wiki-style collaboration on large formalization projects, in particular both the aspect of immediate editing, verification and rendering of formal code, and the aspect of gradual and mutual refactoring and correspondence of the initial informal text and its formalization. Here, we highlight these features within the Flyspeck Wiki.

1 Introduction

The formal development of large parts of mathematics is gradually becoming mainstream. In various proof assistants, large repositories of formal proof have been created, e.g. in Mizar [1], Coq [2], Isabelle [3] and HOL Light [4]. This has led to fully formalized proofs of some impressive results, for example the odd order theorem in Coq [5], the proof of the 4 color theorem in Coq [6] and a significant portion of the proof of the Kepler conjecture [7] in HOL Light.

Even though these results are impressive, it is still quite hard to get a considerable speed-up in the formalization process. If we look at Wikipedia, we observe that due to its distributed nature everyone can and wants to contribute, thus generating a gigantic increase of volume. If we look at the large formalization projects, we see that they are very hierarchically structured, even if they make use of systems like Coq, that very well support a cooperative distributed way of working, supported by a version control system. An important reason is that the *precise* definitions *do* matter in a computer formalised mathematical theory:

J. Carette et al. (Eds.): CICM 2013, LNAI 7961, pp. 152–167, 2013.

some definitions work better than others and the structure of the library impacts the way you work with it.

There are other reasons why formalization is progressing at a much slower rate than, e.g. Wikipedia. One important reason is that it is very hard to get access to a library of formalised mathematics and to reuse it: specific features and notational choices matter a lot and the library consists of such an enormous amount of detailed formal code that it is hard to understand the purpose and use of its ingredients. A formal repository consists of computer code (in the proof assistant's scripting language), and has the same challenges as a programming source code regarding understanding, modularity and documentation. Also, if you want to make a contribution to a library of formalized mathematics, it really has to be all completely verified until the final proof step. And finally, giving formal proofs in a proof assistant is very laborious, requiring a significant amount of training and experience to do effectively.

To remedy this situation we have been developing the Agora platform: wiki technology that supports the development of large coherent repositories of formalised mathematics. We illustrate our work by focusing on the case of a wiki for the Flyspeck project, but the aims of Agora are wider. In short we want to provide proof assistant users with the tools to

1. Document and display their developments for others to be read and studied,
2. Cooperate on formalizations,
3. Speed up the proving by giving them special proof support via AI/ATP tools.

All this is integrated in one web-based framework, which aims at being a "Wiki for Formal Mathematics". In the present paper we highlight and advocate our framework by showing the prototype Flyspeck Wiki. We first elaborate on the three points mentioned above and indicate how we support these in Agora.

Documenting formal proofs. An important challenge is the communication of large formalizations to the various different communities interested in such formalizations: PA users that want to cooperate or want to build further on the development, interested readers who want to understand the precise choices made in the formalization and mathematicians who want to convince themselves that it is really the proper theorem that has been proven. All these communities have their own views on a formalization and the process of creating formalization, giving a diverse input that benefits the field. Nonetheless, communicating a formal proof is hard, just as hard as communicating a computer program.

Agora provides a wiki based approach: Formal proofs are basically program code in a high-level programming language, which needs to be documented to be understandable and maintainable. A proof development of mathematics is special, because there typically is documentation in the form of a mathematical text (a book or an article) that describes the mathematics informally. This is what we call the *informal mathematics* as opposed to the *formal mathematics* which is the mathematics as it lives inside a proof assistant. For software verification efforts, there is no pre-existing documentation, but Agora can be used to provide documentation of the verification as well. These days, informal mathematics consists of LaTeX files and formal mathematics usually consist of a set of text files that are given as input to a proof assistant to be checked for correctness.

In Agora, one can automatically generate HTML files from formal proof developments, where we maintain all linking that is inherently available in the formal development. Also, one can automatically generate files in wiki syntax from a set of LaTeX files. These wiki files can also be rendered as HTML, maintaining the linking inside the LaTeX files, but more importantly, also the linking with the formal proof development. Starting from the other end, one can write a wiki document about mathematics and include snippets of formal proof text via an inclusion mechanism. This allows the dynamic insertion of pieces of formal proof, by referencing the formal object in a repository.

Cooperation on formal proofs. With Agora, we also want to lower the threshold for participating in formalization projects by providing an easy-to-use web interface to a proof assistant [8]. This allows people to cooperate on a project, the files of which are stored on the server.

Proof Support. We provide additional tools for users of proof assistants, like automated proof advice [9]. The proof states resulting from editing HOL Light code in Agora are continuously sent to an online AI/ATP service which is trained in a number of ways on the whole Flyspeck corpus. The service automatically tries to discharge the proof states by using (currently 28) different proof search methods in parallel, and if successful, it attempts to create the corresponding code reconstructing such proofs in the user's HOL Light session.

To summarize, the Agora system now provides the following tooling for HOL Light and Flyspeck:

- a rendering of the informal proof texts, written originaly in LaTeX,
- a hyperlinked, marked up version of the HOL Light and Flyspeck source code, augmented with the information about the proof state after each proof step
- transclusion of snippets of the hyperlinked formal code into the informal text whenever useful
- cross-linking between the informal and formal text based on custom Flyspeck annotations
- an editor to experiment with the sources of the proof by dropping down to HOL Light and doing a formal proof,
- integrated access to a proof advisor for HOL Light that helps (particularly novices) to finish their code while they are writing it, or provide options for improvement, by suggesting lemmas that will solve smaller steps in one go.

Most of these tools are prototypical and occasionally behave in unexpected ways. The wiki pages for Flyspeck can be found at http://mws.cs.ru.nl/agora_cicm/flyspeck. These pages also list the current status of the tooling.

The rest of the paper is structured as follows. Section 2 shows the presentation side of Agora, as experienced by readers. The internal document model of Agora

is described in Section 3, Section 4 explains the interaction with the formal HOL
Light code, and Section 5 describes the inclusion of the informal Flyspeck texts
in Agora. Section 6 concludes and discusses future work.

1.1   Similar Systems

There are some systems that support mashing up informal documentation with
computed information. In particular, Agora shares some similarities with tools
using the OMDoc [10] format, as well as the IPython [11] architecture (and
Sage [12], which uses IPython as an interface to computer algebra functionality).

OMDoc is mainly a mechanization format, but supports workflows that are
similar to Agora's, but differs in execution: OMDoc is a stricter format, requiring
documents to be more structured and detailed. In particular, this requires its
input languages, such as sTeX, to be more structured. On the other hand, Agora
does not define much structure on the files its includes, rather extracting as much
information as possible and fitting it in a generic tree structure. Because Agora
is less strict in its assumptions, it becomes easier to write informal text, freeing
the authors of having to write semantic macros.

The IPython architecture has the concept of a *notebook* which is similar to a
page in Agora: it is a web page that allows an author to specify 'islands' of Python
that are executed on the server, with the results displayed in the notebook. Agora
builds on top of this idea, by having a collection of documents referring to each
other, instead of only allowing the author of a document to define new islands.

2   Presenting Formal and Informal Mathematics in **Agora**

Agora has two kinds of pages: fully *formal* pages, generated from the sources of
the development, and *informal* pages, which include both markup and snippets
of formal text. To give readers, in particular readers not used to reading the
syntax of a proof assistant, insight in a formal development, we believe that it
is not enough to mark up the formal text prettily:

- there is little to no context for an inexperienced reader to quickly understand
  what is being formalized and how: items might be named differently, and in
  a proof script, all used lemmas are presented with equal weight. This makes
  it difficult for a reader to single out what is used for what purpose;
- typically, the level of detail that is used to guide the proof assistant in its
  verification of a proof is too high for a reader to understand the essence of
  that proof: it is typically decorated with commands that are administrative
  in nature, proof steps such as applying a transitivity rule. A reader makes
  these steps implicitly when reading an informal proof, but they must be
  spelled out for a formal system. In the extreme, this means that a proof that
  is 'trivial' in an informal text still requires a few lines of formal code;
- because most proof assistants are programmable, a proof in proof assistant
  syntax can have a different structure than its informal counterpart: proofs
  can be 'packed' by applying proof rules conditionally, or applying a proof
  rule to multiple similar (but not identical) cases.

On the other hand, it is not enough to just give informal text presenting a formalization: without pointers to the location of a proof in the formal development, it is easy for a reader to get lost in the large amount of code. To allow easier navigation by a reader, the informal text should provide *references* to the formal text at the least, and preferably include the portions of formal text that are related to important parts of the informal discussion.

By providing the informal documentation and formal code on a single web platform, we simplify the task of cross-linking informal description to formal text. The formal text is automatically cross-linked, and annotated with proper anchors that can also be referenced from an informal text. Moreover, our system uses this mechanism to provide a second type of cross-reference, which includes a formal entity in an informal text [13]: these references are written like hyperlinks, using a slightly different syntax indicating that an inclusion will be generated. Normal hyperlinks can refer to concepts on the same page, the same repository, or on external pages.

These mechanisms allow an author of an informal text to provide an overview of a formal development that, at the highest level, can give the reader insight in the development and the choices made. Should the reader be interested in more details of the formalization, cross-linking allows further investigation: clicking on links opens the either informal concepts or shows the definition of a formal concept.

The formalization of the Kepler conjecture in the Flyspeck project provides us with an opportunity to display these techniques: not only is it a significant nontrivial formalization, but its informal description in LaTeX [14] contains explicit connections between the informal mathematics and the related formal concepts in the development. We have transformed these sources into the wiki pages available on our Agora system[1]. Parts of one page are shown in Figures 1 and 2.

2.1   Informal Descriptions

The informal text on the page is displayed similarly to the source (Flyspeck) document, from which it is actually generated (see Section 5), keeping the formulae intact to be rendered by the MathJax[2] JavaScript library. The difference to the Flyspeck source document is that the source document contains *references* to formal items (see also Section 5), while the Agora version *includes* the actual text of these formal entities. To prevent the reader from being confused by the formal text, which can be quite long, the formal text is hidden behind a clearly-labeled link (for example the FAN and XOHLED links in Figure 1 which link to the formal definition of *fan* and the formal statement of lemma *fan_cyclic*).

The informal page may additionally *embed* editable pieces of formal code (instead of just including addressable formal entities from other files as done in the demo page). In that case (see Section 4) clicking the 'edit' on these blocks opens up an editor on the page itself, which gives direct feedback by calling

---

[1] http://mws.cs.ru.nl/agora_cicm/flyspeck/doc/fly_demo/

[2] http://mathjax.org

| Informal | Formal |

Definition of [fan, blade] DSKAGVP (fan) [fan ↔ FAN]

Let (V, E) be a pair consisting of a set $V \subset \mathbb{R}^3$ and a set E of unordered pairs of distinct elements of V. The pair is said to be a *fan* if the following properties hold.

1. (CARDINALITY) V is finite and nonempty. [cardinality ↔ fan1]
2. (ORIGIN) $\mathbf{0} \notin V$. [origin ↔ fan2]
3. (NONPARALLEL) If $\{\mathbf{v}, \mathbf{w}\} \in E$, then $\mathbf{v}$ and $\mathbf{w}$ are not parallel. [nonparallel ↔ fan6]
4. (INTERSECTION) For all $\varepsilon, \varepsilon' \in E \cup \{\{\mathbf{v}\} : \mathbf{v} \in V\}$, [intersection ↔ fan7]

$$C(\varepsilon) \cap C(\varepsilon') = C(\varepsilon \cap \varepsilon').$$

When $\varepsilon \in E$, call $C^0(\varepsilon)$ or $C(\varepsilon)$ a *blade* of the fan.

basic properties

The rest of the chapter develops the properties of fans. We begin with a completely trivial consequence of the definition.

| Informal | Formal |

Lemma [] CTVTAQA (subset-fan)

If (V, E) is a fan, then for every $E' \subset E$, (V, E') is also a fan.

Proof

This proof is elementary.

| Informal | Formal |

Lemma [fan cyclic] XOHLED

$[E(v) \leftrightarrow \text{set\_of\_edge}]$ Let (V, E) be a fan. For each $\mathbf{v} \in V$, the set

$$E(\mathbf{v}) = \{\mathbf{w} \in V : \{\mathbf{v}, \mathbf{w}\} \in E\}$$

is cyclic with respect to $(\mathbf{0}, \mathbf{v})$.

Proof

If $\mathbf{w} \in E(\mathbf{v})$, then $\mathbf{v}$ and $\mathbf{w}$ are not parallel. Also, if $\mathbf{w} \neq \mathbf{w}' \in E(\mathbf{v})$, then

Fig. 1. Screenshot of the Agora wiki page presenting a part of the "Fan" chapter of the informal description of the Kepler conjecture formalization. For each formalized section, the user can choose between the informal presentation (shown here) and its formal counterpart (shown on the next screenshot). The complete wikified chapter is available at: http://mws.cs.ru.nl/agora_cicm/flyspeck/doc/fly_demo/.

Informal Formal

```
#DSKAGVP⁷
let FAN=new_definition`FAN(x,V,E) <=> ((UNIONS E) SUBSET V) /\ graph(E) /\ fan1(x,V,E) /\ fan2(x,V
fan6(x,V,E)/\ fan7(x,V,E)`;;
```

basic properties

The rest of the chapter develops the properties of fans. We begin with a completely trivial consequence of the definition.

Informal Formal

```
let CTVTAQA=prove(`!(x:real^3) (V:real^3->bool) (E:(real^3->bool)->bool) (E1:(real^3->bool)->bool)
FAN(x,V,E) /\ E1 SUBSET E
==>
FAN(x,V,E1)`,

REPEAT GEN_TAC
THEN REWRITE_TAC[FAN;fan1;fan2;fan6;fan7;graph]
THEN ASM_SET_TAC[]);;
```

Informal Formal

```
let XOHLED=prove(`!(x:real^3) (V:real^3->bool) (E:(real^3->bool)->bool) (v:real^3).
FAN(x,V,E) /\ v IN V
==> cyclic_set (set_of_edge v V E) x v`,

MESON_TAC[CYCLIC_SET_EDGE_FAN]);;
```

Fig. 2. http://mws.cs.ru.nl/agora_cicm/flyspeck/doc/fly_demo/ (formal)

HOL Light in the background, and displaying the resulting proof assistant state, together with a *proof advice* which uses automated reasoning tools to try to find a solution to the current goal.

2.2 Formal Texts

The formal text of the development, in the proof assistant syntax, is included in Agora as a set of hyperlinked HTML pages that provide *dynamic* access to the proof state, using the Proviola [15] technology we have previously developed: pointing at the commands in the formal text calls the proof assistant and provides the state on the page. The results of this computation are memoized for future requests: this makes it possible for future visitors to obtain these states quickly, while not taking up space unnecessarily.

The pages are hyperlinked (see Section 4.2) to allow a reader to explore the presented formalization. The formalization could be large and, in projects like Flyspeck, produced by a number of collaborators. The current alternatives to hyperlinking are unsatisfactory in such circumstances: it amounts to either memorization by the reader of large parts of the libraries, or mandatory access to a search facility. In HOL Light, this search facility is the system itself: typing in the name of a lemma prints out its statement.

3 Document Structure: Frames and Scenes

The pages in Agora are generated from in-memory *documents*: (Python) objects equipped with methods for rendering and storing the internal files. To cater for multiple proof assistants and document-preparation tools, such as a renderer for wiki syntax, we use the object-inheritance to instantiate documents for different systems, while providing a common interface. This interface consists of a tree-like structure of *frames*, grouped into *scenes*.

Documents in Agora are structured according to our earlier work on a system called Proviola [16], for replaying formal proof: this tool takes a "proof script" and uses a light-weight parser to transform it into a list of separate commands. This list can then be submitted to a proof assistant, storing the responses in the process. This memoization of the proof assistant's responses is stored together with the command, into a data structure we call a *frame*. Frames can store more than just a response and a command, in particular, we assume that all frames in Agora documents store a markup element that contains the HTML markup of the frame's command.

To display a document as a page, it would be enough to display the list of frames in order, rendering the markup of each frame, and this is how the purely formal pages in Agora are rendered. However, we want our tools to be able to display not only flat lists of text, but also combine them in meaningful ways: for example by grouping a lemma with its proof, but also combining multiple lemmas into a self-contained section. For this, we introduced a *scene*: a scene is a grouping of (references to) frames and other scenes, that can combine them in any order. The system will render such a tree structure recursively, displaying the markup of each frame referenced to. The benefit of grouping files into scenes is that it becomes easier to re-mix parts of a document into a new document, such as including formal text into an informal page.

Inclusion. To allow remixing scenes from documents into new content, it is necessary to provide an interface that allows including scenes into pages. In previous work [13], we introduced an interface in the form of syntax: Agora allows users to write narratives in a markup language similar to Wikipedia's, which is extended with the notion of a *reference*. This reference is similar to Isabelle's antiquotation: it is syntax for pointing to formally defined entities on the Web which carry some metadata, which can be automatically provided by a theorem prover. When rendered, the references are resolved into marked up 'islands' of formal text. The rest of the syntax is a markup language allowing mathematical notation and hyperlinks.

These islands are included in the scene structure as references to the marked up scenes. At the moment, we only allow referring to formal scenes from informal text, which is enough to render the Flyspeck text. Having an inclusion syntax fits the Agora philosophy: the documentation workflow can use the formal code, but it should not change it. Instead, writing informal documentation about a development should be similar to writing a LaTeX article, only in a different markup language. However, it is occasionally necessary to add code directly to an informal page, for example to write an illustrative example or a failed attempt; such a code block is not part of the formal development, but benefits from the markup techniques applied to the development.

In the document structure, such code blocks are just scenes, that are marked to be written in a particular language. From the rendered page, it is possible to open an editor for each scene, which requires special functionality to support writing formal proofs.

4   Interaction with Formal **HOL Light** Code

4.1   Parsing and Proving

For HOL Light, adding Proviola support implies adding a parser that can transform a proof script into a list of commands, and adding a layer to communicate with the prover's read-eval-print loop (REPL). This is sufficient, but so far does not create a very illustrative Proviola display: most HOL Light proofs are *packaged* into a single REPL-invocation that introduces and discharges a theorem. Making this into a useful Proviola display is left for future work, but we will sketch how a better display can be implemented using the scene structure of a Proviola document.

To illustrate the workings of the parser and the prover, we use the following example code:

```
(* Example code fragment. *)
g `x=x`;;
e REFL_TAC;;
let t = (* Use top_thm to verify the proof. *)
   top_thm ();;
```

Parser Because HOL Light proofs are written as syntactically correct scripts that are interpreted by the OCaml read-eval-print loop (REPL), the parser separates a proof script into the single commands that can be interpreted by this REPL. These commands are, in the Flyspeck sources, terminated by ';;'[3] and followed by a newline, so our parser splits a proof script into commands by looking for this terminator. Additionally, the proof can contain comments, surrounded by '(*' and '*)': we let the parser only emit a command if the terminator does not occur as part of a comment. Finally, comment blocks that are not within other commands are treated as separate commands. This last decision differs from traditional source-code parsers, which regard comments as white space, because Agora reconstructs the proof script's appearance from the frames in the movie, in order to show the complete proof script if a reader desires it.

The parser does not group the frames into a scene structure: a HOL Light proof is represented as a single scene containing all frames. For our example, the following frames are generated:

---

[3] According to the OCaml reference manual,
   http://caml.inria.fr/pub/docs/manual-ocaml-4.00/manual003.html#toc4

- *(* Example code fragment. *)*
- g 'x=x';;
- e REFL_TAC;;
- **let** t = *(* Use top_thm to verify the proof. *)*
  top_thm ();;

The first comment does not occur within a command, so it is parsed as a separate command, and the second comment occurs inside a command.

Prover. HOL Light is not implemented as a stand-alone program with its own REPL. Instead, it is implemented as a collection OCaml scripts and some parsing functions. This means that the 'prover' instance is actually a regular OCaml REPL instance, which loads the appropriate bootstrap script. The problem of this approach is that these scripts take several minutes to load, a heavy penalty for wanting to edit a proof on the Web. To offset the load time, one can *checkpoint* the OCaml instance after it has bootstrapped HOL Light. Checkpointing software allows the state of a process to be written to disk, and restore this state from the stored image later. We use DMTCP[4] as our checkpointing software: it does not require kernel modifications, and because of that is one of the few checkpointing solutions that works on recent Linux versions.

Communication with the provers is encapsulated by a Python class: creating an instance of the class loads the checkpoint and connects to its standard input and output. The resulting object has a **send** method which writes a provided command to standard input and returns the REPL's response. Beyond this low-level communication mechanism, the object also provides a **send_frame** method. This method takes an entire frame and sends the command stored in it. This method does not only send the text, but also records the number of tactics that the prover has executed so far, by examining the length of the current goalstack. This gives an indication of how far a list of frames is processed, and allows the prover to use HOL Light's undo function to prevent executing too many commands.

After sending the frames generated from our example code, the frames have stack numbers as shown in Table 1.

When the frame with the REFL_TAC invocation is changed, the **send_frame** method will send the HOL Light undo function, b ();; as many times as is necessary to return to state 1. Afterwards, it will send the command of the changed frame.

Table 1. Frames with state numbers

Command	State
(* Example code fragment. *)	0
g 'x=x';;	1
e REFL_TAC;;	2
let t = ...	2

---

[4] http://dmtcp.sourceforge.net

The HOL Light glue does not send all commands equally: the Flyspeck formalization packs its proofs within an OCaml module, which causes the REPL not to give output until the module is closed. Because we want to give state information per command, the gluing code ignores the `module` and `end` commands that signal the opening and closing of modules.

Packaged Proofs. To allow Proviola to record a packaged proof, it needs to break the proof down to its individual commands. To do this, we propose to use the Tactician tool [17]: this is an extension to HOL Light that records a packaged proof as it is executed, and allows the user to retrieve the actual tactics executed, which exposes the tree-like structure of such a proof: some of the tactics in the packaged proof might be applied multiple times, to different subgoals generated during the proof.

We can use the sequential tactic script generated by Tactician directly, rendering it instead of the packaged proof, or do more sophisticated post-processing: we could match up the generated tactics to their occurrence in the packaged proof, and generate a special scene for each packaged proof. This scene would render as the original proof, but execute the Tactician-generated sequence to provide responses. This gives readers a better feel of what is going on in such a packaged proof, but depends on a correct matching of the packaged proof to the sequential proof. We have not yet fully investigated the reach of these possibilities, however, so this remains as future work.

4.2   Hyperlinking

It seems that no proper hyperlinking facility exists so far for HOL-based systems. Such a facility should plug in to the parsing layer of the systems (as done, e.g., for Coq and Mizar), and either export the information about symbols' definitions relative to the original formal text, or directly produce a hyperlinked version of the text: this hyperlinking pass should be fast, so it can be run when a page is loaded in the browser.

For HOL Light (and Flyspeck), we so far did not try to hook into the parsing layer of the system, and only provide a heuristic hyperlinking system. Still, such a hyperlinker can be useful, because relatively few concepts are overloaded in the formalization, and most of the definitions and theorems are introduced using a regular syntax: this means that the hyperlinker can generate an index for file definitions with only a small chance of ambiguity. The hyperlinking proceeds in two broad steps, an indexing step and a rendering step. The indexing is done by a Perl script that generates a symbol index by:

1. collecting the globally defined symbols and theorem names from the formal texts by heuristically matching the most common patterns that introduce them,[5] and
2. optionally adding and removing some symbols based on a predefined list.

---

[5] To help this, we also use the theorem names stored by the HOL Light processing in the "theorems" file, using the mechanisms from the file update_database_**.ml.

The page renderer of Agora then processes the texts again by heuristically to-kenizing the text, looking up tokens and their linking in the generated index. Additionally, the page rendering also uses the index to generate metadata that can be used by the referencing mechanism [13].

The complete hyperlinking of the whole library now takes less than ten sec-onds, and while obviously imperfect, it seems to be already quite useful tool that allowed us to browse and study the library. The generated index of 15,780 Flyspeck entities together with their URLs can be loaded into arbitrary external application, and used for separate heuristic hyperlinking of other texts. This function is used by the script that translates the LaTeX sources of the informal text describing Flyspeck into wiki syntax (Section 5), to link the formally defined concepts to their HOL Light definitions.

4.3   Editing and Proof Advising

Editing. We can directly use the tools that turn text into frames for building the server backend of a (simple) web-based editor: the front end of this editor just gathers the entered text and sends it to the server, the server processes it into a list of frames and post-processes it: both by generating proof assistant (HOL Light) responses and by sending markup information based on the correctness of a part of the text. Because this processing is incremental, information can be returned on demand: after the text has been parsed into frames, the server can give the editor information as it is produced, using the protocols described in [8]. As also described in that paper, it remains an open question on how to properly deal with the impact of the formal text written in the editor, as this might invalidate the entire repository. An example of the editor interaction is shown in Figure 3. It already shows also the proof advising facility.

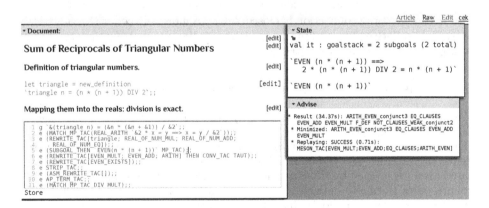

Fig. 3. The interactive editor built in the Wiki with the proof state for the line with the cursor. The screenshot features a section of Harrison's triangular numbers formal-ization. In line 5 the advisor automatically finds a proof that $n(n + 1)$ is even, slightly different from the one used in the edited formalization.

Proof Advising In order to further facilitate the online Wiki authoring using HOL Light, we have added a post-processing step to the editor. For each goal interactively computed by the proof assistant, the editor automatically submits this goal to the AI/ATP proof advisor (HOL(y)Hammer) service [18]. The advisor uses a number of differently parametrized premise-selection methods (based on various machine-learning algorithms) to find the most relevant theorems from the Flyspeck library for a given goal, and passes them (after translation to first-order logic) to automated theorem provers (ATPs) such as Vampire [19], E [20], and Z3 [21]. If an ATP proof is found, it is minimized and reconstructed by a number of reconstruction strategies described in [22]. In parallel to such AI/ATP methods, a number of decision procedures are tried on the goal. The currently used decision procedures are able to solve boolean goals (tautologies), goals that involve naturals (arithmetic), integers, rationals, reals and complex numbers including Gröbner bases. Whenever any of the strategies finds a tactic that solves the goal, all other strategies are stopped and the result of the successful one is transmitted to the Agora users through a window. The users can immediately use the successful results in their proof.

The protocol to communicate with the advisor has been designed to be as simple as possible, in order to enable using it not only as a part of Agora but also via an experimental Emacs interface [18] and from the command line tool in the spirit of old style LCF. A request for advice consists of a single line which is a text representation of a goal to prove. To encode a goalstate as text the goal assumptions need to be separated from the goal conclusion and from each other. We use the ' character as such separator, since the character never appears in normal HOL Light terms as it is used to denote start and end of terms by the Camlp5 preprocessor. When a request for advice is received the server parses the goal assumptions and conclusion together, to allow matching the free variables present in more than one of them and ensure proper typing. The response is also textual and the connection is closed when no more advice for the goalstate is available. Server-side caching is used to handle repeated queries, typically produced by refactoring an existing proof script in the Wiki.

5   Inclusion of the Informal **Flyspeck** Texts

We have used a version of the informal Flyspeck LATEX text that has 309 pages, but only a smaller part has so far been chosen for the experiments: Chapter 5 (Fan). The file fan.tex has 1981 lines. There are 15 definitions (some of them define several concepts) and 36 lemmas. The definitions have the following annotated form (developed by Hales), which already cross-links to some of the formal counterparts (formally defined theorem names like QSRHLXB and MUGGQUF and symbols like azim_fan and is_Moebius_contour):

```
\begin{definition}[polyhedron]\guid{QSRHLXB}
A \newterm{polyhedron} is the
intersection of a finite number of closed half-spaces in
$\ring{R}^n$.
\end{definition}
```

The lemmas are written in a similar style:

```
\begin{lemma}[Krein--Milman]\guid{MUGGQUF}
Every compact convex set $P\subset\ring{R}^n$ is the convex hull
of its set of extreme points.
\end{lemma}
```

The text contains many mappings between informal and formal concepts, e.g.:

```
\formaldef{$\op{azim}(x)$}{azim\_fan}
\formaldef{M\"obius contour}{is\_Moebius\_contour}
\formaldef{half space}{closed\_half\_space, open\_half\_space}
```

There are several systems that can (to various extent) transform LaTeX texts to (X)HTML and similar formats. Examples include LaTeXML[6], PlasTeX[7], xhtmlatex[8], and TeX4ht.[9] Often they are customizable, and some of them can be equipped with custom non-HTML (e.g., wiki) renderers. For the first experiments we have however relied only on MathJaX for rendering mathematics, and custom transformations from LaTeX to wiki syntax that allow us to easily experiment with specific functions for cross-linking and formalization without involving the bigger systems. The price for this is that the resulting wiki pages are more similar to presentations in ProofWiki and Wikipedia than to full-fledged HTML book presentations. We might switch to the larger extendable systems when it is clear what extensions are needed for our use-case.

The transformations are now implemented in about 200 lines of a Perl script (Creolify.pl) translating the Flyspeck LaTeX sources into the enhanced Creole wiki syntax used by Agora. The script is easily extendable, and it now consists mainly of about 30 regular-expression replacements and related functions taking care of the non-mathematical LaTeX syntax and macros. The mathematical text is handled by the (slightly modified) macros taken from Flyspeck (kepmacros.tex) that are prepended to any Agora Flyspeck text and used automatically by Math-JaX. Producing and tuning the transformations took about one to two days of work, and should not be a large time investment for (formal) mathematicians interested in experimenting with Agora. The particular transformations that are now used for Flyspeck include:

- Transformations that handle wiki-specific syntax that is (intentionally or accidentally) used in LaTeX, such as comments, white space, fonts and section markup.
- Transformations that create wiki subsections for various LaTeX blocks, sections, and environments. Each definition, lemma, remark, corollary, and proof environment gets its own wiki subsection, similarly, e.g., to ProofWiki and Wikipedia.

---

[6] http://dlmf.nist.gov/LaTeXML/
[7] http://plastex.sourceforge.net/
[8] http://www.matapp.unimib.it/~ferrario/var/x.html
[9] http://tug.org/tex4ht/

– The transformation that add linking and cross-linking, based on the LaTeX annotations. Each LaTeX label produces a corresponding wiki anchor, and each LaTeX reference produces a link to the anchor. Newly defined terms (introduced with the `newterm` macro) also produce anchors. Formal annotations (introduced with the `guid` and `formaldef` macros) are first looked up in the index of all formal concepts produced by hyperlinking of the formalization (Section 4.2), and if they are found there, such annotations are linked to the corresponding formal definition.

6   Conclusion and Future Work

The platform is still in development, and a number of functions can be improved and added. For example, whole-library editing, guarded by global consistency checking of the formal code that has been already verified (as done for Mizar [23]), is future work. On the other hand, the platform already allows the dual presentation of mathematical texts as both informal and formal, and the interaction between these two aspects. In particular, the platform takes both LaTeX and formal input, cross-links both of them based on simple user-defined macros and on the formal syntax, and allows one to easily browse the formal counterparts of an informal text. It is already possible to add further formal links to the informal concepts, and thus make the informal text more and more explicit. A particular interesting use made possible by the platform is thus an exhaustive collaborative formal annotation of the Flyspeck book. The platform also already includes interactive editing and verification, which allows at any point of the informal text to switch to formal mode, and to add the corresponding formal definitions, theorems, and proofs, which are immediatelly hyperlinked and equipped with detailed proof status information for every step. The editing is complemented by a relatively strong proof advice system for HOL Light. This is especially useful in a Wiki environment, where redundancies and deviations can be discovered automatically. The requests for advice can become grounds for further experiments on strengthening the advice system.

One future direction is to allow even the non-mathematical parts of the wiki pages to be written directly with (extended) LaTeX, as it is done for example in PlanetMath. This could facilitate the presentation of the projects developed in the wiki as standalone LaTeX papers. On the other hand, it is straightforward to provide a simple script that translates the wiki syntax to LaTeX, analogously to the existing script that translates from LaTeX to wiki.

References

1. Grabowski, A., Kornilowicz, A., Naumowicz, A.: Mizar in a nutshell. Journal of Formalized Reasoning 3(2), 153–245 (2010)
2. Bertot, Y., Casteran, P.: Interactive Theorem Proving and Program Development - Coq'Art: The Calculus of Inductive Constructions. Texts in Theoretical Computer Science. Springer (2004)
3. Nipkow, T., Paulson, L.C., Wenzel, M. (eds.): Isabelle/HOL. LNCS, vol. 2283. Springer, Heidelberg (2002)

4. Harrison, J.: HOL Light: An overview. In: Berghofer, S., Nipkow, T., Urban, C., Wenzel, M. (eds.) TPHOLs 2009. LNCS, vol. 5674, pp. 60–66. Springer, Heidelberg (2009)
5. Gonthier, G.: Engineering mathematics: the odd order theorem proof. In: Giacobazzi, R., Cousot, R. (eds.) POPL, pp. 1–2. ACM (2013)
6. Gonthier, G.: The four colour theorem: Engineering of a formal proof. In: Kapur, D. (ed.) ASCM 2007. LNCS (LNAI), vol. 5081, p. 333. Springer, Heidelberg (2008)
7. Hales, T.C., Harrison, J., McLaughlin, S., Nipkow, T., Obua, S., Zumkeller, R.: A revision of the proof of the Kepler conjecture. Discrete & Computational Geometry 44(1), 1–34 (2010)
8. Tankink, C.: Proof in context — web editing with rich, modeless contextual feedback. To appear in Proceedings of UITP 2012 (2012)
9. Kaliszyk, C., Urban, J.: Learning-assisted automated reasoning with Flyspeck. CoRR abs/1211.7012 (2012)
10. Kohlhase, M. (ed.): OMDoc. LNCS (LNAI), vol. 4180. Springer, Heidelberg (2006)
11. Pérez, F., Granger, B.E.: IPython: a System for Interactive Scientific Computing. Comput. Sci. Eng. 9(3), 21–29 (2007)
12. Stein, W.A., et al.: Sage mathematics software (2009)
13. Tankink, C., Lange, C., Urban, J.: Point-and-write – documenting formal mathematics by reference. In: Jeuring, J., Campbell, J.A., Carette, J., Dos Reis, G., Sojka, P., Wenzel, M., Sorge, V. (eds.) CICM 2012. LNCS, vol. 7362, pp. 169–185. Springer, Heidelberg (2012)
14. Hales, T.C.: Dense Sphere Packings - a blueprint for formal proofs. Cambridge University Press (2012)
15. Tankink, C., McKinna, J.: Dynamic proof pages. In: ITP Workshop on Mathematical Wikis (MathWikis). CEUR Workshop Proceedings, vol. 767 (2011)
16. Tankink, C., Geuvers, H., McKinna, J., Wiedijk, F.: Proviola: A tool for proof re-animation. In: [24], pp. 440–454
17. Adams, M., Aspinall, D.: Recording and refactoring HOL Light tactic proofs. In: Proceedings of the IJCAR Workshop on Automated Theory Exploration (2012)
18. Kaliszyk, C., Urban, J.: Automated reasoning service for HOL Light. In: Carette, J., Aspinall, D., Lange, C., Sojka, P., Windsteiger, W. (eds.) CICM 2013. LNCS (LNAI), vol. 7961, pp. 120–135. Springer, Heidelberg (2013)
19. Riazanov, A., Voronkov, A.: The design and implementation of VAMPIRE. AI Commun. 15(2-3), 91–110 (2002)
20. Schulz, S.: E - A Brainiac Theorem Prover. AI Commun. 15(2-3), 111–126 (2002)
21. de Moura, L., Bjørner, N.: Z3: An Efficient SMT Solver. In: Ramakrishnan, C.R., Rehof, J. (eds.) TACAS 2008. LNCS, vol. 4963, pp. 337–340. Springer, Heidelberg (2008)
22. Kaliszyk, C., Urban, J.: PRocH: Proof reconstruction for HOL Light (2013)
23. Urban, J., Alama, J., Rudnicki, P., Geuvers, H.: A wiki for Mizar: Motivation, considerations, and initial prototype. In: [24], pp. 455–469
24. Autexier, S., Calmet, J., Delahaye, D., Ion, P.D.F., Rideau, L., Rioboo, R., Sexton, A.P. (eds.): AISC 2010. LNCS, vol. 6167. Springer, Heidelberg (2010)

Determining Points on Handwritten Mathematical Symbols

Rui Hu and Stephen M. Watt

The University of Western Ontario
London Ontario, Canada N6A 5B7
{rhu8,Stephen.Watt}@uwo.ca

Abstract. In a variety of applications, such as handwritten mathematics and diagram labelling, it is common to have symbols of many different sizes in use and for the writing not to follow simple baselines. In order to understand the scale and relative positioning of individual characters, it is necessary to identify the location of certain expected features. These are typically identified by particular points in the symbols, for example, the baseline of a lower case "p" would be identified by the lowest part of the bowl, ignoring the descender. We investigate how to find these special points automatically so they may be used in a number of problems, such as improving two-dimensional mathematical recognition and in handwriting neatening, while preserving the original style.

Keywords: Handwriting analysis, Handwriting neatening, Mathematical handwriting recognition, Pen computing.

1   Introduction

Many digital ink applications allow handwritten characters in various sizes and in different locations. For example, in mathematics, subscripts and superscripts appear relatively smaller than normal text and are written slightly below or above it. Moreover, these subscripts and superscripts may themselves have subscripts or superscripts. Such notation is easily read and understood. This involves determining the relative baselines and sizes of symbols. This process may present various ambiguities, for example whether a particular symbol is a lower case "p" or an upper case "P" giving the subscripted p_q or the juxtaposed Pq.

In order to find the scale and offset of individual characters, it is necessary to identify the location of certain expected features which are typically defined by particular points. These particular points occur at different locations in different symbols, and the precise location can vary in different handwriting samples of the same symbol. For example, the baseline of lowercase "p" would be identified by the lowest part of the bowl, ignoring the descender. In contrast, the baseline of lowercase "k", would be identified by the toes. In this article we refer to a point such as this, that determines the height of a metric line, as a *determining point*. Knowing the determining points of each symbol can help us solve a number of problems. For example, one can use the determining points to improve two-dimensional mathematical recognition. By comparing the baseline locations

J. Carette et al. (Eds.): CICM 2013, LNAI 7961, pp. 168–183, 2013.

and the sizes of adjacent symbols, one can identify superscripts and subscripts (e.g. S_2, $S2$, S^2) with more confidence. Another application is in handwriting neatening. Since handwritten symbols often come with variations in alignment and size, certain transformations based on determining points can be applied to obtain normalized samples while preserving the original writing style.

Recording determining points for an individual handwritten symbol is easy. One can manually annotate the symbol with the positions of all its determining points. However, finding determining points for all symbols in a collection is much more challenging. First, with a large database the labour for manual annotation would be prohibitively costly. Secondly, applications such as mathematics involve a large variety of symbols derived from a range of alphabets and other sources. In practice, many of them are often poorly written and there is no fixed dictionary of words to aid in disambiguation [1]. This increases the difficulty to find determining points reliably. Meanwhile, each person's handwriting is unique — even identical twins write differently [2]. Even if a training database were to be fully annotated, it is not entirely clear how this should best be used to identify the points of interest in new input. Last, but not least, the usual methods for detecting determining points depend on device resolution significantly. With rapidly evolving technology, this means that new algorithms cannot use archival data directly and therefore must be "re-sampled" (interpolated).

We are interested in the problem of how to automatically find determining points of handwritten mathematical symbols and to use them in a variety of problems. Considerable related work has been conducted, some of which we highlight here. Pechwitz and Märgner [3] proposed an algorithm that can find determining points from symbol skeleton approximated by piecewise linear curve. However, these determining points are only useful in detecting baseline locations. In 2010, Infante Velázquez [4] developed an annotation tool to record determining points manually for handwritten characters represented in InkML [5]. The determining points were later used to neaten new handwriting, making it uniform in size, alignment and slant while preserving writers' particular writing styles. However, this tool recorded each determining point with absolute coordinates and was therefore subject to device resolution and variations in style. As device resolution may vary among different vendors and over generations of technology, this approach is not device-independent. Similar problems exist in [6]. In addition, Zanibbi et al. [7] proposed a technique to automatically improve the legibility of handwriting by gradually translating and scaling individual symbol to closely approximate their relative positions and sizes in a corresponding typeset version. This technique detects baseline locations by comparing symbols' bounding boxes, which leads to troubles with vertical placement and scale. For example, it fails to distinguish between "$x2$" and "x^2". In 2012, Hu and Watt [8] presented an algorithm to find turning points that determine the shape of characters, but that approach lacked the ability to capture the geometric meaning of each determining point and therefore does not provide sufficient information to calculate certain desired symbol metrics, such as the location of baseline. Harouni et al. [9] later proposed a method to find determining points in handwritten Arabic

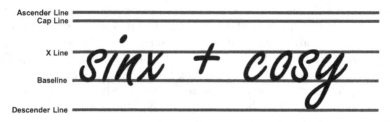

Fig. 1. An example to illustrate the concepts of metric lines

characters. The method consisted of two stages. In the first stage, the raw input data were converted to a standard format using smoothing, normalization and interpolation techniques. In the second stage, each stroke of input characters was split into several pieces. The method calculated the local maximum and minimum of each piece and recorded them as determining points. However, this method is not optimal as it requires extra effort to split strokes and may generate undesired determining points that lack meaning.

In this article, we present an algorithm to find determining points automatically and suggest how they may be used in areas such as improving two-dimensional mathematical recognition and in neatening handwriting. The basic approach is to identify the points of interest on one average instance of each type of symbol, and to use this information to find the corresponding points on newly written symbols. We borrow ideas from typography, where a number of determining points are identified to measure the metrics of different font families, and apply these to handwriting. We consider several types of determining points, which, in turn, determine certain metrics. These include the locations of the five main metric lines, i.e. the baseline, x line, ascender line, cap line, and descender line, as shown in Figure 1, as well as symbol width and slant. To make the determining points device-independent, the algorithm first converts all handwritten symbols into parametric curves approximated by truncated orthogonal series, mapping each symbol to a single point in a low dimensional vector space of series coefficients. We then compute the average symbol for each class by computing the average of the points for the class in the vector space. The determining points of interest are identified on these average symbols. From these, the algorithm can derive corresponding determining points in samples automatically. The beauty of this algorithm is that it is writer-independent. We only need to annotate once, on the average symbols. This reduces cost significantly. Furthermore, the algorithm is device-independent as all symbols are represented in the functional space, which is robust against changes in device resolution.

The remainder of this article is organized as follows. In Section 2, we recall how to represent digital ink using functional approximation. Section 3 discusses several types of determining points that are useful in finding symbol alignment lines. In Section 4, we present the algorithm that can identify determining points in handwritten mathematical symbols automatically. Section 5 evaluates the performance of the algorithm. We then investigate the possible use of the algorithm in a number of problems in Section 6. Section 7 concludes the article.

2 Functional Approximation for Digital Ink

Digital ink is generated by sampling points from a traced curve at a certain rate, and thus is typically given in the form of a series of points, each of which contains x and y values in a rectangular coordinate system at a sequence of times. Since the sampling rate and resolution typically depend on the hardware type, different devices usually result in different numerical point values for the same character. In order to take device differences into account, various *ad hoc* treatments have been developed, such as size normalization and "re-sampling" (interpolation). To make the representation more robust under changes in hardware, we represent handwritten symbols as coefficients for an approximating basis in a function space. This approach has been used in earlier work [10–13].

We consider an ink trace as a segment of a plane curve $(x(s), y(s))$, parameterized by Euclidean arc length

$$s = \int \sqrt{dx^2 + dy^2}.$$

This parameterization has been found to lead to good recognition and is intuitive sense, since it gives curves that look the same the same parameterization [10]. Given a digital ink trace $(x(s), y(s))$ and an approximating basis $\{B_i(s)\}_{i=0,\ldots,d}$, we represent the trace using the coefficients x_i and y_i from

$$x(s) \approx \sum_{i=0}^{d} x_i B_i(s) \qquad y(s) \approx \sum_{i=0}^{d} y_i B_i(s)$$

It is convenient to choose the functions $B_i(s)$ to be orthogonal polynomials, e.g. Chebyshev, Legendre or some other polynomials. By choosing an appropriate family of basis polynomials to high enough degree, the approximating curve can be made arbitrarily close to the original trace.

We have found a Legendre-Sobolev basis allows approximating curves to have the desired shape for relatively low degrees. These may be computed by Gram-Schmidt orthogonalization of the monomials $\{s^i\}$ with respect to the inner product

$$\langle f, g \rangle = \int_a^b f(s)g(s)\mathrm{d}s + \mu \int_a^b f'(s)g'(s)\mathrm{d}s.$$

If a symbol has multiple strokes, we join consecutive strokes by concatenating the point series, which yields a single curve. For more details see [13]. An example of using Legendre-Sobolev polynomials in approximation is shown in Figure 2. After approximation, we may now represent the digital ink trace, or symbol, as the coefficient vector $(x_0, \ldots, x_d, y_0, \ldots, y_d)$. We may standardize the location and size of the character by setting x_0, y_0 to 0 and the norm of the vector to 1.

(a)   (b)

Fig. 2. Approximation using Legendre-Sobolev series. (a) Original.
(b) Approximated using series of degree 12 with $\mu = 1/8$.

3   Handwriting Metrics

In order to understand the scale of individual symbols, it is necessary to identify
the location of certain expected features which are typically defined by a number
of determining points. These determining points have locations that vary from
symbol to symbol, but typically occur where parts of the symbols touch certain
invisible horizontal lines. To discuss this, we adopt concepts from typeface design.
In this article, we consider several types of determining points related to the
following metrics. We concentrate on symbols used in European alphabets. Many
other writing systems would have other metric lines determined in a similar way.

Baseline. Most scripts share the notion of *baseline*. It is a guide line for writing
so that adjacent symbols can retain their horizontal alignment. It is also used as
the reference to obtain other metrics such as x height, ascender height, etc. While
some symbols such as lower case "p" may extend below the baseline, it serves as
the imaginary base for most symbols. Figure 3 shows examples of baselines and
their determining points. As shown in Figure 3(b), the three legs of the lowercase
"m" are not completely aligned. In such case, multiple determining points are
identified and the location of the baseline may be determined by the average y
value of all the determining points.

X Line and Height. The x *line* falls at the top of most lowercase symbols,
such as "a" and "y", and is located over the baseline. Some symbols may extend
above the x line, such as "h" where the x line is located at the top of the shoulder.
The x *height* is the distance between the *baseline* and the x *line*. Figure 4 shows
an example of x line and associated determining points. Certain symbols, such
as lowercase "x", may have multiple determining points to define the x line. In
such a case, the location of the x line is determined by the average of their y
values.

Ascender Line and Height. The part of a lowercase symbol, such as "h" and
"k", that extends above the x *line* is known as an *ascender*. The *ascender line*
is located above the x *line* and is determined by the height of the ascenders.
The *ascender height* is the distance between the baseline and the ascender line.
Figure 5 shows an example of an ascender line and ascender height. The location

Fig. 3. Baseline with (a) one, and (b) multiple determining points.

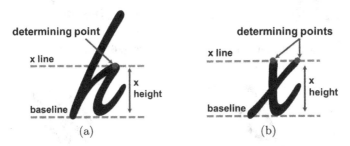

Fig. 4. x line and x height with (a) one, and (b) multiple determining points.

of the ascender line is determined by the determining point shown in red. In the case that there are multiple determining points, the location of the ascender line is given by the average y value of all the relevant determining points.

Cap Line and Height. The *cap line* is used to align uppercase symbols and is usually located below the ascender line, although it is not limited to that position. Indeed, in handwriting it often coincides with the ascender line. The *cap height* is the distance between the baseline and the cap line. Figure 6 shows an example of a cap line and cap height. The location of the cap line is determined by the determining point shown in red. In the case that there are multiple determining points, the location of the cap line may be taken as the average y value of all the determining points.

Descender Line and Height. The *descender line* is located below the baseline. It is used to align descenders, which are the parts of symbols that extend below the baseline. Figure 7 shows an example of a descender line and descender height. If there are multiple determining points, the location of the descender line is given by the average y value of all the determining points.

Slant and Width. In some handwriting styles, symbols are written with inclination either to the left or to the right. The degree of inclination is referred to as the *slant*. The *width* of a symbol is given by the horizontal distance from the left-bounding and right-bounding lines with the given slant. Figure 8 shows an example of symbol width and slant.

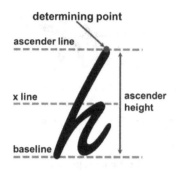

Fig. 5. Ascender line and height

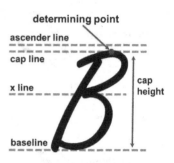

Fig. 6. Cap line and height

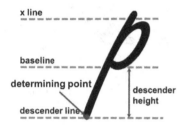

Fig. 7. Descender line and height

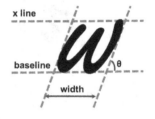

Fig. 8. Slant (θ) and width

4   Algorithm

In this section, we present an algorithm to find automatically the determining points for newly written symbols. The algorithm derives determining points for a new symbol from the known determining points of an annotated average symbol of the same type.

Average Symbols

We classify symbols so that symbols that are written the same way and could be interpreted the same way are in the same class. So, for example, there may be several classes for the numeral "8", depending on whether the symbol is written with one continuous stroke or two separate strokes, which stroke is written first and the direction of writing. On the other hand, a Latin letter "O" and the numeral "0" could belong to the same class.

Taking each sample as a point in the functional approximation space, it has been found in earlier work that classes of points are almost completely pairwise separable by single hyperplanes. Thus the convex hulls of the class point sets are to a good approximation non-intersecting. Any point on a line segment between two sample points of the same class falls within that class. It is therefore meaningful to compute the average of a set of known samples for a class as the average point in the function space

(a)   (b)

Fig. 9. (a) Samples provided by different writers. (b) The average symbol.

$$\bar{C} = \sum_{i=1}^{n} C_i/n,$$

where n is the number of the samples and C_i is the coefficient vector for the ith sample. Figure 9(a) shows a set of samples provided by different writers and Figure 9(b) shows the average symbol.

Deriving Determining Points from Average Symbols

Our algorithm is based on the observation that the average symbols typically look similar to the samples of the same class. Within a given class, the features present in one sample should be present in other samples and at a similar location. We can take the location to be the arc length along the ink trace to the defining point of the feature. We assume that, if two symbols are sufficiently similar, the locations of corresponding determining points will be similar (given by distance along the curve).

This suggests that we can find the determining points of a new symbol by taking the known locations on an annotated symbol and making an adjustment. In more detail, to detect the determining points in a sample, we start with an annotated sample in the same class. For now, this will be the average of the training samples, annotated with its determining points. Each annotation consists of the location (as arc length), the type of determining point (e.g. baseline, x line, etc) and whether it is located at a local minimum or local maximum of y value.

For each determining point of the annotated sample, we guess that the corresponding determining point on the new sample will be near the same arc length location. So we take the point at that location in the new sample and follow the trace upward or downward, depending on whether that determining point is supposed to be at a local minimum or local maximum. This can be easily done using a number of numerical methods. In our implementation, we applied Newton's method to solve $y'(s) = 0$. A formal algorithm is given in Algorithm 1.

Figure 10 shows examples of using Algorithm 1. Figure 10(a) shows the determining points annotated on the average symbol "η". This is the reference symbol A in the algorithm. Figures 10(b1) and 10(c1) show two example input samples S

Algorithm 1. LocateDeterminingPoints

Input : A, the coefficient vector for a reference symbol.

$D_A = [(s_1, T_1, K_1), \ldots (s_n, T_n, K_n)]$, a vector of determining points. For each, the position is given as arc length s_i on the curve of A, the value T_i states which type of metric line is being defined, and the value K_i states whether the metric line is given by a local minimum or local maximum at $y_A(s_i)$.

S, the coefficient vector for the input sample whose determining points are to be found.

Output: $D_S = [(\ell_1, T_1, K_1), \ldots, (\ell_n, T_n, K_n)]$, giving the locations, ℓ_i, and types of the determining points of S.
The value of ℓ_i along S corresponds to the value s_i along A.

1. Let $x_A(s)$, $y_A(s)$, $x_S(s)$, $y_S(s)$ be the coordinate functions defined by the coefficient vectors A and S.
2. **for** $i \in 1..n$ **do**
  | **if** $K_i = \boldsymbol{max}$ **then**
  | └ $f \longleftarrow -y_S$
  | **else if** $K_i = \boldsymbol{min}$ **then**
  | └ $f \longleftarrow y_S$
  | $\ell_i \longleftarrow$ local minimum of $f(s)$ nearest $s = s_i$.
  |
  | *Note this local minimum or maximum is of a real univariate polynomial and*
  | *any standard method may be used. For example, we use Newton's method to*
  | └ *solve $f'(s) = 0$ with initial point $s = s_i$.*
3. Return$[(\ell_1, T_1, K_1), \ldots, (\ell_n, T_n, K_n)]$

---

with initial approximate locations s_i for the determining points. Figures 10(b2) and 10(c2) show the determining points found at locations ℓ_i. Figure 11 shows several examples of determining points found for samples of "π".

5   Experiments and Testing

We developed a software tool to annotate handwriting samples with their determining points. Figure 12 shows the user interface. By selecting a nearby location, the tool is able to find the target determining point automatically. The locations of all the metric lines discussed in Section 3 can be detected. Multiple determining points may exist for certain metrics lines. In such circumstances, the location of the corresponding metric line is determined by the average of the values given by all the determining points of that kind. Symbol slant can also be recorded by adjusting a spinner. Symbol width is automatically detected with slant considered.

To evaluate the performance, we have tested the algorithm against a large handwriting dataset. The handwriting dataset we used contained altogether

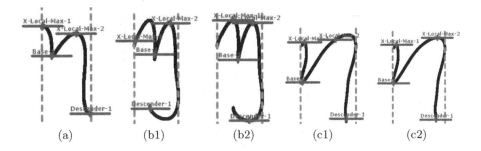

(a)   (b1)   (b2)   (c1)   (c2)

Fig. 10. Automatically finding determining points. (a) Average symbol "η".
(b1) Sample 1 initial approximations and (b2) with determining points found.
(c1) Sample 2 initial approximations and (c2) with determining points found.

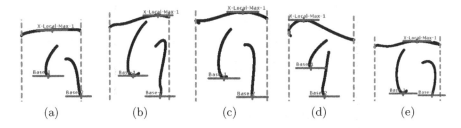

(a)   (b)   (c)   (d)   (e)

Fig. 11. Automatically finding determining points. (a) Average symbol "π".
(b-e) Determining points derived from the average symbol.

64944 samples of 240 different symbols. Most of the samples are Latin and Greek
letters, digits, operators, or other mathematical symbols provided by various
writers. All of these samples had been classified in advance. As some symbols
were written in different styles (e.g. completely different forms, different numbers
of strokes, or strokes in different orders), a total of 382 classes were examined. We
first computed the average symbol for each class, in which determining points
were identified using the software tool shown in Figure 12.

We then computed determining points for all the samples using Algorithm 1.
The number of determining points varied from 2 to 5, according to the sample.
If any of the determining points were mis-positioned, we considered it as incorrect. We chose up to 30 samples randomly from each class and examined their
correctness visually. In total, we examined 8119 samples, of which 421 samples
have at least one mis-positioned determining point. This gave a measured error
rate of 5.2%.

We found the error was introduced mainly from two sources. The first was
mis-classified samples in the original data set. These were either mis-labelled (e.g.
"e" of style 1, instead of "e" of style 2), or had strokes given in a different order
from the usual. In this latter case, we have the option of defining a new style or
normalizing the order of the strokes. The second source was that some samples
are significantly different from the average symbol. As a result, the determining

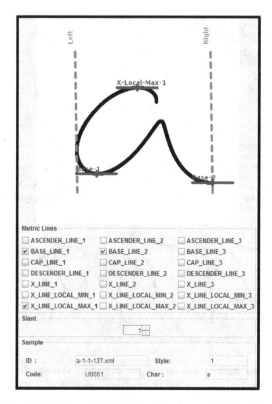

Fig. 12. Software tool to identify determining points.

points in the average symbol may not be mapped correctly to those dissimilar samples.

As misclassified samples were errors in the training data, rather than errors by the algorithm, we excluded those samples from the experiment. We further added 39 new classes (giving 421 classes in total) to split out those samples with different stroke orders. After these corrections, the measured error rate decreased to 2.0% (9593 samples reviewed, of which 189 samples had at least one mis-positioned determining point).

To address the second issue, that of points mis-positioned because the sample was far from the average shape, we used a homotopy between the average and the test sample in a multi-step method. Recall that, in the function space, a line from the average symbol to the test sample lies entirely within the class. By dividing this line into several equal steps, we may apply Algorithm 1 several times to follow the determining points through the homotopy. If $\bar{C}$ is the average symbol for the class and C_{targ} is the input sample, then the line joining the two points in the function space is given by $C(t) = (1 - t)\bar{C} + tC_{targ}$, with t ranging from 0 to 1. The determining points should move smoothly as the character is deformed by the homotopy, and we can choose a step size. Figure 13 shows an

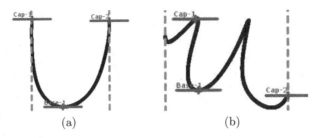

Fig. 13. Failure example: (a) average symbol, (b) target with one point misplaced

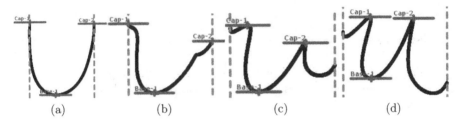

Fig. 14. Success in 3 steps: (a) average (b) step 1 (c) step 2 (d) step 3 = target

Table 1. Error rates of the multi-step method on 9593 samples

Steps	1	2	3	4	6	8	10	20
Failed Samples	189	69	36	28	25	25	24	24
Error Rates	2.0%	0.72%	0.38%	0.29%	0.26%	0.26%	0.25%	0.25%

example where Algorithm 1 fails to identify one of the determining points when applied naively. However, when applied in a 3 step homotopy, it succeeded, as shown in Figure 14.

We have tested the multi-step method against the same handwriting dataset. We chose up to 30 samples randomly from each class and examined their correctness visually. The measured error rates are reported in Table 1. The samples that failed in the 10-Step and 20-Step methods typically either had slants that interfered with the strategy of using local minimum or maximum y value to find determining points or that were very badly written. For these samples, our algorithm was able to identify some determining points correctly but not all of them, as shown in Figure 15. Note that the points found would in any case be sufficient for most applications.

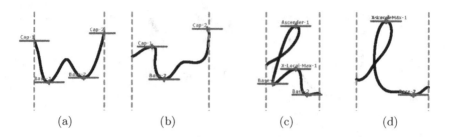

Fig. 15. Multi-step failures: (a) Average, (b) target. (c) Average, (d) target.

6   Use Cases

Determining points can be used in a variety of digital ink applications to solve different problems. Here we describe two scenarios in which determining points have been found useful.

Handwriting Recognition

Juxtaposition ambiguity is common in mathematical handwriting recognition. This is usually caused by symbols that are next to each other are written in different sizes and at different heights. Figure 16 shows an example with several relative positionings of two characters. The first character can in each case be a "P" or "p" and the second can be interpreted as a "q" or "9". Together there could be a variety of possible interpretations:

$$P^9 \; P9 \; P_9 \qquad p^9 \; p9 \; p_9$$
$$P^q \; Pq \; P_q \qquad p^q \; pq \; p_q$$

However, by comparing symbols' baseline locations and sizes, we can predict each expression with more confidence. This is because the baselines of subscripts and superscripts are typically placed slightly below or above the normal line of text and their sizes are relatively smaller. Note that to determine the relative position, it is definitely not sufficient to compare the baselines of the symbol bounding boxes. This is seen in Figure 17(d). Similarly, having an imputed baseline determined by symbol class (such as at 50% height for "q") is insufficient. We thus find it is important to find and use the symbol's determining points.

Handwriting Neatening

Handwriting neatening is becoming possible in some digital ink applications. It is used to transform handwriting to obtain visually appealing output while preserving the original writing style. Figure 18 shows an example. By identifying

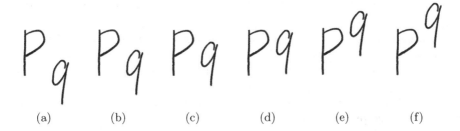

Fig. 16. Juxtaposition ambiguity.

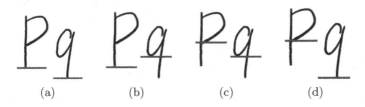

Fig. 17. Disambiguation by baselines. (a) P9 (b) Pq (c) pq (d) p9

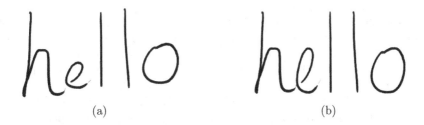

Fig. 18. Neatening using determining points. (a) original, (b) neatened.

$$a_1 x^2 + a_2 \qquad a_1 x^2 + a_2$$

        (a)   (b)

Fig. 19. Neatening using determining points. (a) original, (b) neatened.

the determining points of each character, we can shift and scale these characters to make corresponding metrics lines aligned properly, as shown in Figure 18(b). Figure 19 shows a second example. In this case, all characters including the superscripts and subscripts were adjusted in order to obtain a normalized output. Transforming the function $y(s)$ for each symbol is the simplest approach to neatening. A more aggressive approach is to replace each input symbol with the appropriately scaled version of the average of like symbols seen by the same writer, and further transformations can be employed. However, this is beyond the scope of the present article.

7   Conclusion and Future Work

We have presented an algorithm to identify automatically the determining points in handwritten symbols. Identifying these determining points helps us better understand the scale of individual characters as well as find the locations of certain desired features. In contrast to existing methods, which treat digital ink traces as a collection of discrete points, this algorithm relies on interpreting ink traces as single points in a functional space. This allows device independence, on one hand, and a simple formulation of homotopic deformation, on the other.

Various features can be recorded by using the determining point algorithm. The nature of the determining points depends on the symbol set used. In our case, the symbols were based mostly on those of European languages and mathematical operators, so the baseline, x line, ascender line, descender line and cap line were used.

To evaluate the performance of the algorithm, we have tested it against a database of handwritten mathematical characters. The experiments showed promising results. To demonstrate possible use of determining points, we have described two scenarios: handwriting recognition and handwriting neatening, in both of which determining points have been found useful.

There are a few directions we would like to pursue in the future. First, we wish to include determining points in our handwriting recognizer. It is expected that, combined with ambient baseline information, this will improve the recognition rate. Secondly, we would like to investigate using rotation- and slant-invariant techniques [14, 15] in conjunction with the present methods. At a more detailed-level, we would like to annotate all samples in our database using a supervised multi-step method. This will allow us to perform a more satisfying statistical analysis of the effectiveness of our method. Finally, before incorporating these techniques in our recognition framework, we would like to investigate the correlation between the model-sample distance and the number of steps required for low error rates, and how the number of required steps varies by class.

We would like to thank Isaac Watt for helping to organize the handwriting dataset used in the experiments.

References

1. Smirnova, E., Watt, S.M.: A context for pen-based mathematical computing. In: Proceedings of the 2005 Maple Summer Workshop, Waterloo, Canada, pp. 409–422 (2005)
2. Srihari, S., Huang, C., Srinivasan, H., Srihari, S., Huang, C., Srinivasan, H.: On the discriminability of the handwriting of twins. J. For. Sci. Journal of Forensic Identification 126 53, 430–446 (2008)
3. Pechwitz, M., Margner, V.: Baseline estimation for arabic handwritten words. In: Proceedings of Eighth International Workshop on Frontiers in Handwriting Recognition, pp. 479–484 (2002)
4. Infante Velázquez, M.T.: Metrics and neatening of handwritten characters. Master's thesis, The University of Western Ontario, Canada (2010)
5. Watt, S.M., Underhill, T. (eds.): Ink Markup Language (InkML) W3C Recommendation (September 2011), http://www.w3.org/TR/InkML/
6. Connell, S.D., Jain, A.K.: Template-based online character recognition. Pattern Recognition 34, 1–14 (1999)
7. Zanibbi, R., Novins, K., Arvo, J., Zanibbi, K.: Aiding manipulation of handwritten mathematical expressions through style-preserving morphs. In: Proceedings of Graphics Interface 2001, pp. 127–134 (2001)
8. Hu, R., Watt, S.: Optimization of point selection on digital ink curves. In: Proceedings of 2012 International Conference on Frontiers in Handwriting Recognition (ICFHR), pp. 527–532 (September 2012)
9. Harouni, M., Mohamad, D., Rasouli, A.: Deductive method for recognition of online handwritten persian/arabic characters. In: Proceedings of 2010 The 2nd International Conference on Computer and Automation Engineering (ICCAE), vol. 5, pp. 791–795 (February 2010)
10. Watt, S.M.: Polynomial approximation in handwriting recognition. In: Proceedings of the 2011 International Workshop on Symbolic-Numeric Computation, SNC 2011, pp. 3–7. ACM (2011)
11. Golubitsky, O., Watt, S.M.: Online stroke modeling for handwriting recognition. In: Proceedings of the 2008 Conference of the Center for Advanced Studies on Collaborative Research: Meeting of Minds, CASCON 2008, pp. 6:72–6:80. ACM (2008)
12. Char, B.W., Watt, S.M.: Representing and characterizing handwritten mathematical symbols through succinct functional approximation. In: Proceedings of the Ninth International Conference on Document Analysis and Recognition, ICDAR 2007, vol. 02, pp. 1198–1202. IEEE Computer Society (2007)
13. Golubitsky, O., Watt, S.M.: Distance-based classification of handwritten symbols. Int. J. Doc. Anal. Recognit. 13(2), 133–146 (2010)
14. Golubitsky, O., Mazalov, V., Watt, S.M.: Orientation-independent recognition of handwritten characters with integral invariants. In: Proceedings of Joint Conference of ASCM 2009 and MACIS 2009: Asian Symposium of Computer Mathematics and Mathematical Aspects of Computer and Information Sciences, ASCM 2009, pp. 252–261 (2009)
15. Golubitsky, O., Mazalov, V., Watt, S.M.: Toward affine recognition of handwritten mathematical characters. In: Proceedings of the 9th IAPR International Workshop on Document Analysis Systems, DAS 2010, pp. 35–42. ACM (2010)

Capturing Hiproofs in HOL Light

Steven Obua, Mark Adams, and David Aspinall

School of Informatics, University of Edinburgh

Abstract. Hierarchical proof trees (hiproofs for short) add structure to ordinary proof trees, by allowing portions of trees to be hierarchically nested. The additional structure can be used to abstract away from details, or to label particular portions to explain their purpose.

In this paper we present two complementary methods for capturing hiproofs in HOL Light, along with a tool to produce web-based visualisations. The first method uses *tactic recording*, by modifying tactics to record their arguments and construct a hierarchical tree; this allows a tactic proof script to be modified. The second method uses *proof recording*, which extends the HOL Light kernel to record hierachical proof trees alongside theorems. This method is less invasive, but requires care to manage the size of the recorded objects. We have implemented both methods, resulting in two systems: *Tactician* and *HipCam*.

1 Overview

Proofs constructed by an interactive theorem proving system can be extremely complex. This complexity is reflected in the size of input proof scripts needed for large verifications, which may amount to hundreds of thousands of lines of proof script. It is also reflected in attempts to demonstrate the overall result of a proof development as a proof tree: real proof trees rapidly become large and unwieldy and the debate over their utility continues.

A common way of managing complexity is by introducing hierarchy. This can be done in the input proof script language: an example is the Isar proof language [17]. Isar uses block structure to induce a hierarchy; new blocks are introduced for proof constructs like induction and case distinction.

Hierarchy can also be used to tame large proof trees, which is our focus in this paper. We employ a notion of hierarchical proof known as *hiproofs* [11,6]. The hope is that by providing mechanisms to add hierarchy to proofs as they are constructed, we may build proof trees that can be more easily managed and exploited in useful ways. With good interfaces, users may be able to navigate proof trees comfortably, to zoom in on some detail about how a proof proceeded, or to gain an oversight without having expert knowledge of the source language. Automatic tools may be provided that take advantage of hierarchy and labelling, for example, allowing operations for querying and transforming proofs (such as the prototype query language in [7]) or providing inputs for machine learning to investigate patterns in proof.

Block structured proof scripts and hierarchical proofs fit well together; the latter can provide a semantics for the former [19]. Here we chose to start work from

J. Carette et al. (Eds.): CICM 2013, LNAI 7961, pp. 184–199, 2013.

HOL Light [1], which does not have a hierarchical input language. Proofs are constructed by composing tactics in the meta-language OCaml. So we need other ways of introducing hierarchy. This is possible by several means: by transforming a previously produced proof tree, by modifying standard tactics to produce nested labelled proofs, or by introducing dedicated user-level tactics. We use the tactic based mechanisms here.

Outline. We will first give a quick introduction to hiproofs. We then describe two methods for obtaining hierarchical proofs in HOL Light. Both work by instrumenting the HOL Light theorem prover, but they work on different levels of atomicity. The *Tactician* tool works at the layer of *tactics* by modifying them so that proof information is recorded in a goal tree. The *HipCam* tool works at the layer of *inference rules* and modifies the HOL Light kernel so that hiproofs are recorded in the theorem data structure, thereby extending the proof recording approach described in [16] to also record hierarchy. The two approaches have complementary advantages and disadvantages; further discussion follows as they are introduced and in the concluding section.

2   Hierarchical Proofs

As an introductory example, Figure 1 shows the proof of the HOL Light theorem `TRANSITIVE_STEPWISE_LT_EQ`. HOL Light is written in OCaml and therefore we use a prettified OCaml notation in this paper. Figure 2 shows the hierarchical proof generated from this proof by Tactician (a similar visualisation can be generated via HipCam), and Figure 3 shows the expanded version of the `<==` box. All boxes have been introduced automatically during the generation of the hiproof, with the exception of the box labelled "Prepare induction hypothesis", which has been introduced by an explicit labelling command.

```
let TRANSITIVE_STEPWISE_LT_EQ = prove
 ('!R. (!x y z. R x y /\ R y z ==> R x z)
 ==> ((!m n. m < n ==> R m n) <=> (!n. R n (SUC n)))',
     REPEAT STRIP_TAC THEN EQ_TAC THEN ASM_SIMP_TAC[LT] THEN
     DISCH_TAC THEN SIMP_TAC[LT_EXISTS; LEFT_IMP_EXISTS_THM] THEN
     GEN_TAC THEN ONCE_REWRITE_TAC[SWAP_FORALL_THM] THEN
     REWRITE_TAC[LEFT_FORALL_IMP_THM; EXISTS_REFL; ADD_CLAUSES] THEN
     INDUCT_TAC THEN REWRITE_TAC[ADD_CLAUSES] THEN ASM_MESON_TAC[]);;
```

Fig. 1. Tactic-style proof of TRANSITIVE_STEPWISE_LT_EQ

Hiproofs were introduced by Denney et al [11], as a uniform formalisation of ideas that had been experimented with in several proof development systems. Denotationally, hiproofs are described as a forest of trees with a nesting relation. A syntactic formulation was added later [6]; adapted to the purposes of this paper, this syntax can be represented as a datatype as follows:

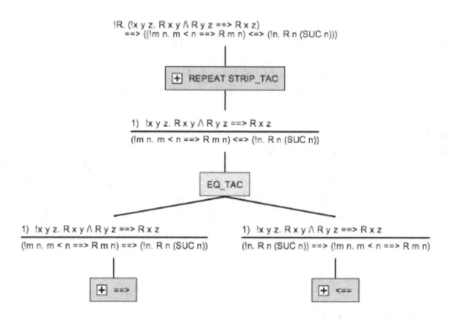

Fig. 2. Hierarchical proof of TRANSITIVE_STEPWISE_LT_EQ

```
type hiproof =
          Atomic of label × goal × int
        | Sequence of hiproof list
        | Tensor of hiproof list
        | Box of label × hiproof
```

Here $\text{Atomic}\,(l, g, n)$ represents the application of an atomic tactic labelled l to a goal g yielding n subgoals (Fig. 4) whereas Sequence and Tensor are used to build more complex proofs. The left hiproof in Figure 5 illustrates this, the picture shown corresponds to the hiproof expression g defined by

$$g = \text{Sequence}\,[\text{Atomic}\,(T_1, A, 2), \text{Tensor}\,[\text{Atomic}\,(T_2, B, 1), \text{Atomic}\,(T_3, C, 2)]].$$

The basic idea of hierarchical proofs is now that tactics are not necessarily atomic but that it is possible to "look inside" a tactic by representing its inside as a hiproof, too. The expression $\text{Box}\,(l, h)$ allows this and denotes a tactic labelled l with an inner hiproof. The right hiproof in Figure 5 boxes up the hiproof g to its left and is written in our notation as $\text{Box}\,(\text{``Tactic''}, g)$.

Labels are arbitrary and can be used for different purposes; they can contain simple names for tactics or proof methods as we show in examples, or could, for example, contain references into the source code of the proof.

We are only interested in *well-formed* hiproofs. In well-formed hiproofs sequences and tensors are at least two elements long. To check further requirements, let the function $\text{IN} : \text{hiproof} \to \text{int}$ be defined via

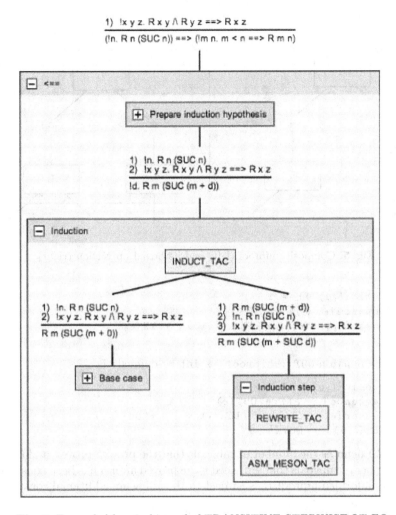

Fig. 3. Expanded box in hiproof of TRANSITIVE_STEPWISE_LT_EQ

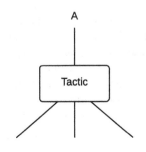

Fig. 4. Atomic (Tactic, A, 3)

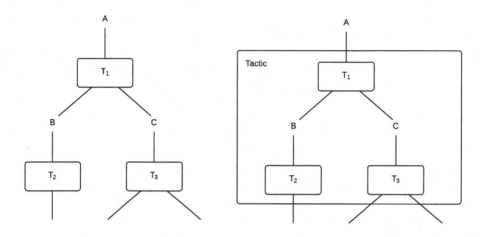

Fig. 5. Composite hiproof (left) and its boxed up version (right)

```
IN(Atomic(l,g,n)) = 1
IN(Sequence[e,...]) = IN(e)
IN(Tensor[e₁,...,eₙ]) = IN(e₁) + ... + IN(eₙ)
IN(Box(l,h)) = IN(h)
```

and let the function OUT : hiproof → int be defined via

```
OUT(Atomic(l,g,n)) = n
OUT(Sequence[...,e]) = OUT(e)
OUT(Tensor[e₁,...,eₙ]) = OUT(e₁) + ... + OUT(eₙ)
OUT(Box(l,h)) = OUT(h)
```

Now $IN(h)$ denotes the number of subgoals that the proof h proves, and $OUT(h)$ is the number of subgoals that still need to be proved after h has been considered. Then a well-formed hiproof H is subject to the following additional constraints:

- Every hiproof contained in H is well-formed, too.
- $H = \text{Box}(l, h)$ implies $IN(h) = 1$.
- $H = \text{Sequence}[e_1, \ldots, e_n]$ implies $OUT(e_i) = IN(e_{i+1})$ for $1 \leq i < n$.

The well-formedness constraints ensure that hiproofs are "plugged together" correctly, and serve as (informal) invariants maintained in our software.

Our implementations include a module (based on Javascript and HTML5 Canvas) that displays well-formed hiproofs in a web browser as shown in Figure 2 and Figure 3. Boxes can be collapsed so that they display only their label and not their inner hiproof. The display of intermediate goals can be toggled individually.

In the next two sections we will present two methods for capturing hiproofs of HOL Light theorems.

label	flattened proof
	g '!R. (!x y z. R x y /\ R y z ==> R x z) ==> ((!m n. m < n ==> R m n) <=> (!n. R n (SUC n)))';;
A	e (REPEAT STRIP_TAC);;
B	e (EQ_TAC);;
	(* *** Subgoal 1 *** *)
C	e (ASM_SIMP_TAC [LT]);;
	(* *** Subgoal 2 *** *)
D	e (ASM_SIMP_TAC [LT]);;
E	e (DISCH_TAC);;
F	e (SIMP_TAC [LT_EXISTS; LEFT_IMP_EXISTS_THM]);;
G	e (GEN_TAC);;
H	e (ONCE_REWRITE_TAC [SWAP_FORALL_THM]);;
I	e (REWRITE_TAC [LEFT_FORALL_IMP_THM; EXISTS_REFL; ADD_CLAUSES]);;
J	e (INDUCT_TAC);;
	(* *** Subgoal 2.1 *** *)
K	e (REWRITE_TAC [ADD_CLAUSES]);;
L	e (ASM_MESON_TAC []);;
	(* *** Subgoal 2.2 *** *)
M	e (REWRITE_TAC [ADD_CLAUSES]);;
N	e (ASM_MESON_TAC []);;

Fig. 6. Flattened proof of TRANSITIVE_STEPWISE_LT_EQ

3   Tactician

Tactician is a productivity tool for refactoring individual HOL Light tactic proof scripts. It supports two main refactoring operations: packaging up a series of tactic steps into a single compound tactic joined by THEN and THENL tacticals, and the reverse operation, for flattening out a packaged-up tactic into a series of tactic steps. It is aimed at helping experts maintain their proof scripts, and helping beginners learn from existing proof scripts. It can be obtained from [4].

Behind the scenes, Tactician uses a representation of the recorded tactic proof tree which is close to a hiproof; recording hierachical proofs was one of its original design goals.

3.1   Example

A typical packaged up proof has already been presented in Fig. 1. The result of flattening out this proof is shown in Fig. 6. The following hiproof can be directly read off from the flattened proof (we omit the goals in Atomic):

```
Sequence[
    Atomic(A,1),Atomic(B,2),
    Tensor[
        Atomic(C,0),
        Sequence[
            Atomic(D,1),Atomic(E,1),Atomic(F,1),
```

```
Atomic(G,1),Atomic(H,1),Atomic(J,2),
Tensor[
    Sequence[Atomic(K,1),Atomic(L,0)],
    Sequence[Atomic(M,1),Atomic(N,0)]]]]]]
```

This hiproof corresponds (after introduction of several Boxes) to the visualisation shown in Fig. 2 and Fig. 3.

3.2 Tactic Recording

It helps to recall how a tactic proof is constructed in HOL Light. The user starts with a single main goal, which gets broken down over a series of tactic steps into hopefully simpler-to-prove subgoals. The user works on each subgoal in turn. The proof is complete when the last subgoal has been proved. Behind the scenes, the standard subgoal package maintains a proof state that consists of a list of current proof goals and a justification function for constructing the formal proof of a goal from the formal proofs of its subgoals. Tactics are functions that take a goal and return a subgoal list plus a justification function. The subgoal package state is updated every time a tactic is applied, incorporating the tactic's resulting subgoals and justification function.

Tactician works by recording such a tactic-style proof in a proof tree, where each node in the tree corresponds to a goal in the proof. When a user wants to refactor the proof, the proof tree is abstracted to a hiproof, which then gets refactored accordingly before being emitted as an ML tactic proof script. We give a brief overview of the recording mechanism here; more details are in [5].

The proof tree gets initialised when a tactic proof is started, and is added to as tactics are executed. Tactics are modified so that they work on a modified, or "promoted", goal datatype called xgoal (Fig. 7). Each xgoal carries a unique goal id, which corresponds to a node in the proof tree. A modified tactic has type xtactic. It takes an xgoal input, strips away its id, applies the original unmodified tactic, and generates new ids for each of the resulting goals. Information about the tactic step, including an abstraction of the text of the tactic as it would appear in the proof script, is then inserted into the proof tree at the node indicated by the input's id. An index of ids and references to their corresponding nodes is maintained to enable nodes to be located.

Boxes around tactics can be introduced manually via a function

```
val hilabel : label → xtactic → xtactic
```

```
type goalid = int
type xgoal = goal × goalid
type xgoalstate = (term list × instantiation) ×
                  xgoal list × justification
type xtactic = xgoal → xgoalstate
```

Fig. 7. Modifications of HOL Light's datatypes

so `hilabel`(l, t) sets up a new box with the label l around the tactic t. The input of t becomes the input of the box, and the subgoals which result from applying t become the outputs.

Apart from basic tactics, there are tactic-producing functions which depend on additional arguments like terms, theorems, or other tactics. We also need to modify these more complicated tactic forms. Because each tactic form has a fixed ML type signature, a generic wrapper function can be written for performing this modification for each such form. About 20 such wrappers need to be written to cover the commonly used tactic forms in the HOL Light base system.

3.3 Capturing Hiproofs with Tactician

Based on its tactic recording mechanism, Tactician can generate hiproofs from tactic style proofs by a straightforward transformation of the tactic proof tree to hiproofs. Because the proof tree corresponds naturally to the user's actual proof script, so does the hiproof. Hierarchical boxes can optionally be introduced by the various wrapper functions. This method of generating hiproofs works also for proofs that have not been completed yet, and can therefore potentially be used to visualise the current proof state during interactive proof as a hiproof.

Tactician outputs a proof at the user level, i.e., involving the same atomic ML tactics, rules and theorems as occur in the original proof script. Low-level information of the proof is not retained. This has the advantage that hierarchical proofs are maintained at a level meaningful to the user, and the overhead of recording is kept low.

One fundamental limitation of Tactician is that tactics that take functions as arguments cannot be "promoted" if the function itself does not return a promotable datatype (the only common instance of this in HOL Light is `PART_MATCH`, which takes a term transformation function as an argument). Another is that ML type annotations in the proof script need to mention promoted, rather than unpromoted, ML datatypes.

With Tactician version 2.2, proof script files involving several hundred lines of ML will typically encounter one or two occurrences of such limitations. It is possible to get around these problems by making hand edits to the proof script, but highly-automated processing of very large bodies of proof is not currently feasible.

4 HipCam

The basic idea of HipCam is to modify the HOL Light kernel, instead of modifying the higher-level tactic-layer like Tactician does. While Tactician relies on *tactic recording*, HipCam instead uses a *proof recording* approach closely related to that pioneered in [16] . HipCam can be downloaded from [15].

HipCam is minimally invasive; any theorem proven using the original HOL Light kernel can be proven using the HipCam-modified kernel and modification of proof scripts is not needed, except to add explicit hierarchy and labelling, incrementally as desired. However, HipCam does not allow recovering proof scripts from recorded proofs; it is intended primarily as a tool to construct large hiproofs for inspection, rather than replay or refactor proof scripts.

4.1 Proof Recording

HipCam does not alter the signature of the HOL Light kernel except to add two functions. To extract a hierarchical proof from such a theorem in the modified kernel, one applies the new kernel function

```
val hiproof : thm → hiproof
```

to the theorem. This is made possible by changing the definition for the type `thm` from

```
type thm = Sequent of term list × term
```

which stores the assumptions and conclusion of a theorem to

```
type thm = Sequent of hiproof × term list × term
```

which in addition stores the hiproof of a theorem. This change in the implementation of `thm` is visible outside the kernel in only one way, by using native ML equality to compare theorems. Fortunately, after proof recording was introduced to HOL Light, native ML equality is not used to compare theorems anymore. To test two theorems for equality, the function `equals_thm` is called; it only compares assumptions and conclusions.

All kernel primitives which produce values of type `thm` are modified to also produce corresponding theorem-internal hiproofs. None of those primitives introduce hierarchy, though. To produce hiproofs with an actual hierarchy we need another new kernel function, `hilabel`, which will draw a box around an existing hierarchical proof. What could be the signature of such a function? Our first guess might be

```
val hilabel₁ : label → hiproof → hiproof
```

which can simply be defined via

```
let hilabel₁ l h = Box(l,h)
```

The obvious problem with this is that still for no theorem t will $\mathtt{hiproof}(t)$ contain any boxes. This is simply because `hilabel₁` does not allow any change of the internal hiproofs of theorems.

Our next guess might therefore be to rectify this problem as follows:

```
val hilabel_thm : label → thm → thm
let hilabel_thm l (Sequent(h,asms,concl)) =
        Sequent(Box(l,h),asms,concl)
```

Unfortunately, `hilabel_thm` does not allow us to create boxes as flexibly as we want to. This is because for any theorem $t = \mathtt{Sequent}(h,asms,concl)$ the invariants $\mathtt{IN}(h) = 1$ and $\mathtt{OUT}(h) = 0$ hold. In other words, no sub-goals can be exported from nested boxes and we could only construct fully nested trees.

So a sub-hiproof H like the ones from Fig. 5 could not be contained in any of the hiproofs created via $\texttt{hilabel}_{\texttt{thm}}$ because $\texttt{OUT}(H) = 3$ holds.

What we need when drawing a box onto a hiproof is the ability to specify which part of the hiproof should become part of the box, and which part should stay outside of the box. We gain this ability by drawing boxes around *rules* instead of just theorems:

```
type rule = thm list → thm
val hilabel : label → rule → rule
```

We will see in the next section how `hilabel` works and how it can be implemented. Meanwhile, we can see that `hilabel` will satisfy all of our boxing needs. It is still trivial to box theorems:

```
let hilabelthm l t = hilabel l (fun _ → t) []
```

It is also straightforward how to label tactics (with a reminder of the types):

```
type goalstate = (term list × instantiation) ×
                 goal list × justification
type tactic = goal → goalstate
val hilabeltac : label → tactic → tactic
let hilabeltac l t g =
  let (inst, gls, j) = t g in
  let k inst = hilabel l (j inst)
  in (inst, gls, k)
```

The above code reduces labelling a tactic to labelling the justification function that is obtained as the result of applying the tactic to a goal.

4.2 Implementing `hilabel`

Let us examine how we want `hilabel` to behave. Assume we have a rule

```
val rule : thm list → thm
```

and three theorems α_1, α_2 and α_3 such that

$\texttt{rule}\,[\alpha_1,\alpha_2,\alpha_3]$

yields a new theorem as the result of applying `rule` to these theorems. The hiproof of this new theorem will then in some way depend on the hiproofs of the α_i, e.g. like depicted on the left in Fig. 8. Now, if instead of applying the original rule, we apply the labelled rule

$(\texttt{hilabel}\ \texttt{"rule"}\ \texttt{rule})\,[\alpha_1,\alpha_2,\alpha_3]$

then we'd like the hiproof of the resulting theorem to look like depicted on the right in Fig. 8. Note that all we guarantee for the boxed hiproof is that the hiproofs of the α_i will appear (if at all) outside of the box. In particular, there is no predetermined order in which the hiproofs of the α_i will appear. It might even be the case that some of these hiproofs are not used at all, or are used more than once. This situation is shown in Fig. 9: here the proof of $\texttt{rule}\,[\alpha_1,\alpha_2,\alpha_3]$

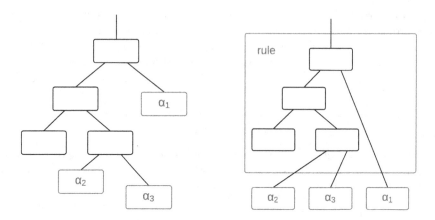

Fig. 8. `rule[`α_1`,`α_2`,`α_3`]` vs. `(hilabel "rule" rule)[`α_1`,`α_2`,`α_3`]`

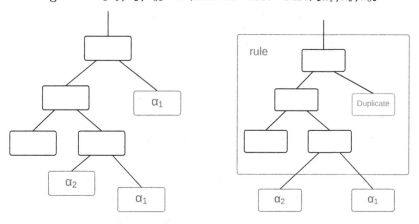

Fig. 9. Unused α_3 and duplicate α_1

does not make use of α_3, and uses α_1 twice. We detect multiple appearances of the same α_i and treat only the first occurrence normally. The other occurrences are marked as being duplicate instances of goals proven elsewhere.

Our design of `hilabel` is driven by trying to make the collapsing and expanding of boxes in a visualised hiproof straightforward. One can imagine other ways of dealing with reordered, duplicate, or missing dependencies. For example, we could introduce a new hiproof constructor for boxes which rewire the outputs of their inner hiproofs such that externally, the outputs of the box correspond 1-to-1 and in the right order to the arguments of the rule the box is supposed to represent (a *swap* primitive can be used for this purpose; see [18]).

To implement `hilabel`, we first introduce three kinds of labels: the identity label L_{id}, the duplicate label L_{dup} and a family of variable labels L_{var}^{name} where *name* is from some infinite set V of variable names. We then define

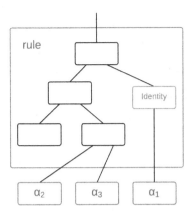

Fig. 10. Explicit display of identity tactic

$$\text{Identity}\,(g) \;\equiv\; \text{Atomic}\,(L_{id}\,,g\,,1)$$
$$\text{Duplicate}\,(g) \;\equiv\; \text{Atomic}\,(L_{dup}\,,g\,,0)$$
$$\text{Variable}\,(name\,,g) \;\equiv\; \text{Atomic}\,(L_{var}^{name}\,,g\,,0)$$

to serve us as identity tactic, duplicate marker, and hiproof variable, respectively. We need the identity tactic because without it we could not represent the right hand side hiproof in Fig. 8 (which is just a prettification of the hiproof shown in Fig. 10). We have already motivated why we need duplicate markers (Fig. 9). And we need variables so that we can track how the hiproofs of α_i are being used in constructing the hiproof for $\text{rule}\,[\alpha_1,\ldots,\alpha_k]$. The details of how this tracking is achieved are shown in Fig. 11. There the notation α/h is used to represent the theorem resulting from replacing the hiproof of the theorem α with the hiproof h. The heavy lifting in hilabel is done by the function turnvars.

A major challenge in the actual implementation of hilabel and turnvars is that recorded proof trees quickly grow to be enormous. Their representations in memory exploit sharing, but repeatedly traversing such trees depth-first to compute or update them is impractical. More sophisticated data structures could help with this, but we use the simple fix of adapting the described algorithms so that all important properties of a hiproof are computed (and then cached for shared reuse) during the construction of the hiproof, so later traversals are unnecessary. One such property of a hiproof we have introduced is its *shallow size* $\text{SS}(h)$ which measures the size of a hiproof h as if all boxes it contained were replaced by atomics instead:

$$\text{SS}\,(\text{Atomic}\,(l\,,g\,,n)) \;=\; 1$$
$$\text{SS}\,(\text{Sequence}\,[e_1,\ldots,e_n]) \;=\; 1 \;+\; \text{SS}\,(e_1) \;+\; \ldots \;+\; \text{SS}\,(e_n)$$
$$\text{SS}\,(\text{Tensor}\,[e_1,\ldots,e_n]) \;=\; 1 \;+\; \text{SS}\,(e_1) \;+\; \ldots \;+\; \text{SS}\,(e_n)$$
$$\text{SS}\,(\text{Box}\,(l\,,h)) \;=\; 1$$

We can use the shallow size of a hiproof h to adjust h to the needs of the hiproof consumer. For visualisation, for example, we are not interested in hiproofs that

```
val hilabel : label → rule → thm list → thm
let hilabel l rule [α₁,...,αₖ] =
  let N be a set {n₁,...,nₖ} of k fresh names in
  let α'ᵢ = αᵢ/(Variable(nᵢ, goal(αᵢ))), i=1...k, in
  let β = rule[α'₁,...,α'ₖ] in
  let (names, h) = turnvars N (hiproof(β)) in
  let H be a function such that H(nᵢ) = hiproof(αᵢ) in
  let b = Box(l,h) in
  let h' =
    match (map H names) with
       [] → b
    | [a] → Sequence[b,a]
    | ā → Sequence[b,Tensor(ā)]
  in β/h'


val turnvars : V set → hiproof → V list × hiproof
let turnvars N h =
  let h' =
      (replace all occurrences of Var(n,g) in h where n ∈ N
       either with Identity(g) or with Duplicate(g)
       and massage the result so that it is well-formed)
  in let names =
      (the list of variable names which correspond
       to the outputs of h')
  in (names,h')
```

Fig. 11. Description of hilabel and turnvars

have a shallow size larger than a certain threshold τ, say $\tau = 1000$. Therefore we replace subexpressions of the form $\text{Box}(l,h)$ with an atomic whenever $\text{SS}(h) > \tau$. Note that such a replacement requires that we introduce a property for atomics which keeps track of the variables that the replaced box contained.

Another way to cut-out uninteresting detail is to look at the label of a box. For example, when l indicates that the box corresponds to a standard HOL Light inference rule, we could also elide the detail. Therefore HipCam has two modes: a *max-detail* mode, which does not replace boxes corresponding to standard inference rules, and a *high-level* mode which does.

4.3   Capturing Hiproofs with HipCam

We have applied HipCam to several formalisations that ship with HOL Light. Fig. 12 displays how much time and space were needed for each formalisation, without HipCam, with HipCam in max-detail mode, and with HipCam in high-level mode. The results are quite surprising: using HipCam incurs only a modest speed penalty of a maximum factor of not more than 1.5. HipCam's memory

Proof	Original	Max-Detail	High-Level
`#use "hol.ml";;`	4 min 30 sec / 0.1 GB	7 min / 2.2 GB	6 min / 0.7 GB
Gödel 1	10 min 30 sec / 0.1 GB	15 min / 3.1GB	13 min / 0.9 GB
e is transcendental	13 min / 0.2 GB	19 min / 4.2 GB	16 min / 1.5 GB
Jordan curve theorem	31 min / 0.4 GB	out of memory	45 min / 6.5 GB

Fig. 12. HipCam statistics, using a MacBook Pro, Quad Core 2.4GHz, 16GB RAM

usage is more taxing, using several gigabytes for large formalisations. The max-detail mode needs about three times as much memory as the high-level mode.

The memory usage of HipCam's max-detail mode is higher, but similar to the memory used by the standard proof recording approach, allowing some inflation for the extra information used by HipCam like the shallow size. But depending on the use of the recorded proof we can dramatically undercut these memory requirements as the high-level mode shows; this is not possible in a simple proof recording approach where larger examples would fail outright.

5   Related Work

In Section 1 we explained the difference between hierarchical structure of stored proof trees and the hierachical structure of proof script input languages (provided by languages such as Isar [17]). The later may be manipulated by text-based interfaces, for example, to fold (temporarily hide) sub-sections.

The urge to present proofs in two dimensions is widespread. Some systems have taken a tree-like approach from the start, using interfaces that present proofs in a nested hierarchical form as they are developed, enforcing structure rigidly, or using a GUI to build trees. One early example is Nuprl's tactic trees [12]; another is the Tecton system which introduced proof forests and allowed to print out their graphical representation [14]. The more recent ProofWeb system [13] allows both "flag" style proof development as well as tree-style, connecting each style back to source Coq code. An interesting mix of proof scripts and a graphical representation also appears in Hyperproof and its methodology of heterogenous reasoning [9,8]. Proofscape [2] is a recently launched project which aims to become a visual library of mathematics. It displays proofs with an adjustable level of detail which corresponds to our notion of hiproof boxes which are either collapsed or expanded in their visualisation.

A complete survey of proof visualisation tools for proof is out of scope, but we mention one example that inspired the visualisation work here: the Prooftree tool by Tews [3] displays proof trees for Coq in Proof General (in turn itself inspired by the similar feature provided in PVS). Contrary to our current visualisation software, Prooftree supports interactive visualisation during a proof, but it does not yet include hierarchical proof trees.

On a different strand, a main purpose of Tactician is to serve as a refactoring tool that can convert between "flat" and "packaged up" proofs. There is related work on proof refactoring, including some approaches designed based on hiproof semantics by Whiteside [19,18], as well as tools that have been implemented such as the conversion between procedural and declarative proof scripts inside ProofWeb [13] and the Levity tool [10] which allows moving lemmas between different theories. Generally such tools are still in the early days.

6   Conclusions

Hierarchical proofs as invented in [11] have so far been mostly theoretical constructs. Our work is directed towards gaining hands-on experience with "real-world" hiproofs. With Tactician and with HipCam we have laid the technical foundations for such an undertaking, we are able now to take existing bodies of formalisations, represent them as hiproofs, and study them as such.

Tactician can present individual proofs at the level of detail given by the user, but because of its mentioned limitations, it is less suitable to automatically obtain hiproof representations of existing large formalisations, or to delve arbitrarily deep to understand the results of a complicated tactic; this is what HipCam was designed for. Tactician recovers user-level proof steps lost to HipCam, but hierarchy doesn't appear for free in either case. Both tools allow the user to annotate tactics to automatically add labels (for example, boxing up an induction or simplification tactic), or add labels manually in particular proofs.

It is natural to ask whether the approaches can be combined. The disadvantage with Tactician is the need to modify scripts pervasively, but this issue arises mainly because of its aim to record proof script input fully to allow refactoring; without this the wrapper functions are much simpler. Conversely, HipCam could be provided with a modified subgoal package like Tactician's that records user-level proof steps and triggers only high-level capturing mode between steps.

More crucially, looking at hiproofs generated automatically via HipCam from theorems like the Jordan Curve Theorem is not very illuminating, because there isn't enough hierarchy yet. The problem is that it is hard to distinguish between those parts of the proof which convey its meaning, and those parts which exist for mostly technical reasons. So the challenge is how to "box up" the technical parts of a proof, so that its meaningful parts are emphasised, and do this in a hierarchical way. In future work we plan to investigate semi-intelligent ways of transforming a hiproof to introduce structure, as well as using some manual labelling on some case study large developments.

Acknowledgements. We're grateful to the CICM referees for providing good suggestions to improve the paper and to members of the Mathematical Reasoning Group at Edinburgh for feedback and discussions. Our research was supported by funding from UK EPSRC grant EP/J001058/1 and from the Laboratory for Foundations of Computer Science, University of Edinburgh.

References

1. HOL light, http://www.cl.cam.ac.uk/~jrh13/hol-light
2. Proofscape, http://proofscape.org

3. Prooftree, http://askra.de/software/prooftree/
4. Adams, M.: Tactician (2012), http://www.proof-technologies.com/tactician
5. Adams, M., Aspinall, D.: Recording and refactoring HOL light tactic proofs. In: Workshop on Automated Theory eXploration (2012)
6. Aspinall, D., Denney, E., Lüth, C.: Tactics for hierarchical proof. Mathematics in Computer Science 3(3), 309–330 (2010)
7. Aspinall, D., Denney, E., Lüth, C.: Querying proofs. In: Bjørner, N., Voronkov, A. (eds.) LPAR-18. LNCS, vol. 7180, pp. 92–106. Springer, Heidelberg (2012)
8. Barker-Plummer, D., Etchemendy, J.: A computational architecture for heterogeneous reasoning. Journal of Experimental & Theoretical Artificial Intelligence 19(3), 195–225 (2007)
9. Barwise, J.: Heterogeneous reasoning. In: Mineau, G.W., Moulin, B., Sowa, J.F. (eds.) ICCS 1993. LNCS, vol. 699, pp. 64–74. Springer, Heidelberg (1993)
10. Bourke, T., Daum, M., Klein, G., Kolanski, R.: Challenges and experiences in managing large-scale proofs. In: Jeuring, J., Campbell, J.A., Carette, J., Dos Reis, G., Sojka, P., Wenzel, M., Sorge, V. (eds.) CICM 2012. LNCS, vol. 7362, pp. 32–48. Springer, Heidelberg (2012)
11. Denney, E., Power, J., Tourlas, K.: Hiproofs: A hierarchical notion of proof tree. Electronic Notes in Theoretical Computer Science 155 (2006)
12. Griffin, T.: Notational definition and top-down refinement for interactive proof development systems. PhD thesis, Cornell University (1988)
13. Kaliszyk, C., Wiedijk, F.: Merging procedural and declarative proof. In: Berardi, S., Damiani, F., de'Liguoro, U. (eds.) TYPES 2008. LNCS, vol. 5497, pp. 203–219. Springer, Heidelberg (2009)
14. Kapur, D., Nie, X., Musser, D.: An overview of the tecton proof system. Theoretical Computer Science 133(2), 307–339 (1994)
15. Obua, S.: HipCam (2013), http://github.com/phlegmaticprogrammer/hipcam
16. Obua, S., Skalberg, S.: Importing HOL into Isabelle/HOL. In: Furbach, U., Shankar, N. (eds.) IJCAR 2006. LNCS (LNAI), vol. 4130, pp. 298–302. Springer, Heidelberg (2006)
17. Wenzel, M.: Isar - A generic interpretative approach to readable formal proof documents. In: Bertot, Y., Dowek, G., Hirschowitz, A., Paulin, C., Théry, L. (eds.) TPHOLs 1999. LNCS, vol. 1690, pp. 167–183. Springer, Heidelberg (1999)
18. Whiteside, I.: Refactoring Proofs. PhD thesis, University of Edinburgh (2013)
19. Whiteside, I., Aspinall, D., Dixon, L., Grov, G.: Towards formal proof script refactoring. In: Davenport, J.H., Farmer, W.M., Urban, J., Rabe, F. (eds.) Calculemus/MKM 2011. LNCS, vol. 6824, pp. 260–275. Springer, Heidelberg (2011)

A Qualitative Comparison
of the Suitability of Four Theorem Provers
for Basic Auction Theory*

Christoph Lange[1], Marco B. Caminati[2], Manfred Kerber[1], Till Mossakowski[3],
Colin Rowat[4], Makarius Wenzel[5], and Wolfgang Windsteiger[6]

[1] Computer Science, University of Birmingham, UK
[2] http://caminati.net.tf, Italy
[3] University of Bremen and DFKI GmbH Bremen, Germany
[4] Economics, University of Birmingham, UK
[5] Univ. Paris-Sud, Laboratoire LRI, UMR8623, Orsay, F-91405, France
[6] RISC, Johannes Kepler University Linz (JKU), Austria
http://www.cs.bham.ac.uk/research/projects/formare/code/auction-theory/

Abstract. Novel auction schemes are constantly being designed. Their
design has significant consequences for the allocation of goods and the
revenues generated. But how to tell whether a new design has the desired
properties, such as efficiency, i.e. allocating goods to those bidders who
value them most? We say: by formal, machine-checked proofs. We investigated the suitability of the Isabelle, Theorema, Mizar, and Hets/CASL/
TPTP theorem provers for reproducing a key result of auction theory:
Vickrey's 1961 theorem on the properties of second-price auctions. Based
on our formalisation experience, taking an auction designer's perspective,
we give recommendations on what system to use for formalising auctions,
and outline further steps towards a complete auction theory toolbox.

1   Motivation: Why Formalise Auction Theory?

Auctions are a widely used mechanism for allocating goods and services[1], perhaps second in importance only to markets. They are used to allocate electro-
magnetic spectrum, airplane landing slots, oil fields, bankrupt firms, works of
art, eBay items, and to establish exchange rates, treasury bill yields, and stock
exchange opening prices. Novel auction schemes are constantly being designed,
aiming to maximise the auctioneer's revenue, foster competition in subsequent
markets, and to efficiently allocate resources.

Auction design can have significant consequences. Klemperer attributed the
low revenues gained in some government auctions of the 3G radio spectrum in

---

* This work has been supported by EPSRC grant EP/J007498/1. We would like to
thank Peter Cramton and Elizabeth Baldwin for sharing their auction designer's
point, and Christian Maeder for his recent improvements to Hets.

[1] For the US, the National Auctioneers Association reported $268.5 billion for 2008 [2].

J. Carette et al. (Eds.): CICM 2013, LNAI 7961, pp. 200–215, 2013.

2000 (€20 per capita vs. €600 in other countries) to bad design [18]. Design practice outstrips theory, especially for complex modern auctions such as combinatorial ones, which accept bids on subsets of items (e.g. collections of spectrum). Designing a revenue-maximising auction is *NP*-complete [6] even with a single bidder. Important auctions often run 'in the wild' with few formal results [19]. We aim at convincing auction designers that investing into formalisation pays off with machine-checked proofs and a deeper understanding of the theory. To this end, we want to provide them with a toolbox of basic auction theory formalisations, on top of which they can formalise and verify their own auction designs – which typically combine standard building blocks, e.g. an ascending auction converting to a sealed-bid auction when the number of remaining bidders equals the number of items available. Given the ubiquity of specialist support across a range of service sectors, we conjecture that auction designers might be supported by formalisation experts, creating a niche for specially trained experts at the interface of the core mechanised reasoning community and auction designers.

Our ForMaRE project (formal mathematical reasoning in economics [22]) seeks to increase confidence in economics' theoretical results, to aid in discovering new results, and to foster interest in formal methods within economics. To formal methods, we seek to contribute new challenge problems and user experience feedback from new audiences. Auctions are representative of practically relevant fields of economics that have hardly been formalised so far.[2] Economics has been formalised before [15], particularly social choice theory (cf. §5 and [10]) and game theory (cf. [37] and our own work [16]). However, none of these formalisations involved economists. Formalising (mathematical) theories and applying mechanised reasoning tools remain novel to economics.[3]

§2 establishes requirements for the Auction Theory Toolbox (ATT); §3 explains our approach to building it. §4 is our main contribution: a qualitative comparison of how well four different theorem provers satisfy our requirements. §5 reviews related work, and §6 concludes and provides an outlook.

2   Requirements for an Auction Theory Toolbox

Conversations with auction designers established ATT requirements as follows:

D1. Formalise ready-to-use basic auction concepts, including their definitions and essential properties.
D2. Allow for extension and application to custom-designed auctions without requiring expert knowledge of the underlying mechanised reasoning system.

From a computer scientist's technical perspective, these translate to:

---

[2] Even code verification is typically not considered, although Leese, who worked on the UK's spectrum auctions, has called for auction software to be added to the Verified Software Repository at http://vsr.sourceforge.net [47].
[3] There is a field 'computational economics'; however, it is mainly concerned with the *numerical* computation of solutions or simulations (cf., e.g., [13]).

C1. Identify the right language to formalise auction theory. This language should (a) be sufficiently expressive for concisely capturing complex concepts, while supporting efficient proofs for the majority of problems, (b) be learnable for economists used to mathematical textbook notation, and (c) provide libraries of the mathematical foundations underlying auctions.

C2. Identify a mechanised reasoning system (a) that assists with cost-effective development of formalisations, (b) that facilitates reuse of formalisations already existing in the toolbox, (c) that creates comprehensible output to help users understand, e.g., why a proof attempt failed, or what knowledge was used in proving a goal, and (d) whose community is supportive towards users with little specific technical and theoretical background.

Note the conflicts of interest: a single language might not meet requirement C1a, and if it did, it might not be supported by a user-friendly system.

3   Approach to Building the Auction Theory Toolbox

To avoid a chicken-and-egg problem, we identify relevant domain problems in parallel to identifying languages and systems suitable for formalisation.

3.1   The Domain Problem: Vickrey's Theorem and Beyond

We started with Vickrey's 1961 theorem on the properties of second-price auctions of a single, indivisible good, whose bidders' private values are not publicly known. Each participant submits a sealed bid; one of the highest bidders wins, and pays the highest *remaining* bid; the losers pay nothing. Vickrey proved that 'truth-telling' – submitting a bid equal to one's actual valuation of the good – was a *weakly dominant* strategy, i.e. that no bidder can do strictly better by bidding above or below their valuation *whatever* the other bidders do. Thus, the auction is also *efficient*, allocating the item to the bidder with the highest valuation. Bidders only have to know their own valuations; in particular they need no information about others' valuations or the distributions these are drawn from.

As variants of Vickrey auctions are widely used (e.g. by eBay, Google and Yahoo! [45]), this formalisation will enable us to prove properties of contemporary auctions as well. The underlying theory is straightforward to understand even for non-economists and can be formalised with reasonable effort. Finally, formalising Vickrey provides a good introduction for domain experts to mechanised reasoning technology by serving as a small, self-contained showcase of a widely known result, helping to build trust in this new technology.

Maskin collected 13 theorems, including Vickrey's, in a review [24] of an influential auction theory textbook [25]. This sets the roadmap for building the ATT – a collaborative effort, to which we welcome community contributions [23].

3.2   Paper Elaboration to Prepare the Machine Formalisation

To prepare the machine formalisation, we refined the original paper source, aware that current mechanised reasoning systems typically require much more explicit statements than commonly found on paper: automated provers must find proofs without running out of search space, whereas proof checkers require proofs at a certain level of detail, which in turn requires detailed statements. Maskin states Vickrey's theorem in two sentences and proves it in another six sentences [24, Proposition 1].[4] Our elaboration uses eight definitions specific to the domain problem plus an auxiliary one about maximum components of vectors, as follows:

$N = \{1, \ldots, n\}$ is a set of *participants*, often indexed by i. An *allocation* is a vector $\boldsymbol{x} \in \{0, 1\}^n$ where $x_i = 1$ denotes participant i's award of the indivisible good to be auctioned (i.e. 'i wins'), and $x_j = 0$ otherwise. An *outcome* $(\boldsymbol{x}, \boldsymbol{p})$ specifies an allocation and a vector of payments, $\boldsymbol{p} \in \mathbb{R}^n$, made by each participant i. Participant i's *payoff* is $u_i \equiv v_i x_i - p_i$, where $v_i \in \mathbb{R}_+$ is i's valuation of the good. A *strategy profile* is a vector $\boldsymbol{b} \in \mathbb{R}^n$, where $b_i \geq 0$ is called i's *bid*.[5] For an n-vector $\boldsymbol{y} = (y_1, \ldots, y_n) \in \mathbb{R}^n$, let $\bar{y} \equiv \max_{j \in N} y_j$ and $\bar{y}_{-i} \equiv \max_{j \in N \setminus \{i\}} y_j$.

Definition 1 (Second-Price Auction). *Given* $M \equiv \{i \in N : b_i = \bar{b}\}$, *a second-price auction is an outcome* $(\boldsymbol{x}, \boldsymbol{p})$ *satisfying:*
1. $\forall j \in N \setminus M, x_j = p_j = 0$; *and*
2. *for one*[6] $i \in M$, $x_i = 1$ *and* $p_i = \bar{b}_{-i}$, *while,* $\forall j \in M \setminus \{i\}, x_j = p_j = 0$.

Definition 2 (Efficiency). *An efficient auction maximises* $\sum_{i \in N} v_i x_i$ *for a given v, i.e., for a single good,* $x_i = 1 \Rightarrow v_i = \bar{v}$.

Definition 3 (Weakly Dominant Strategy). *Given some auction, a strategy profile* $\boldsymbol{b}$ *supports an equilibrium in weakly dominant strategies if, for each* $i \in N$ *and any* $\hat{\boldsymbol{b}} \in \mathbb{R}^n$ *with* $\hat{b}_i \neq b_i$, $u_i \left(\hat{b}_1, \ldots, \hat{b}_{i-1}, b_i, \hat{b}_{i+1}, \ldots, \hat{b}_n \right) \geq u_i \left(\hat{\boldsymbol{b}} \right)$.[7] *I.e., whatever others do, i will not be better off by deviating from the original bid* b_i.

Theorem 1 (Vickrey 1961; Milgrom 2.1). *In a second-price auction, the strategy profile* $\boldsymbol{b} = \boldsymbol{v}$ *supports an equilibrium in weakly dominant strategies. Furthermore, the auction is efficient.*

The attempt to be close to a paper formalisation may introduce artefacts that unnecessarily complicate machine formalisation. E.g., the contiguous numeric participant indexing is merely a convention: formally any relation between participants' valuation, bid, allocation, and payment vectors suffices. Similarly, the product $v_i x_i$ recalls the general divisible good case ($x_i \in [0, 1]$) and works around the lack of an easy and compact 'if–then–else' textbook notation.[8]

---

[4] The high level of Maskin's text is owed to its summative nature. Original proofs in auction theory are typically more thorough.

[5] This simplification is sufficient for proving the theorem. More precisely, all participants know that each v_i is an independent realisation of a random variable with distribution density f. A participant's *strategy* is a mapping g_i such that $b_i = g_i(v_i, f)$.

[6] When running an auction in practice, this i may be selected randomly, but this circumstance does not matter for the proof of Vickrey's theorem.

[7] The notation $u_i(\boldsymbol{b})$ is standard in economics but formally misleading. A more careful notation is $u_i(x_i, v_i, p_i)$, where x_i and p_i depend on $\boldsymbol{b}$ and the auction type.

[8] Case distinctions with curly braces consume at least two lines.

Proof. Suppose participant i bids $b_i = v_i$, whatever $\hat{b}_j$ the others bid. Let $\hat{b}^{i \leftarrow v}$ abbreviate the overall vector $(\hat{b}_1, \ldots, \hat{b}_{i-1}, v_i, \hat{b}_{i+1}, \ldots, \hat{b}_n)$. There are two cases[9]:

1. i wins. This implies $b_i = v_i = \overline{\hat{b}^{i \leftarrow v}}$, $p_i = \overline{\hat{b}^{i \leftarrow v}}_{-i}$, and $u_i(\hat{b}^{i \leftarrow v}) = v_i - p_i = \overline{\hat{b}^{i \leftarrow v}} - \overline{\hat{b}^{i \leftarrow v}}_{-i} \geq 0$. Now consider i submitting an arbitrary bid $\hat{b}_i \neq b_i$, i.e. assume an overall bid vector $\hat{b}$. This has two sub-cases:
   (a) i wins with the other bid, i.e. $u_i(\hat{b}) = u_i(\hat{b}^{i \leftarrow v})$, as the second highest bid has not changed.
   (b) i loses with the other bid, i.e. $u_i(\hat{b}) = 0 \leq u_i(\hat{b}^{i \leftarrow v})$.

2. i loses. This implies $p_i = 0$, $u_i(\hat{b}^{i \leftarrow v}) = 0$, and $b_i \leq \overline{\hat{b}^{i \leftarrow v}}_{-i}$; otherwise i would have won. This yields again two cases for i's alternative bid :
   (a) i wins, i.e. $u_i(\hat{b}) = v_i - \overline{\hat{b}}_{-i} = b_i - \overline{\hat{b}^{i \leftarrow v}}_{-i} \leq 0 = u_i(\hat{b}^{i \leftarrow v})$.
   (b) i loses, i.e. $u_i(\hat{b}) = 0 = u_i(\hat{b}^{i \leftarrow v})$.

By analogy for all i, $b = v$ supports an equilibrium in weakly dominant strategies. Efficiency is immediate: the highest bidder has the highest valuation. □

3.3 Choosing Language and System

In terms of *logic*, it is not immediately obvious whether Vickrey's theorem is inherently higher-order. Defining the maximum operator on arbitrarily sized finite sets of real-valued bids and proving its essential properties requires induction and thus exceeds first-order logic (FOL): similarly for the finiteness of a set[10] and a formalisation of real numbers.[11] However, if one takes real vectors and a maximum operation on them for granted, and explicitly requires the maximum to exist, FOL suffices to formalise the relevant domain concepts: single good auctions, second-price auctions, and the theorem statement.[12]

In terms of *syntax*, we assume that auction designers will prefer a language that is close to the textbook mathematics they are used to, rather than having a programming language flavour. We assume that at least optional type annotations support intuitive modelling of domain concepts (e.g. an auction as a function that takes bids and returns an allocation and payments) and prevent formalisation mistakes by cheap early checks (cf. [21]).

In terms of *user experience*, we study two paradigms: *automated provers* try, given a theorem and a knowledge base, to automatically find a proof, potentially appealing to our audience if the user just has to push a button (as with model checkers). *Interactive provers* interactively check a proof written by the user, which may be convenient when a paper proof already exists.

---

[9] Our initial elaboration of Maskin's proof, which distinguishes cases on the basis of participants' bids, resulted in nine leaf cases. Straightforward on paper, we found them tedious to formalise in Isabelle, which triggered the rearrangement shown here.

[10] Finiteness matters: the set $\{b_i = 1 - \frac{1}{i} : i = 1, 2, 3, \ldots\}$ has no maximum.

[11] Real numbers are not usually required for running auctions in *practice*. Even financial exchanges that allow 'sub-pennying' have a minimal discrete quantum of currency.

[12] For instance, our Mizar proof never invokes any second-order *scheme* directly. Two proof steps use the fact that a finite set of numbers includes its maximum, which is proved in the Mizar Mathematical Library (MML) using the induction scheme.

4   Qualitative Comparison of the Languages and Systems

We have formalised Vickrey's theorem in four systems, which differ in logic, syntax and user experience: Isabelle, followed by Mizar, CASL and Theorema. For each system at least one author has in-depth knowledge. The purpose of redoing formalisations from scratch is to understand the specific advantages and disadvantages of the systems and to obtain as idiomatic a formalisation as possible. The formalisations and instructions for using them are available from the ATT homepage [23]. Tab. 1 compares the features of the systems and their languages and shows the state of our work. The following subsections assess the languages and systems w.r.t. the technical requirements C* of §2. Tab. 2 at the end of this section summarises our findings to underpin our final recommendations.

4.1   Level of Detail and Explicitness Required (req. C1a)

All systems required greater detail and explicitness than the paper elaboration of §3.2. The Isabelle formalisation needs 3 additional definitions and 7 auxiliary lemmas. Guiding the automated provers of Theorema and Hets and Mizar's proof checker required similar numbers of auxiliary statements, plus, in Theorema and Hets, further ones to emulate proof steps (cf. §4.2). However, first steps beyond Vickrey's theorem suggests that these auxiliaries make it easier to formalise *further* notions. As our work involved beginners and experts[13], we can only approximately quantify the formalisation effort beyond the paper elaboration. The 'de Bruijn factor' [40], the formalisation size divided by the size of an informal TEX source, measured after stripping comments and xz compression, is around 1.5 for all formalisations[14] except Theorema[15]. This observation suggests that machine formalisation is generally still harder than elaboration on paper.

Even while explicit machine formalisation imposes tedious work on the author, it can also prove beneficial. On paper, it was neither immediately obvious that exactly one participant wins a second-price auction, nor that the outcome is a function of the bids. While obvious that at least two participants are required to define the 'second highest bid', the standard literature largely overlooks this, but formalisation forced us to choose whether to allow it (by, e.g., defining $\max \emptyset \equiv 0$) or to explicitly require $n \geq 2$.

4.2   Expressiveness vs. Efficiency (req. C1a)

As discussed in §3.2, we did not strictly take the elaborated paper source as a specification for the formalisation, but wrote idiomatic formalisations. In Isabelle and Mizar, we, e.g., avoided specific intervals $\{1, \ldots, n\}$ as sets of auction

---

[13] The Mizar formalisation was, e.g., completely written by an expert (Caminati), whereas the Isabelle formalisation was initially written by a first-time user with a general logic background (Lange), then largely rewritten by an expert (Wenzel).

[14] A typical average is 4, but our paper proof is particularly detailed.

[15] Determining a de Bruijn factor for Theorema does not make sense: single keystrokes or clicks may yield complex inputs, Mathematica notebooks store layout and maintenance information, and Theorema caches proofs in the notebook (cf. §4.6).

Table 1. Languages and systems we compared; state of our formalisations

Language	Logic	Prover	User Interface	Licence	Formalisation
Isabelle/HOL 2013 [14]	HOL (simply-typed set theory)	interactive[a]	document-oriented IDE (Isabelle/jEdit [39]) or programmer's text editor (Proof General Emacs [1])	BSD/LGPL/GPL	complete incl. proof
Theorema 2.0 [46]	FOL + set theory[b]	automated[c]	textbook-style documents, proof management GUI (add-on for Mathematica CAS)	GPL[d]	statements complete, no proof[e]
Mizar 8.1.01 [11]	FOL[f] + set theory	batch verifier	CLI[g]; programmer's text editor (Emacs add-on)	freeware/GPL+ CC-BY-SA[h]	complete incl. proof
CASL/TPTP[i] [5]	sorted FOL[i]	automated[j]	progr.'s text editor (Emacs add-on), proof mgmt. GUI+ CLI (Hets 0.98[k] [27]), web service (System on TPTP [34])	GPL	complete incl. proof

[a] Isabelle integrates internal and external automated provers.

[b] Theorema actually supports HOL. We, however, just needed FOL besides the built-in sets, tuples, and the max operator.

[c] For each goal, the prover can be configured individually.

[d] Theorema is under GPL but needs the commercial, closed-source Mathematica. Economists tend to be pragmatic about that.

[e] Theorema is in transition to the new 2.0. Its architecture, inference engine, and user interface are fully implemented, but its collection of *inference rules* is still incomplete. Therefore, the proof does not yet work.

[f] *Schemes* permit a limited degree of higher-order reasoning.

[g] The *verifier* produces a list of numerical errors codes and their source file positions. The ancillary utilities *errflag* and *addfmsg* decorate source files with this information, and optionally append terse textual explanations of the relevant error codes.

[h] The Mizar proof checker is closed-source; the MML is free.

[i] Common Algebraic Specification Language. 'CASL/TPTP' denotes our use of CASL as an input language for automated FOL provers (here: SPASS, E, Darwin) using the TPTP [32] exchange language. CASL features some second-order features, e.g. inductive datatypes.

[j] The proof is largely automatic. However, Vickrey's theorem is too complex to for automated proving in one step. Thus, the proof script introduces auxiliary lemmas and selects suitable axioms and provers for proving them. Proof times range from fractions of seconds if the exact list of axioms used is known beforehand to hours if not. However, once a proof is found, the prover can output the list of axioms used and thus speed up subsequent replays of the proof.

[k] Heterogeneous tool set; gives access to a wide range of automated theorem provers. We use FOL provers, most of which share the unsorted TPTP FOF [32] as a common input format. Hets translates CASL to FOF by introducing auxiliary predicates for sorts.

participants: arbitrary (finite) sets of natural numbers simplify the formalisation, and the highest and second highest bids are determined using library set operations. In contrast, Theorema naturally indexes its built-in tuples from 1 to n and allows for restricting quantified variables to such ranges, e.g. $\forall_{i=1,\dots,n}$.

The CASL formalisation confirms the assumption of §3.3 that FOL suffices for expressing and proving the essence of Vickrey's theorem. For many FOL provers, CASL's (sub)sorts[16] are mere syntactic sugar but allow us to stay close to the domain language, speaking, e.g., of 'valuation vectors', each of which also is a valid 'bid vector'. Note that we have avoided using partial functions (e.g., for modelling out-of-scope vector indices) because of the complex logic translations required for coding them out.

Isabelle and Mizar process the proof in a few seconds on a 2.5 GHz dual-core processor; Hets/TPTP need about an hour; in Theorema it is not yet complete. We used rather weak HOL features, e.g., no synthesisation of functions. Coinciding with earlier, general observations on HOL [8], the low processing time suggests that there is no disadvantage in choosing a rich logic, which allows for expressing relevant concepts (such as maxima of finite sets of real numbers) naturally. Our formalisations' small size (less than 5 K after compression) does not yet warrant a precise quantitative judgement of time efficiency. Particularly for FOL there exist highly optimised automated provers. They are conveniently accessible in Hets, via the System on TPTP [34] web service (accepting TPTP input that Hets can generate), but also from Isabelle/HOL via the Sledgehammer interface (see §4.3). Still, we observed a source of inefficiency in formalising for automated provers: the high share of preconditions with long conjunctions in our CASL formalisation makes it hard for the automated FOL provers to identify applicable axioms. Such conjunctions result from the absence of structured proofs in CASL. This requires, whenever a theorem is too complex for automated proving, to 'emulate' proofs steps via auxiliary lemmas, whose antecedents are conjunctions of all relevant assumptions in the current branch of the proof tree. Performance improvements by guiding provers through the search space can, however, be achieved with the extra effort of grouping frequently occurring conjunctions of assumptions into single abstract predicates, as in the following concrete case for the proof of Vickrey's theorem: $spaWithTruthfulOrOtherBid(n, x, p, v, \hat{\boldsymbol{b}}, i, \boldsymbol{b}) \Leftrightarrow secondPriceAuction(n, x, p) \wedge |v| = |\hat{\boldsymbol{b}}| = n \wedge inRange(n, i) \wedge \hat{\boldsymbol{b}}_i \neq v_i \wedge \boldsymbol{b} = \hat{\boldsymbol{b}}[i \leftarrow v]$.

4.3   Proof Development and Management (req. C2a)

The systems we studied offer different ways of invoking automated provers and keeping track of proof efforts in progress. The 'apparent' difference between automated and interactive theorem proving blurs at a closer look. The interactive prover Isabelle features various automated proof methods; furthermore Sledgehammer gives access to E, SPASS, and TPTP provers. One can configure the facts they should take into account (e.g. local assumptions and conclusions).

---

[16] TPTP's typed first-order form (TFF [33]) is sorted, but without subsorts. We have not used it, as Hets cannot currently produce it from CASL.

For Mizar, there are also automated external tools (MPTP, MoMM, MizAR) [31]. Theorema's automated proving workflow is conceptually similar: specifying the knowledge to be used, then configuring the prover.[17] Hets users can select axioms and previously proved theorems to be sent to an automated prover but have little control beyond. Isabelle's prover configuration is editable within the formalisation source. Theorema stores it in hidden fields within the formalisation and exposes it via a dedicated GUI. Configuring proof tools in Hets is separate from the formalisation: the proof management GUI does not currently store settings persistently; however one can write scripts to be processed on the command line.

Just as Isabelle requires complex statements to be proved in multiple steps, involving different proof methods, the automated provers of Theorema[18] and Hets also require guidance by explicit configuration at times, as can be seen from the *.hpf proof scripts in our Hets formalisation [23]. Often, a theorem $c : A \Rightarrow C$ was too complex for automated proving, whereas the job could be done by a script that first proved auxiliary lemmas $a : A \Rightarrow B$ and $b : B \Rightarrow C$, possibly with different provers, and then proved c providing only a and b as axioms. This is conceptually the same as in Isabelle but has four significant user experience differences: 1. Each additional 'proof step' has to be stated as a lemma with full assumptions on the left hand side (similar to the example in §4.2), 2. CASL, originally a specification rather than a prover language, does not syntactically distinguish theorems from lemmas, 3. the scripts have to be maintained separately from the formalisation, and 4. a multi-step proof takes many seconds longer, as Hets translates the input theory from CASL to the respective prover's native language before each proof.[19] This gives a clear incentive to eliminate unnecessary proof steps from a CASL formalisation. This experience also influenced our Isabelle formalisation, where writing multi-step proofs is comparatively painless. There, one lemma had a three-step proof, until experiments with the CASL formalisation made us attempt an automated proof. Thus we realised that we could reduce the Isabelle proof to a single step.[20]

Mizar differs by focusing, instead of built-in tactics and automated proof methods, on a natural deduction style which 'tries to "keep a low profile" in its logical foundations' and aims at 'clarity, human readability and closeness to standard mathematical proofs' [38]. Influenced by Mizar, the Isar language ('intelligible semi-automated reasoning') replaced Isabelle's original tactic interface. In the name of its readability focus, Mizar deliberately prevents users from extending the verifier's power [38, §2.1], often forcing them to justify trivial passages. Mizar's *registrations* do allow for custom automation [4]; however, these at times involute exploits often push registrations beyond their intended scope [20] and may result in implicit inferences and less readable proofs.

Particularly in developing the proof of a theorem as complex as Vickrey's top-down, it is useful to defer proofs of lemmas or proof steps, as to use them

---

[17] For Theorema, a prover is a *collection of inference rules* applied in a certain *strategy*.

[18] This assessment relies on experience with Theorema 1.

[19] This is necessary as, by default, each successful proof adds one theorem to the theory.

[20] As it makes use of one definition and two lemmas, this was not obvious a priori.

in a larger proof without the workaround of temporarily declaring them as axioms. Theorema proofs can use unproved theorems as knowledge. Isabelle's `sorry` keyword creates a fake proof. CASL theorems are formulas with the annotation `%implied`. When imported into a theory, (open) theorems become axioms, and Hets can use them without proof, but the open proof obligation is still visible in the imported theory. Mizar's verifier offers top-down proving for free by marking unaccepted inferences as errors *and then proceeding*. This results in a formal proof *sketch*, 'very close to informal mathematical English' but still close to a fully formalised proof [41]. Furthermore, one can prefix the keyword `proof` with `@` to expressly and silently skip a proof, or disable the verifier on arbitrary code portions using pragmas. Mizar's Emacs mode exposes these as one-touch macros, which speeds up the verification process and improves interaction [38].

4.4 Library Coverage and Searchability (reqs. C1c, C2b)

To a varying degree we have been able to reuse mathematical foundations from the systems' libraries. Isabelle can *find* reusable material by `find_theorems` queries; Sledgehammer helps to extract a sufficient set of lemmas from the library, which is then minimised towards a necessary set. MML Query is a search engine for the MML [3]. CASL's library is searchable as plain text; Theorema's is not.

Theorema has a built-in tuple type, including a maximum operation, we used it to formalise bid vectors. The CASL library provides inductive datatypes such as arrays [29] but no n-argument maximum operation. The Isabelle/HOL library provides a *Max* operation on finite sets, and various Cartesian product types suitable for representing bids. Given Isabelle's functional programming syntax we found it, however, most intuitive to model our own vectors as functions $\mathbb{N} \to \mathbb{R}$ evaluated up to a given n. Wrappers make the set maximum operator work on these vectors and prove the properties required subsequently. Our Mizar formalisation draws on generic relations and functions, which the MML richly covers. Thus, we only had to add a few interfacing lemmas.

4.5 Term Input Syntax (req. C1b)

Conversations with auction designers suggest that they find Theorema's term input syntax most accessible. The two-dimensional notation in Mathematica notebooks is similar to textbook notation, and our target audience is largely familiar with Mathematica. The syntax of Isabelle and CASL is closer to programming languages. Isabelle's functional type syntax $f : A \Rightarrow B \Rightarrow C$ looks less closely related to textbook notation than CASL's $f : A * B \to C$. Isabelle, CASL and Mizar allow for defining custom 'mixfix' operator notations. Isabelle provides rich translation mechanisms beyond that, but the layout remains one-dimensional, e.g. $\forall x \in A.\ B(x)$ instead of Theorema's $\underset{x \in A}{\forall} B[x]$ for bounded quantification. Isabelle Proof General and Isabelle/jEdit approximate textbook notation by Unicode symbols. Isabelle, Mizar and Hets can export LaTeX. Mizar uses ASCII; its lack of binders makes mathematical concepts such as limits and sums cumbersome to denote [43]. A major reason for us not to cover the TPTP language is its technical, non-extensible ASCII syntax (using, e.g., !/? for $\forall/\exists$).

Theorema, CASL and Mizar support sharing common quantified variables across multiple statements, corresponding to the practice of starting a textbook section even of several axioms like 'let n, the number of participants, be a natural number ≥ 1'. This helps to avoid redundancy but is prone to copy/paste errors. For example, our CASL formalisation has sections with global quantifiers $\forall i, j$ (e.g. to accommodate the maximum and second-price auction definitions of §3.2), but these include axioms that only use i. Literally pasting into this axiom an expression using j does not cause an error, as j is bound in the current scope as well, but changes the semantics of the axiom in a way hard to detect.

4.6 Comprehensibility and Trustability of the Output (req. C2c)

Machine proofs may 'succeed' for unintended reasons, e.g. accidentally stating a tautology such as an implication with an unsatisfiable antecedent. Or they succeed as intended, but the user cannot follow the (automated) deduction. In such situations the prover's *output* is crucial. Isabelle provides tracing facilities for simplification rules and introduction and elimination rules used in standard reasoning steps. Its inference kernel can produce a full record (usually large and unreadable) of the internal reasoning of automated tools via explicit proof terms, e.g. for independent checking. By default the kernel relies on static ML type-discipline to achieve correctness by construction, without explicit proof terms. Theorema's proof data structure captures the entire proof generation according to the rules and strategy selected. It can be displayed as a structured textbook-style proof with configurable verbosity, and visualised as a browsable tree that distinguishes successful from failed branches. Mizar 'just' verifies what the user wrote according to natural deduction rules, hence he is unlikely to doubt the result. On the other hand, for the same reason, Mizar has no way to detect proofs succeeding for unintended reasons, and offers little help to a user clueless about a failing step. A correct Mizar proof can be improved by enhancer utilities [11, §4.6]: some report useful additional information (e.g., unneeded statements referred in a step, unneeded library files, unneeded lemmas); others cut steps that a human might want to see, impacting readability and possibly the original confidence the user had in the proof. Hets uniformly displays the success of a proof and the list of axioms used; however the latter output is only informative with SPASS. Otherwise, the raw technical output of the prover is displayed, which strongly differs across provers. E.g., SPASS uses resolution calculus, which looks different from a textbook proof. Similarly, System on TPTP outputs performance measures and the status of the given problem (e.g. 'Theorem' or 'Unsatisfiable'), but otherwise the raw prover output.

When a proof attempt fails because the statement was wrong, studying a counterexample may help. Isabelle has the Nitpick counterexample finder built in. Hets integrates several ones (Darwin is supported best [28]) and also employs them for consistency checking, as importing a theory whose axioms have no model results in vacuous truth. Both Isabelle and Hets can attempt a proof or otherwise try to find a counterexample in the same run. Theorema and Mizar do not support counterexamples.

Before proving, all systems check whether the input is syntactically well-formed and well-typed. Isabelle/jEdit performs parsing, type checking and proof processing during editing, and attaches warnings and error messages like modern IDEs. The other systems require the user to explicitly initiate checking. Mizar and Hets check complete files, whereas in Theorema (which only checks syntax), one can individually check each notebook cell (typically containing one to a few statements). Mizar's verifier is particularly error resilient: it seldom aborts before the last input line, thus reporting errors for the whole file.

4.7   Online Community Support and Documentation (req. C2d)

Community support and documentation are major prerequisites for system adoption. We assume that users with little previous mechanised reasoning and formalisation knowledge will seek low-threshold support from tutorial documents or mailing lists rather than attending community meetings – which, in theorem proving, so far focus on scientific/technical aspects rather than applications.

We compare the community sizes, assuming that large communities are responsive even to non-experts: Isabelle is developed at multiple institutions; its user mailing list gets more than 100 posts a month, with over 1000 different authors since 2000. CASL, an international standard, has been subject of hundreds of publications but does not currently have a mailing list. Hets is mainly developed and used within a single institution; its user mailing list receives less than 10 posts a month. Recalling that Hets is an integrative environment, users can also request help from the communities of TPTP (subject of more than 1000 publications, no mailing list) and individual provers. Theorema is developed within a single institution and will not have a mailing list before the 2.0 release. Mizar is developed at one institution by a team that provides dedicated email user assistance: the 'Mizar User Service'. MML grows by 30–60 articles a year, with 241 contributors so far. The mailing list gets around 10 posts a month.

Isabelle and CASL feature comprehensive tutorials and reference manuals, Hets has a user guide, Mizar offers tutorials [26]. Theorema has partial built-in help texts and is documented in a few publications.

5   Related Work

§1 mentioned earlier efforts to *formalise economics*. Particularly Arrow's impossibility theorem, one of the most striking results in theoretical economics, has been a focus for formalisation efforts, including Nipkow's Isabelle and Wiedijk's Mizar formalisation [30, 42]. As in our case (cf. §3.2), they required initial paper elaboration; additionally, it helped them to identify omissions in their source [9]. This source states three alternative proofs, but Tang's/Lin's fourth, induction-based proof, allowed for obtaining insights on the general structure of social choice impossibility results using computer support [36].

Table 2. Performance (as far as results were comparable)

System/ Language	Proof speed	Textbook closeness		Top-down proofs	Library		Output			Commu- nity	Documen- tation	de Bruijn factor
		PI[a]	TI[a]		LC[a]	LS[a]	PO[a]	CE[a]	WF[a]			
	§4.2	§4.3		§4.5	§4.4		§4.6			§4.7		§4.1
Isabelle/HOL	++[b]	++	+	++	++	++	○	++	++	++	++	1.3
Theorema	?	n/a[c]	++	++	+	—	++	n/a	–	—	–	n/a
Mizar	++	++	–	++	++	+	○	n/a	++	+	○	1.7
CASL/TPTP	○[d]	–	+	++	+	–	○	+	+	○	+	1.5

[a] PI/TI = proof/term input; LC/LS = library coverage/search; PO = proof output; CE = counterexamples (incl. consistency checks); WF = well-formedness check.   [b] scores from very bad (—) to very good (++)   [c] fully GUI-based   [d] automated provers

The formal verification technique of model checking has been applied to *auctions*. Tadjouddine et al. proved the strategy statement of Vickrey's theorem via two abstractions to reduce the model checker's search space: program slicing to remove variables irrelevant w.r.t. the property, and discretising bid values (e.g. 'higher than someone's valuation v_i') [35]. Our formalisation is, to the best of our knowledge, the first for *theorem provers*; in the more expressive languages it has the comprehensibility advantage of preserving the structure of the original domain problem. From earlier economics formalisation efforts cited above, it differs in its goal to (ultimately) help economists to use formal methods themselves.

Our focus thus lies on *comparing* different provers by full parallel formalisation. Wiedijk compared Isabelle/HOL, Mizar, Theorema, and 14 other provers by general, technical criteria, studying the code resulting from experts formalising a pure mathematics theorem ($\sqrt{2} \notin \mathbb{Q}$), and comparing it to a detailed paper proof [44]. We complement this with the end user's perspective: our observations, e.g., on the closeness of the input syntax to textbook notation or the comprehensibility of the output are general, but we emphasised these criteria as they are important to auction designers. Griffioen's/Huisman's 1998 PVS and Isabelle/HOL comparison is, like Wiedijk's, independent from a specific application but closer to ours in its look at systems' weaknesses from a user's perspective [12]. Like us, they rate proof management and user support, but go into more detail up to the 'time it takes to fix a bug'. Their *findings* on user interfaces have been obsoleted by progress in developing textbook-like proof languages and editors with random access and asynchronous validation.

6   Conclusion and Outlook

Auctions allocate trillions of dollars in goods and services every year, but their design is still 'far less a science than an art' [24]. We aim at making it a science by enabling auction designers to verify their designs. By parallel formalisation of the first major theorem in a toolbox for basic auction theory (ATT), we have investigated the suitability of four different theorem provers for this job, taking the perspective not only of experienced formalisers but also of our end users. Our contribution is 2×2-fold: 1. to auction designers we provide (a) a growing library to build their formalisations on, and (b) guidelines on what systems to use; 2. to the CICM community we provide (a) challenge problems[21] and (b) user experience feedback from a new audience. This paper focuses on 1b and 2b.

For a concrete application, our findings confirm the widespread intuitions that formalisation benefits from an initial paper elaboration, that the 'automated vs. interactive' distinction proves of little importance in practice, and that no single system satisfies all requirements. For now, our comparison results in Tab. 2 guide auction designers in choosing a system, given their formalisation requirements and experience. The ideal theorem proving environment would feature a *library* as versatile as in Isabelle or Mizar, a *prover* as efficient as those of Isabelle or Mizar, giving *error messages* as informative as in Isabelle/jEdit, further a *proof input language* as close to textbook style as those of Isabelle or Mizar, or an *interface to explore* automated proofs as informative as Theorema's, a *textbook-like term syntax* as Theorema's, an integration of diverse *tools* as in Isabelle or Hets, and a *community* as lively as Isabelle's. We have not yet exploited all strengths of the systems evaluated: maintaining a growing ATT with increasingly complex dependencies will benefit from stronger modularisation, as supported by Isabelle and even more so by the theory graph management of Hets/CASL. Regarding auction *practice*, we are working towards ways to check that formal definitions of auctions are well-defined functions ('for each admissible bid input there is a unique outcome, modulo some randomness'). Given a constructive proof of this property, it should be possible to obtain verified program code that determines the outcome of an auction given the bids. This may work using Isabelle's code generator, but we will also explore provers based on constructive type theory.

Broader conclusions about auction theory require further research. Bidding typically requires forming conjectures of others' beliefs, involving integration over conditional density functions (cf., e.g., Proposition 13 in Maskin's review [24]). We expect that much of the required foundations should already be available in the libraries of Isabelle and Mizar. Maskin limits his review to single good auctions, noting that few general results exist for multi-unit and combinatorial auctions.[22] Such auctions are often more economically critical (e.g. spectrum auctions, monetary policy [19]) but also more complicated. The real challenge for mechanised reasoning will be to demonstrate its use in this domain.[23]

---

[21] Our problems are not currently challenging systems' performance but the promises of their languages and libraries.

[22] The last two chapters of [25] address multi-unit auctions; multi-unit and combinatorial auctions are the focus of [7].

[23] Even more ambitiously, many results in auction theory are simplified or extended by explicit application of mechanism design; cf. [17].

References

1. Aspinall, D.: Proof General: A Generic Tool for Proof Development. In: Graf, S. (ed.) TACAS 2000. LNCS, vol. 1785, pp. 38–43. Springer, Heidelberg (2000)
2. Auctions: The Past, Present and Future, http://realestateauctionglobalnetwork.blogspot.co.uk/2011/11/auctions-past-present-and-future.html
3. Bancerek, G.: Information Retrieval and Rendering with MML Query. In: Borwein, J.M., Farmer, W.M. (eds.) MKM 2006. LNCS (LNAI), vol. 4108, pp. 266–279. Springer, Heidelberg (2006)
4. Caminati, M.B., Rosolini, G.: Custom automations in Mizar. Automated Reasoning 50(2) (2013)
5. CASL, http://informatik.uni-bremen.de/cofi/wiki/index.php/CASL
6. Conitzer, V., Sandholm, T.: Self-interested automated mechanism design and implications for optimal combinatorial auctions. In: Conference on Electronic Commerce. ACM (2004)
7. Cramton, P., Shoham, Y., Steinberg, R. (eds.): Combinatorial auctions. MIT Press (2006)
8. Farmer, W.M.: The seven virtues of simple type theory. Applied Logic 6(3) (2008)
9. Geanakoplos, J.D.: Three brief proofs of Arrow's impossibility theorem. Discussion Paper 1123RRR. Cowles Foundation (2001)
10. Geist, C., Endriss, U.: Automated search for impossibility theorems in social choice theory: ranking sets of objects. Artificial Intelligence Research 40 (2011)
11. Grabowski, A., Korniłowicz, A., Naumowicz, A.: Mizar in a Nutshell. Formalized Reasoning 3(2) (2010)
12. Griffioen, D., Huisman, M.: A comparison of PVS and isabelle/HOL. In: Grundy, J., Newey, M. (eds.) TPHOLs 1998. LNCS, vol. 1479, pp. 123–142. Springer, Heidelberg (1998)
13. Initiative for Computational Economics, http://ice.uchicago.edu
14. Isabelle, http://isabelle.in.tum.de
15. Kerber, M., Lange, C., Rowat, C.: An economist's guide to mechanized reasoning (2012), http://cs.bham.ac.uk/research/projects/formare/
16. Kerber, M., Rowat, C., Windsteiger, W.: Using *Theorema* in the Formalization of Theoretical Economics. In: Davenport, J.H., Farmer, W.M., Urban, J., Rabe, F. (eds.) Calculemus/MKM 2011. LNCS (LNAI), vol. 6824, pp. 58–73. Springer, Heidelberg (2011)
17. Kirkegaard, R.: A Mechanism Design Approach to Ranking Asymmetric Auctions. Econometrica 80(5) (2012)
18. Klemperer, P.: Auctions: theory and practice. Princeton Univ. Press (2004)
19. Klemperer, P.: The product-mix auction: a new auction design for differentiated goods. European Economic Association Journal 8(2-3) (2010)
20. Korniłowicz, A.: On Rewriting Rules in Mizar. Automated Reasoning 50(2) (2013)
21. Lamport, L., Paulson, L.C.: Should your specification language be typed? ACM TOPLAS 21(3) (1999)
22. Lange, C., Rowat, C., Kerber, M.: The ForMaRE Project – Formal Mathematical Reasoning in Economics. In: Carette, J., Aspinall, D., Lange, C., Sojka, P., Windsteiger, W. (eds.) CICM 2013. LNCS (LNAI), vol. 7961, pp. 330–334. Springer, Heidelberg (2013)
23. Lange, C., et al.: Auction Theory Toolbox (2013), http://cs.bham.ac.uk/research/projects/formare/code/auction-theory/

24. Maskin, E.: The unity of auction theory: Milgrom's master class. Economic Literature 42(4) (2004)

25. Milgrom, P.: Putting auction theory to work. Cambridge Univ. Press (2004)

26. Mizar manuals (2011), http://mizar.org/project/bibliography.html

27. Mossakowski, T.: Hets: the Heterogeneous Tool Set, http://dfki.de/cps/hets

28. Mossakowski, T., Maeder, C., Codescu, M.: Hets User Guide. Tech. rep. Version 0.98. DFKI Bremen (2013), http://informatik.uni-bremen.de/agbkb/forschung/formal_methods/CoFI/hets/UserGuide.pdf

29. Mosses, P.D. (ed.): CASL Reference Manual. LNCS, vol. 2960. Springer, Heidelberg (2004)

30. Nipkow, T.: Social choice theory in HOL: Arrow and Gibbard-Satterthwaite. Automated Reasoning 43(3) (2009)

31. Rudnicki, P., Urban, J., et al.: Escape to ATP for Mizar. In: Workshop Proof eXchange for Theorem Proving (2011)

32. Sutcliffe, G.: The TPTP Problem Library and Associated Infrastructure: The FOF and CNF Parts, v3.5.0. Automated Reasoning 43(4) (2009)

33. Sutcliffe, G., Schulz, S., Claessen, K., Baumgartner, P.: The TPTP Typed First-order Form with Arithmetic. In: Bjørner, N., Voronkov, A. (eds.) LPAR-18. LNCS (LNAI), vol. 7180, pp. 406–419. Springer, Heidelberg (2012)

34. System on TPTP, http://cs.miami.edu/~tptp/cgi-bin/SystemOnTPTP

35. Tadjouddine, E.M., Guerin, F., Vasconcelos, W.: Abstracting and Verifying Strategy-Proofness for Auction Mechanisms. In: Baldoni, M., Son, T.C., van Riemsdijk, M.B., Winikoff, M. (eds.) DALT 2008. LNCS (LNAI), vol. 5397, pp. 197–214. Springer, Heidelberg (2009)

36. Tang, P., Lin, F.: Computer-aided proofs of Arrow's and other impossibility theorems. Artificial Intelligence 173(11) (2009)

37. Tang, P., Lin, F.: Discovering theorems in game theory: two-person games with unique pure Nash equilibrium payoffs. Artificial Intelligence 175(14-15) (2011)

38. Urban, J.: MizarMode—an integrated proof assistance tool for the Mizar way of formalizing mathematics. Applied Logic 4(4) (2006)

39. Wenzel, M.: Isabelle/jEdit – a Prover IDE within the PIDE framework. In: Jeuring, J., Campbell, J.A., Carette, J., Dos Reis, G., Sojka, P., Wenzel, M., Sorge, V. (eds.) CICM 2012. LNCS (LNAI), vol. 7362, pp. 468–471. Springer, Heidelberg (2012)

40. Wiedijk, F.: De Bruijn factor, http://cs.ru.nl/~freek/factor/

41. Wiedijk, F.: Formal proof sketches. In: Berardi, S., Coppo, M., Damiani, F. (eds.) TYPES 2003. LNCS, vol. 3085, pp. 378–393. Springer, Heidelberg (2004)

42. Wiedijk, F.: Formalizing Arrow's theorem. Sādhanā 34(1) (2009)

43. Wiedijk, F.: The QED Manifesto Revisited. Studies in Logic, Grammar and Rhetoric 10(23) (2007)

44. Wiedijk, F. (ed.): The Seventeen Provers of the World. LNCS (LNAI), vol. 3600. Springer, Heidelberg (2006)

45. Wikipedia (ed.): Vickrey auction (2012), http://en.wikipedia.org/w/index.php?title=Vickrey_auction&oldid=523230741

46. Windsteiger, W.: Theorema 2.0: A Graphical User Interface for a Mathematical Assistant System. In: UITP Workshop at CICM (2012)

47. Woodcock, J., et al.: Formal method: practice and experience. ACM Computing Surveys 41(4) (2009)

Students' Comparison of Their Trigonometric Answers with the Answers of a Computer Algebra System

Eno Tonisson

Institute of Computer Science, University of Tartu,
2 J. Liivi Tartu, 50409, Estonia
eno.tonisson@ut.ee

Abstract. Comparison of answers offered by a computer algebra system (CAS) with answers derived by a student without a CAS is relevant, for instance, in the context of computer-aided assessment (CAA). The issues of identity, equivalence and correctness emerge in different ways and are important for CAA. These issues are also interesting if a student is charged with the task of comparing the answers. What will happen when students themselves are encouraged to analyse differences, equivalence and correctness of their own answers and CAS answers? What differences do they notice foremost? Would they recognise equivalence/non-equivalence? How do they explain equivalence/non-equivalence? The paper discusses these questions on the basis of lessons where the students solved trigonometric equations. Ten equations were chosen with the aim to ensure that the expected school answer and the CAS answer would differ in various ways. Three of them are discussed more thoroughly in this paper.

Keywords: Computer Algebra Systems, Teaching and Learning Mathematics, Equivalence, Trigonometric Equations.

1 Introduction

The answers offered by a computer algebra system (CAS) are evaluated (in literature) mainly from a professional user's point of view. CAS users could compare their (or others') answers with the answers of a CAS for various reasons. For example, a mathematical researcher could use a CAS to confirm hand-derived solutions (see [1]). Some mathematicians use a CAS in order to check the solutions of students (see [2]).

There are also broader overviews, for example, in [3] where hundreds of answers are evaluated in case of several CAS. A comparison of answers offered by a computer algebra system with answers derived by a student could be also used. One fruitful area is computer-aided assessment (CAA) where a student's answer is assessed automatically with the help of ("invisible") CAS. The capability of a CAA system depends on the capability of the CAS. It is necessary for comparison and assessment to explore the issues of identity, equivalence and correctness, and not only in the sense of classical mathematics (see [4] and [5]).

J. Carette et al. (Eds.): CICM 2013, LNAI 7961, pp. 216–229, 2013.
© Springer-Verlag Berlin Heidelberg 2013

Further interesting issues arise if a student (not a computer program) is charged with the task of comparing the answers. This paper focuses on the following questions: What will happen when students themselves are encouraged to analyse differences, equivalence and correctness of their own answers and CAS answers? What differences do they notice foremost? Would they recognise equivalence/non-equivalence? How do they explain equivalence/non-equivalence? As different systems could present answers in different ways, a particular CAS was prescribed to initiate an "intrigue" and obtain information about the effect of different representations.

The paper is based on lessons where first-year students solved trigonometric equations. Solving trigonometric equations is an interesting topic in this context because of the variety of possible presentations of solutions, units of measurement, general and particular solutions, etc.

The students worked in pairs and their discussions were audio-taped. The students had worksheets with equations and tasks (see Sect. 4). The order of solvable equations was prescribed and was different for different pairs. The students first solved an equation (correctly or not) without and then with a particular CAS. The systems used were Maxima [6], Wiris [7], and WolframAlpha [8]. A specific CAS was prescribed for the equation to attain the expected difference between the students' answers and the CAS answer.

Data of more than 100 instances of equation-solving (29 pairs of students) were collected. Three equations from ten are more interesting from the MKM point of view and they were chosen for deeper analyses in this paper (47 instances of equation-solving, 26 pairs of students).

Before comparing their own answers with the answers of a CAS, students should read and understand CAS answers. Generally, the issue of readability of a CAS answer is multi-faceted. On the one hand, it is connected to sophisticated mathematical reasoning, for example, branch cut and cylindrical algebraic decomposition (CAD) (see [9], [10]). On the other hand, sometimes a rather simple difference (for example, in notation) between a CAS answer and a school-like answer could be confusing for the student (see [11]). The issue of readability of CAS answers is important for this study.

The broader perspectives of the topic could be described in teaching-learning context and in research context. The further purpose is to suggest a new method of using CAS for teaching and learning mathematics where students' discussion, critical thinking and deeper insight into important issues (such as equivalence) should be brought out. The readability of a CAS answer is as challenging as the learning of the CAS syntax. Moreover, the black box nature of a CAS reveals issues that can go as deep as university-grade mathematics (see [2]). However, C. Buteau et al. note in [12]: *Although practitioners have to deal with unusual or unexpected behaviour of CAS, this was occasionally shown to provide pedagogical opportunities.* A thorough study is needed for the method and this paper is a part of it. Besides direct teaching-learning context, perspectives in the research context are also important. Such an analysis of students' worksheets and discussions could provide new opportunities for studying their thinking and learning.

A. Sfard (in [13]) compares the possibility to analyse conversation even to a microscope that gives new power perspectives to 17th century scientists.

Accordingly, the study also includes a search for preliminary answers to questions about suitability of the method for teaching-learning and research.

Section 2 gives a brief overview of related works. The choice of equations is described in Sect. 3. An overview of the lessons is provided in Sect. 4. The main examples are discussed in Sect. 5, 6, 7. Section 8 concludes the paper.

2 Related Work

This study is related to a number of different research areas. For example, comparisons of the different CAS, like [3], [14] and [15] are notable. Such reviews do not focus on pedagogical aspects. However, M. Wester mentioned: *One could invoke mindset* (`Elementary_math_student`) *to initially declare all variables to be real, make* $\sqrt{-1}$ *undefined, etc., for example* [3]. The adequacy of CAS answers is under consideration in [16] and [11], for instance. The paper [11], based on the experiments of P. Drijvers, is focused on parameters, but also defines a more universal list of obstacles. In addition, he suggests that an obstacle could be an opportunity. The CAS answers are observed from a school-oriented point of view in [17] and [18].

The papers [4] and [5] were written from the background of CAA. CAA systems could be connected with a CAS. For example, the STACK system uses Maxima [19]. The issues of identity, equivalence and correctness are very important. They help to distinguish between mathematical, pedagogical and aesthetic correctness. Their study is focused on automated assessment. In our study the students were charged with the task of comparing the answers themselves. The issue of the "right answer" is also very important in this case. Our study is also related to the analysis of discourses, audio recordings, etc., but these topics are too far removed from the main focus of this paper. Furthermore, we do not deal here with the theoretical background of checking equivalence (for example [20]) where trigonometry has a somewhat problematic status. The topic of trigonometric equations is considered in the next section.

3 Choice of Equations and CAS

Our study is focused on trigonometric equations because of the variety of their answers. It is quite usual for a trigonometric equation to have several reasonable representations of the correct answer. Different solution strategies may lead to different-looking but still equivalent answers. A classroom discourse in case of the equation

$$2 + \cos^2 2x = (2 - \sin^2 x)^2 \tag{1}$$

is presented in [21]. Four different answers were under consideration.

The variety of answers is actually even larger as the circumstances involved go beyond a pure solving strategy. For example, one could prefer radians or degrees. General solutions can be sought in some and particular solutions in other instances. Some basic formulae could be slightly different in different regions. For instance, the solution for $\sin x = m$, could be expressed as

$$x = \arcsin m + 2n\pi, \ n \in \mathbb{Z}$$
$$x = \pi - \arcsin m + 2n\pi, \ n \in \mathbb{Z} \tag{2}$$

or (as in Estonian textbooks, for example)

$$x = (-1)^n \arcsin m + n\pi, \ n \in \mathbb{Z} \ . \tag{3}$$

If we use a CAS, the variety is likely to increase because of the peculiarities of the CAS. For example, some notation issues could arise. Different treatments of the (default) number domain can also have an impact. Nevertheless, the issues of general and particular solutions or "regional" differences could be relevant. In this study, 10 equations were chosen for the class. Some of them were from regular school textbooks, others from books where trigonometry is handled at a somewhat advanced level. We analyse three of them in more depth in the paper. These equations seemed to be more suitable for this research track, as the focus is primarily on different representations of the answers and not so much on extraneous roots, complex domain, etc., (like in case of some other equations). The equations were chosen to attain a specific type of difference between the expected answer of the students and the answer of the particular CAS. (Actually, as students solved the equations themselves, they also made mistakes and the comparison was made between their actual answers and the CAS answers. It is more thoroughly explained in the next section.)

The first example is the equation $\sin(4x + 2) = \frac{\sqrt{3}}{2}$, where the students use Formula 3 (as taught in Estonian schools) but WolframAlpha expresses series separately (see Sect. 5). (The issue of branches in CAS is also discussed in [22]). The second example is $\tan^3 x = \tan x$, where students give general solutions but Wiris gives particular solutions (see Sect. 6). The third example is $\cos\left(x - \frac{\pi}{6}\right) = 0.5$, where Maxima uses its own notation with union and %z (see Sect. 7). The other equations with more specific nuances (extraneous roots, issues of domain, indeterminacy, etc.) are not discussed in this paper but are listed for the sake of completeness: $2\sin 2x \cos 2z + \cos 2x = 0$, $\dfrac{\tan^2 x}{\tan x} = 0$, $\tan(x + \dfrac{\pi}{4}) = 2\cot x - 1$, $2\cos^2 x + 4\cos x = 3\sin^2 x$, $\sin x - \sin^2 x = 1 + \cos^2 x$, $\dfrac{1 - \cos x}{\sin x} = 0$, $1 - \cos x = \sqrt{3}\sin x$.

4   In Class

This section gives a brief overview of the lessons in the course "Elementary mathematics", which is a somewhat repetitious course of school mathematics for the first-year university students. The students had quite diverging skill levels in solving trigonometric equations. As the advanced students were dismissed from the course (on the basis of a preliminary test), the proportion of wrong answers probably increased. The students had very few experiences with CAS. CAS were not used in other lessons of the course.

The lesson in question was taught by the author (who was not a regular teacher of the course). The lesson lasted for 90 minutes and consisted of an introduction, a period of equation-solving (ca 70 minutes), and closing (saving and copying data). The introduction gave an overview of the lesson, the aims of the study, etc. The computer algebra systems were not specially introduced but the students were warned that the answers of a CAS could differ from human answers and could also be incorrect. The types of possible differences were not explained. The order of solvable equations was prescribed and was different for different pairs of students in order to collect data about different equations. The students solved the equations in pairs and the discussions were audio-taped in order to obtain a deeper overview beyond the notes on paper. The students first had to solve the trigonometric equations by themselves and then with a particular CAS. They were encouraged to analyse differences, equivalence and correctness of their own answers and CAS answers. The worksheet included the following tasks (in the case of the first example):

- Solve an equation $\sin(4x + 2) = \frac{\sqrt{3}}{2}$ (without the computer at first).

- How confident are you in the correctness of your answer?

- Solve the equation with the CAS WolframAlfa using the command solve.

Fig. 1. WolframAlpha input

- How unexpected is the CAS answer at first view?

- Analyse the accordance of your answer with the CAS answer! If you want to complement/correct your solution, please use the green pen.

- What are the differences between your answer and the CAS answer?

- How are your answer and the CAS answer related (analyse equivalence/non-equivalence, particular solutions/general solutions)?

- Rate the correctness of your (possibly corrected) answer.

- Rate the correctness of the CAS answer.

Some of the issues are discussed thoroughly in this paper, others are mentioned only in the conclusion where further work is described.

The student papers and audio-tapes were analysed and the results in case of three examples are presented in the next sections. Each presentation begins with a brief introduction of the example, including reasons for selecting the example, a possible school answer, and a snapshot of the CAS answer. Next, the equivalence/non-equivalence of the students' answers with the CAS answers is discussed. It is based on mathematical reasoning by the author (*Math.* in tables). The second dimension is the students' opinion about the equivalence/non-equivalence that is based on an analysis of paper and audio data (*Stud.* in tables). The tables are also presented. The discussion concludes with some pedagogical comments.

5   Different Forms of General Solution

The first example is the equation where the CAS answer is particularly unexpected for those who use the $(-1)^n$ formula (Formula 3) in case of $\sin x = m$ (as is common for Estonian students). The possible school answer for the equation

$$\sin(4x + 2) = \frac{\sqrt{3}}{2} \tag{4}$$

is

$$x = -\frac{1}{2} + (-1)^n \frac{\pi}{12} + \frac{n\pi}{4}, \quad n \in \mathbb{Z} . \tag{5}$$

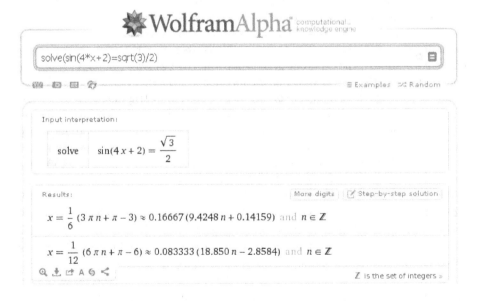

Fig. 2. $\sin(4x + 2) = \frac{\sqrt{3}}{2}$. WolframAlpha

WolframAlpha gives two series of solutions (see Fig. 2). The answers are actually equivalent. The students did not receive any specific information about the CAS answer.

As our textbooks and teachers use mainly the $(-1)^n$ form, the students' answers and the CAS answer seemed quite different at least for this reason. (Twelve pairs (of 17) used $(-1)^n$ form, 4 gave particular solution. One pair gave initially particular solution and after correction $(-1)^n$ form.)

As several pairs made mistakes, one could count 11 cases of equivalence with the CAS answer and 6 cases non-equivalence. Four pairs (of equivalent cases) used both degrees and radians in the same answer, for example:

$$x = 15° \cdot (-1)^n + 45° \cdot n - \frac{1}{2}, \; n \in \mathbb{Z} \; . \tag{6}$$

Our main focus in the paper is to observe how students compare their own and CAS answers. In many cases, their opinion about the equivalence is ascertainable, sometimes not. The results are presented in Table 1.

Table 1. $\sin(4x + 2) = \frac{\sqrt{3}}{2}$. Equivalence/non-equivalence

	Stud. Equivalent	Stud. Non-equivalent	Abstruse
Math. Equivalent	4	5	2
Math. Non-equivalent	3	3	

The depth of discussions about the comparison varied between the student pairs. For example, 3 pairs identified actual equivalence through reasonable discussion, while one pair simply presumed it. There were also 3 pairs whose answer was not equivalent with the CAS answer, but they counted them as equivalent without any real discussion. Seven pairs did not recognize that the answers were equivalent. Mainly, they did not grasp that n in their answer (like in Formula 3) and n in the CAS answer (see Fig. 2) was not the same. This points to an automated (and correct) habit of solving the algorithm of trigonometric equation without exhaustive understanding of the solution. Three pairs identified the non-equivalence of their answer and the CAS answer. Their answers were remarkably different from the CAS answer.

It seems that the different representations of the same answer, like in this example, could initiate instructive discussion. It could also point to a possible superficial treatment of the fairly important issue of the meaning of n. A simpler equation, like $\sin 4x = \frac{\sqrt{3}}{2}$, could probably be a more straightforward means for clarifying the phenomenon. The example is suitable if the students use the $(-1)^n$ formula. This is also an issue of different traditions. For example, it is usual to find solutions such as

$$x = (-1)^n \frac{\pi}{6} + n\pi, \; n \in \mathbb{Z} \tag{7}$$

(being the solution of $\sin x = \frac{1}{2}$) in the textbooks of some countries, such as Estonia, whereas we are aware that many others do not.

6   If the CAS Gives Only Particular Solutions

As the second example, we have chosen a situation where a CAS gives only particular solutions, but the students were asked to present general solutions. Wiris has its own rules for the presentation of solutions to trigonometric equations. In case of $\sin x = a$, for example, $\arcsin a$ and $\pi - \arcsin a$ are presented.

The students should frame the CAS solutions up to their own general solutions. In case of the equation

$$\tan^3 x = \tan x \tag{8}$$

the human answer could be

$$x = n\pi, n \in \mathbb{Z}$$
$$x = \pm\frac{\pi}{4} + n\pi, n \in \mathbb{Z} \tag{9}$$

or

$$x = n\pi, n \in \mathbb{Z}$$
$$x = \frac{\pi}{4} + n\pi, n \in \mathbb{Z} \tag{10}$$
$$x = -\frac{\pi}{4} + n\pi, n \in \mathbb{Z} \ .$$

Wiris gives the particular solutions (Fig. 3).

$$\text{solve}(\tan(x)^3 = \tan(x)) \ \rightarrow \ \left\{ \{x=0\}, \{x=\pi\}, \left\{x=\frac{\pi}{4}\right\}, \left\{x=\frac{3\cdot\pi}{4}\right\}, \left\{x=\frac{5\cdot\pi}{4}\right\}, \left\{x=-\frac{\pi}{4}\right\} \right\}$$

Fig. 3. $\tan^3 x = \tan x$. Wiris

We count these answers as equivalent in the sense that all series are presented by 2 instances. Certainly, $n\pi$ and $\{0; \pi\}$ are not equivalent in the usual mathematical sense. The order of solutions is quite confusing as instances of the series of solutions are not always side by side (for example, $\frac{\pi}{4}$ and $\frac{3\pi}{4}$ are not from same "club"). The students did not receive any specific information about the CAS answer. Many of the student pairs (9 of 14) gave the right answer and they also figured out (after smaller or larger effort and discussion) the relationship between their and CAS answer (see Table 2). One pair could not frame π, $\frac{3\pi}{4}$ and $\frac{5\pi}{4}$ up to their right answer. Again, the meaning of n in the formula seemed to be incoherent for them. The cases where students omitted some solutions were very interesting. One such pair corrected their mistake and finally found the right answer. They added to

$$\pi + \pi n$$
$$\frac{\pi}{4} + \pi n \tag{11}$$

missing

$$-\frac{\pi}{4} + \pi n \ . \tag{12}$$

We do not focus on emotions in this paper but their joy after the correction was remarkable. The other pair (initially only $n\pi$ solution) had a member who already diagnosed their mistake. The third pair did not analyse the CAS solutions thoroughly enough and did not notice that their answer was incomplete. It is impossible to give a thorough overview of the discussion of the pair that got an incomplete answer and also considered it as non-equivalent with the CAS answer, as their discussion was very laconic. It seems that the representation

Table 2. $\tan^3 x = \tan x$. Equivalence/non-equivalence

	Stud. Equivalent	Stud. Non-equivalent
Math. Equivalent	9	1
Math. Non-equivalent	2	1
Non-equivalent → Equivalent	1	

of the answer is generally accomplishable in this case. The possible corrective virtue is also notable. The standard of representation of answers to trigonometric equations could provide more instructive examples, as the choice of a particular solutions is not always as transparent.

7   Unusual Form of Arbitrary Integer

The third example is related to CAS notation. The CAS answer is actually very similar to a normal human answer but with some CAS-specific peculiarity. The human answer to the equation

$$\cos\left(x - \frac{\pi}{6}\right) = 0.5 \tag{13}$$

could be

$$\begin{aligned} x &= -\frac{\pi}{6} + 2n\pi, n \in \mathbb{Z} \\ x &= \frac{\pi}{2} + 2n\pi, n \in \mathbb{Z} \ . \end{aligned} \tag{14}$$

Maxima gives the same answer in a somewhat distinctive way (Fig. 4). The package `to_poly_solve` is used for solving trigonometric equations. We cite the Maxima manual for clarity: *Especially for trigonometric equations, the solver sometimes needs to introduce an arbitrary integer. These arbitrary integers have the form %zXXX, where XXX is an integer* [23].

The meaning of %z was also an important issue for solving the equation with Maxima. The students did not receive any specific information about the CAS answer, but they had additional brief paper manuals (3 pages) on using different CAS where %z was explained. Only two pairs found the info about %z from

(%i6) %solve(cos(x-%pi/6)=0.5,x);

$$(\%o6) \ \%union\left(\left[x=2\ \pi\ \%z6 -\frac{\pi}{6}\right],\left[x=2\ \pi\ \%z8 +\frac{\pi}{2}\right]\right)$$

Fig. 4. $\cos\left(x - \frac{\pi}{6}\right) = 0.5$. Maxima

this manual. Almost all pairs mentioned the %z as a remarkable difference from their own answer. An explanation was given if the students asked about it. Nevertheless, two pairs remained confused and could not understand the CAS answer. The meaning of such a notation could be more clearly indicated in the CAS user-interface. For example, tooltips could be used. Eight pairs (of 16) got the right answer (see Table 3). Five of these pairs quite easily found the CAS answer to be equivalent. Three pairs had an answer equivalent with the CAS answer but their opinion about equivalence was abstruse. One of these pairs could not understand the CAS answer because of %z. The second pair did not observe the CAS answer sufficiently and did not notice the relation between the CAS answer and their own (not fully simplified) answer. The third pair's discussion was too laconic. One pair corrected their mistake and finally found the right answer, from

$$\cdots$$
$$x - 30° = \arccos\frac{1}{2} + 2\pi n \tag{15}$$
$$\cdots$$

to

$$\cdots$$
$$x - 30° = \pm\arccos\frac{1}{2} + 2\pi n \tag{16}$$
$$\cdots\ \cdot$$

Three pairs saw equivalence that really did not exist. There were also four pairs who considered their wrong answers as non-equivalent with the CAS answer. One of these pairs could not understand the meaning of %z correctly. Two pairs tried to find their mistakes, one pair had evidently a different answer.

It seems that the different notation can cause major trouble for some people, while it can be easily acceptable for others. It should be mentioned that the students used Maxima for the first time and many issues would probably be resolved in the course of further use.

Table 3. $\cos\left(x - \frac{\pi}{6}\right) = 0.5$. Equivalence/non-equivalence

	Stud. Equivalent	Stud. Non-equivalent	Abstruse
Math. Equivalent	5		3
Math. Non-equivalent	3	4	
Non-equivalent → Equivalent	1		

Table 4. Adequate identification of equivalence/non-equivalence

Section	Equation	Adequate identification	
Different Forms of General Solution	$\sin(4x + 2) = \frac{\sqrt{3}}{2}$	7 (of 17)	41%
If CAS gives only particular solutions	$\tan^3 x = \tan x$	11 (of 14)	79%
Unusual Form of Arbitrary Integer	$\cos\left(x - \frac{\pi}{6}\right) = 0.5$	10 (of 16)	62%

8   Conclusion

The study focused on a lesson where students solved trigonometric equations at first without a CAS and then with a CAS. The main task was to compare the answers. Recognition of equivalence can be a difficult task for students. Even those students who solve trigonometric equations quickly and correctly can find it hard to correctly compare their answer with the CAS answer. The three examples presented in the paper helped to highlight different aspects of this situation.

If we look at the findings in Sections 5, 6 and 7, it is possible to single out the cases where students identified the equivalence/non-equivalence of their answer and the CAS answer adequately. The proportions of these cases are presented in Table 4. The cases where the non-equivalent answer was changed to equivalent in the light of the CAS answer are also included.

There seem to be different "hindrances" to identification of equivalence/non-equivalence in case of different equations. Probably, the proportion of adequate identifications of equivalence/non-equivalence could be increased by drawing special attention to the problematic issues before solving or in the worksheets. It is important to decide what issues are adequate and useful for the students. For example, the meaning of n in the answers of trigonometric equations is relevant and useful for mathematical insight. (The %z6 topic is also connected to this issue.) It is probably possible to increase the proportion of students' answers that are equivalent to CAS answers. For example, it is possible to use simpler equations or give more hints about the solution. Principally, it is possible to give whole solutions but then the students would have a weaker connection with the exercise.

These questions could be studied in further experiments. Actually, there are various ideas for further work. Data from the lessons (see worksheet in the Sect. 4) include information about students' pen and paper solutions, confidence, rating of correctness of their answers and CAS answers, etc. The data could be analysed with the topic discussed in this paper. Of course, the seven equations not discussed in this paper should be included in the study. It is notable that some of these equations have CAS answers that are non-equivalent to school answers.

As a teacher, the author could argue (so far without scientific proof) that the lessons were successful. (It was also confirmed by the actual teachers of these groups.) It seems that the task of comparing their own answers and CAS answers was interesting to the students. Generally, they became accustomed to the style of the lesson and actively discussed the topic of trigonometry. The method seems to be fruitful in research context as well. The paper data and audio-tapes complement each other and give a good overview about the students' activities during the solving process.

Coming back to the issue of readability of the answer, it should be mentioned that %z6 form could be confusing for some students, but it seems to be easily explainable. However, the change of %z6 form could be considered as a possible suggestion to CAS developers. In addition, it is possible to improve the order of particular solutions in Wiris.

It would be quite useful if a CAS would have the possibility to choose a mode according to a particular style of presenting the answers. For example, one could choose whether a general solution would be in $(-1)^n$ form or in the form of two series. On the one hand, it is good if the CAS answers are very school-like. On the other hand, the moderate difference between the students' and CAS answer could also be challenging and useful. Specification of such moderation is one of the most challenging issues of further work. It opens more questions. For example, how would such a specification look like? Would it be possible to work out indicators that qualify the type of answers?

One could even say that having different answers compared to school solutions is a part of the charm of CAS. It is possible to propose various lesson scenarios other than those used in the lessons considered here. A discussion where all students would participate could be very useful. The discussion could take place during the same lesson after solving and comparing, but it is also possible to arrange the concluding discussion during the next lesson. In any case, the concluding part is desirable, as students need feedback.

It is also possible to direct students to use CAS tools in the comparison of answers. For example, they could try to substitute a solution into the equation, simplify the difference of answers with the help of the CAS. Of course, it is possible that students compare their own answers with CAS answers as they did in these lessons. Another possible task for students could be a comparison of the answers of different computer algebra systems. In addition, one and the same CAS could offer different answers with different commands or assumptions and these answers could be also compared.

We can conclude that the method of asking students to compare their own answers with CAS answers seems to have potential in the context of learning as well as research, but further work is certainly needed. This style of comparison could contribute of the usage of computer-based tools for doing mathematics in different ways. On the one hand, the students see that calculations can be performed faster and easier. On the other hand, one should understand that evaluation of a CAS answer may not be so fast and easy. The abilities of critical thinking (particularly, with respect to computer algebra systems) are likely to be developed by the exercises.

Acknowledgements. This research was funded by Estonian target funding project SF0180008s12.

References

1. Bunt, A., Terry, M., Lank, E.: Challenges and Opportunities for Mathematics Software in Expert Problem Solving. Human-Computer Interaction 28(3), 222–264 (2013)
2. Marshall, N., Buteau, C., Jarvis, D.H., Lavicza, Z.: Do Mathematicians Integrate Computer Algebra Systems in University Teaching? Comparing a Literature Review to an International Survey Study. Computers & Education 58(1), 423–434 (2012)
3. Wester, M.J.: A Critique of the Mathematical Abilities of CA Systems. In: Wester, M. (ed.) Computer Algebra Systems. A Practical Guide, pp. 25–60. J. Wiley & Sons (1999)
4. Bradford, R., Davenport, J.H., Sangwin, C.J.: A Comparison of Equality in Computer Algebra and Correctness in Mathematical Pedagogy. In: Carette, J., Dixon, L., Coen, C.S., Watt, S.M. (eds.) Calculemus/MKM 2009. LNCS, vol. 5625, pp. 75–89. Springer, Heidelberg (2009)
5. Bradford, R., Davenport, J.H., Sangwin, C.J.: A Comparison of Equality in Computer Algebra and Correctness in Mathematical Pedagogy (II). International Journal for Technology in Mathematics Education 17(2), 93–98 (2010)
6. Maxima, a Computer Algebra System, http://maxima.sourceforge.net/
7. Wiris, the Global Solution for Mathematics Education, http://www.wiris.net/demo/wiris/en/index.html
8. WolframAlpha, Computational Knowledge Engine, http://www.wolframalpha.com/
9. Bradford, R., Davenport, J.H.: Towards Better Simplification of Elementary Functions. In: Proceedings of the 2002 International Symposium on Symbolic and Algebraic Computation, pp. 16–22. ACM, New York (2002)
10. Phisanbut, N.: Practical Simplification of Elementary Functions using Cylindrical Algebraic Decomposition. PhD thesis, University of Bath (2011), http://opus.bath.ac.uk/28838/1/UnivBath_PhD_2011_N.Phisanbut.pdf
11. Drijvers, P.: Learning Mathematics in a Computer Algebra Environment: Obstacles Are Opportunities. ZDM 34(5), 221–228 (2002)
12. Buteau, C., Marshall, N., Jarvis, D.H., Lavicza, Z.: Integrating Computer Algebra Systems in Post-Secondary Mathematics Education: Preliminary Results of a Literature Review. International Journal for Technology in Mathematics Education 17(2), 57–68 (2010)
13. Sfard, A.: Thinking as Communicating: Human Development, the Growth of Discourses, and Mathematizing. Cambridge University Press (2008)
14. Bernardin, L.: A Review of Symbolic Solvers. SIGSAM Bulletin 30(1), 9–20 (1996)
15. Bernardin, L.: A Review of Symbolic solvers. In: Wester, M. (ed.) Computer Algebra Systems. A Practical Guide, pp. 101–120. J. Wiley & Sons (1999)
16. Stoutemyer, D.R.: Crimes and Misdemeanours in the Computer Algebra trade. Notices of the American Mathematical Society 38(7), 778–785 (1991)
17. Tonisson, E.: A School-oriented Review of Computer Algebra Systems for Solving Equations and simplifications. Issues of domain. In: Bohm, J. (ed.) Proceedings TIME-2008, bk teachware (2008)

18. Tonisson, E.: Unexpected Answers Offered by Computer Algebra Systems to School Equations. The Electronic Journal of Mathematics and Technology 5(1), 44–63 (2011)
19. Sangwin, C.J.: STACK: Making Many Fine Judgements Rapidly. In: CAME 2007– The Fifth CAME Symposium, pp. 1–15 (2007)
20. Richardson, D.: Some Unsolvable Problems Involving Elementary Functions of a Real Variable. Journal of Symbolic Logic 33, 514–520 (1968)
21. Abramovich, S.: How to 'Check the Result'? Discourse Revisited. International Journal of Mathematical Education in Science and Technology 36(4), 414–423 (2005)
22. Tonisson, E.: Branch Completeness in School Mathematics and in Computer Algebra Systems. The Electronic Journal on Mathematics and Technology 1(3), 257–270 (2007)
23. Maxima Manual, http://maxima.sourceforge.net/docs/manual/en/maxima.pdf

Mathematics and the World Wide Web

Patrick D.F. Ion[1,2,3,*]

[1] Mathematical Reviews, American Mathematical Society
[2] W3C Mathematics Working Group
[3] Department of Mathematics, University of Michigan

Abstract. Mathematics is an ancient and honorable study. It has been called The Queen and The Language of Science. The World Wide Web is something brand-new that started only about a quarter of a century ago. But the World Wide Web is having a considerable effect on the practice of mathematics, is modifying its image and role in society, and can be said to have changed some of its content. This paper explores some of the issues this raises.

1 Introduction

Mathematics, which is presumably the interest most common to readers of this piece, has always been a somewhat abstruse subject to most of the world. How it can be communicated has played a role in how it has been perceived by the public, but is also very important to its practitioners. Mathematicians have generally been using the standard ways to communicate all along[1].

Very roughly speaking we have gone from spoken language, to marks on rocks, bones and wood, cuneiform tablets, papyrus and parchment, on to paper and finally, in a great breakthrough, to printing.

That last development, printing, has been the modern thing for about 500 years. It incorporated special mathematical symbols and textual layouts developed over many years, and increasingly allowed additional aids such as diagrams and illustrations. However, the advent of electronic computing machines did not just make it much easier to produce the penalty copy[2]of printed mathematics, as is now commonly done with desktop publishing. The networked aspect of the

---

[*] My thanks are due to many colleagues with whom I have discussed these things over the years, especially from Mathematics Reviews and the W3C Math Working Group. I am also particularly grateful to Andrei Iacob for his continuing to provide new ideas and pointers to intellectual stimulation, and to Wolfram Sperber of Zentralblatt für Mathematik.
[1] The use by the Inca culture for numerical records of quipu, a system of knotting in color-coded cords, could be considered an exception, but it does make use of ordinary materials and is apparently both in use today and similar to seemingly independent systems elsewhere [Gullberg:2007]. It can be argued though, that numerals were themselves initially on-standard.
[2] Material to be printed which carries extra cost due to the presence of special symbols or layout requirements.

telegraph and telephone, important communication developments from the 19th century, has not effected a sea-change in mathematics (and science), though it did make a great difference to society and popular culture (as did radio and later television).

In 1990 Tim Berners-Lee, with Robert Caillau, working at CERN, capitalized on the networking of computing (internet) to introduce what became the World Wide Web: in simplest terms this is basically hypertext on an internet. That is the change we wish to examine here[3].

Berners-Lee was, and is still, an idealist. He intended the new communication medium he knew they were starting to be a force for public understanding and good. It can be seen to have been that in many ways. One of them is the improved access to a great deal of knowledge. That includes mathematics. For the special case of mathematical knowledge there have had to be some small targeted improvements in the technology, which are still continuing. Wider public access to mathematics may be changing the numeracy level of the population, and the public is becoming aware that mathematics lies behind such desirable things as search services and coding for CDs, and maybe also encryption. Politicians, even in the US where there are significant anti-scientific movements, are getting more of the scientific picture, to the benefit of mathematics I would argue. Mathematics is changing its position in society as a result of the advent of the World Wide Web.

Mathematics itself, as a subject of enquiry, has benefitted from the technology which has been facilitated by the World Wide Web. The fact that mathematical notions can be communicated so much more rapidly and widely than before, using the networks of computers across the globe has led to a real increase in mathematical knowledge.

It is largely email and the sharing of files produced by desktop publishing that can be seen to have started the new age, but it is the World Wide Web that has carried it through to a fully new era. Sharing libraries of papers, books, notes, datasets, software and simulations makes a difference to how we learn more of our subject, whether as a student or as a researcher.

Perhaps this is similar to the change in society produced by the dissemination of knowledge in print material, which devalued the real person of a teacher and facilitated studying while not at someone's feet. This new communication medium, which really is world wide and almost instantaneous, is changing how we learn. There are disadvantages, as there were to print's introduction, notably the information overload we are subject to. We are just beginning to learn how to use our new tools. In addition to the personal communication aspects, there are the effects of globally networked computer power; examples where this has changed matters for mathematics are projects like GIMP for number theory, and the enormous algebra calculations associated with, say, root systems for exceptional Lie algebras. There are also various numerical or symbolic computation services that are offered, such as ones associated with AMPL, or Mathematica and Maple.

---

[3] The first Web page has just been restored to public view at [CERN:WWW-1].

There have been developments within mathematics that are readily traced back to the needs of the developing technology. There has been a great deal of new work associated with networks as mathematical objects, of which study of the physical internet is a fine example; but then networks began to be investigated all over from biologically to sociologically. It's not entirely the case that the World Wide Web network sparked the first interest in such matters, but the realization of the importance for us of that real network has meant an increased consciousness of networks wherever they are to be found.

All sorts of developments to do with signal processing, switching theory, encoding and encryption come out of the technology that lies behind the internet which supports the World Wide Web, and which the World Wide Web needs developed. The advent of large datasets was, and their study and usefulness are, encouraged by the World Wide Web. Next we are going on to the Semantic Web, where there are attempts to encode, transmit and work automatically with more of the meanings we impute to strings of characters that fly about our internet and make up our World Wide Web.

There are disadvantages to this new technology and communication medium. As usual, there are unexpected possibilities opened for miscreants, fraudsters, the misguided but well-intentioned, for social control and crime, and for erosion of values previously held dear by, in this case, the mathematical community. [4] It is incumbent on us to be aware that maybe, as Alan Kay has put it, "The best way to predict the future is to invent it".[5]

The exigencies of time and space available mean that the range of subjects touched on in the introduction cannot be expanded on here. Below there's discussion of mathematical communication in history and of a modern web technology, MathML, the author has been closely involved with. Another such example would be the newly revised Mathematical Subject Classification, MSC2010, now in SKOS form for the new Semantic Web world of Linked Datasets — for description of this see the references at [MSC2010:info]. There's also the matter of the development of new types of mathematical documents such as those with visualizations and manipulatives)for personal examples see [Ion:2011], [Ion:2003-], and [MSC:TiddlyWiki]; for more striking, and important examples see [Arnold:2007], [Wolfram:mathdemos] or [Wolfram:Alpha], and also millions of personal web sites and blogs, some notable ones with serious mathematical reach at the research level [Tao:blog], [Gowers:blog], [Baez:site] and crowd-sourcing efforts [Polymath:Project], [Nielsen:blog], [mathoverflow], [PlanetMath]. The matters of ethical behavior and social media are much discussed and there are questions as to whether the culture of mathematics makes other subjects experiences less

---

[4] Similar things happened upon the introduction of the telegraph, for instance [Standage:2007].

[5] Alan Kay claims to have uttered the phrase at a meeting in 1971 involving people from PARC, *Palo Alto Research Center* and Xerox planners.[Smalltalk:website]. More recently he has apparently considered evolving it to "The best way to predict the future is to {invent, prevent} it" with alternating blinking words.[Windley:blogsite]

relevant [AnaEtAl:2013], [ResnikEtAl:2013], [AMS:Social Media], [BikEtAl:2013]. This list could readily be extended. An attempt to do a more thorough job can be found at the author's [Ion:UM-Home].

Finally I would like to mention the fact that this contribution might be deprecated as nothing more than a cut-and-paste job, which the new technology and Web access have made so much easier. Is it a bit like a rap remix with lots of included samples? Perhaps that is what the Web is facilitating, and that is part of the change in knowledge creation that is going on. What is there that's original about such a piece? Does it matter? Is it not likely that the writer does not know what is original in what's written, and indeed we often fool ourselves into thinking we have come up with something new when we should know we have not? Research seems to show this [Sugimori:2013].

2   A Potted History of Mathematical Communication

2.1   Pre-history

Mathematical notions we can imagine go back to the beginnings of human culture. But the traces of early forms of number though they might be subtly encoded in language and teased out by linguistic research are not obvious until they are recorded on something that persists physically. One of the earliest examples of an artefact that can be pointed to as something mathematical seems to be the Ishango Bone, maybe 22,000 years old from the Upper Paleolithic era, unearthed in 1960 [Huylebrouck:2008] [Cornelissen et al.:2008] [Crèvecoeur:2008] [Ishango:website] [Ishango:Wikipedia] [Ishango:YouTube], and now held at the Royal Belgian Institute for Natural Sciences in Brussels [KBI:website] (see Fig. 1). It shows tally markings that have been claimed to show prime numbers, or perhaps to be associated with a calendar, but may simply be expressed in base 12. There is also a Lebombo Bone dated at 37,000 years old [Bogoshi et al.:1987] [Lebombo:website], as well as a Czech contender for oldest object, and a newly studied second Ishango Bone.

2.2   Ancient Times

Let us move on now quickly to ancient times in Babylonia where organized scripts for setting down language were being developed. Writing on tablets of mud that baked dry has the advantage that it has persisted, or at least some of it has, through the ages to the present. There are examples of bullae bearing cuneiform writing that scholars can interpret to show clearly mathematical calculations dating back to over 2000 BCE. There is one, for instance, that seems to show a table of the powers of 70 times 2, up to 16,470,860,000,000 (2×70^7) from 2050BCE, and another from 1800BCE with the values of $N^3 + 3N^2 + N = N[(N+1)(N+2) - 1]$ for $N = 1..50$ [Sumer:Clay].

Another tablet from these early days, known as MS 3051, shows a geometrical figure [Friberg:2005] (on the Web see [Friberg: page 190]).

Fig. 1. The Ishango Bone; from 4 sides and a close-up

A different medium which has shown its persistence is papyrus, and this example again shows the use of non-textual aids for mathematics, in this case a triangle diagram in an Egyptian fragment known as Problem 59 from ca. 1850 BCE [Egypt:Papyri].

2.3   Historical Times

The Archimedes Palimpsest — a couple of thousand years later in the early part of our modern era, around 1000 CE — is the most recently famous piece of the mathematical record rediscovered. It was recently auctioned for a record price after being lost for many decades. The palimpsest is a series of parchment pages on which a text ascribed to Archimedes was erased to be replaced by a Christian religious text which was thought more valuable at the time[Archimedes:website].

Leibniz (1646–1716) was a towering figure of his time, a truly versatile diplomat and scholar, who was a librarian, lawyer and philosopher. As a librarian he can be said to have introduced a system of indexes (author, title, keywords), timelines, and a classification scheme with a decimal code aspect like that most common today. He even used 500 for Science, as nowadays. Of course, he's widely known for his work on binary notation and his conception of a *calculus ratiocinator* which is the origin of the name of one of the parts of CICM, *Calculemus!*, as well as for his form of the calculus which was seen as a competitor to Newton's. He was a prodigious worker and wrote about 40,000 letters and much other material, so that there are presently six sites undertaking digitization of his works to make it available on the Web.

Pasigraphy is not well-known today, but as a conception it plays a role in our story. In 1897, at the First International Congress of Mathematicians in Zürich there were not many talks. Striking among their titles is "Pasigraphie, ihren jetzigen Zustand, und die pasigraphische Bewegung in Italien"[6]. The presenter, the logician Schröder from Karlsruhe[7] began his talk by saying that if there were any topic that really belonged at an International Conference of Mathematicians, then it was *pasigraphy*. He was sure that pasigraphy would take its rightful place on the agenda of all succeeding such conferences. Pasigraphy is the study of universal languages of symbols intended to encapsulate semantics and to provide a basis for calculational ratiocination, i.e. reasoning by manipulation of symbols. This is in line with Leibniz's conception.

Schröder then went on to disagree with the distinguished chairman of the session, Giuseppe Peano[8], by saying that he did not think that Leibniz's problem of providing an *algebra universalis*, a symbolic calculus for mathematics had been solved. Peano had just published, in 1894, his "Formulaire des Mathématiques" [Peano:1894], in which he felt he had essentially provided just that. It is in fact there that Peano's axioms for the natural numbers are to be found, along with axiomatizations and highly symbolic representations for much of arithmetic, algebra, geometry and calculus.

Schröder goes on to offer some of his own considerations on the topic of universal symbolics for math, including some remarks that can be seen as rather prescient about, for instance, the complexity of formulas[9].

Paul Otlet and his friend Henri La Fontaine, about 1895, conceived the idea of a universal library that would aggregate all knowledge and make it retrievable through targeted searches carried out on a well-organized index system. They designed many subsystems with great care, and were able to obtain funding from the Belgian Government to start realizing their dream. La Fontaine applied the money he received as a Nobel Peace Prize largely to the project of building a Universal Bibliographic Repertoire.

Otlet designed a highly advanced index card machine: "a moving desk shaped like a wheel, powered by a network of hinged spokes beneath a series of moving surfaces. The machine would let users search, read and write their way through a vast mechanical database stored on millions of 3×5 index cards. This new research environment would do more than just let users retrieve documents; it would also let them annotate the relationships between one another, the connections each [document] has with all other [documents], forming from them what might be called the Universal Book."

They negotiated with John Dewey and obtained his permission, under some terms on non-competitiveness, to extend his Dewey Decimal System of classification to a Universal Decimal Classification (UDC), also still in use today. They started cataloging and collecting, and even responding to remote queries not just received through the post, but made using the new telegraph and telephone

---

[6] "Pasigraphy, its present state, and the pasigraphic movement in Italy"

[7] Perhaps best known today from the so-called Schröder-Bernstein theorem.

[8] The Italian mathematician who can be considered leader of the pasigraphic movement in Italy, and who was also active in developing universal natural language Interlingua [Interlingua:Wikipedia].

[9] In fact Galois in an open letter had brought that up already. [Neumann:2011]

systems. Indeed their vision included a distributed information system that is very modern in conception. A big problem in implementing this was that the records were on slips in card files and the querying agents were library assistants scurrying from one box to another and gathering notes. Nonetheless a start of the system called the Mundaneum, in Mons, was demonstrated and ran for a while until larger forces supervened. With the world-wide downturn of economic life around 1930 public funding dried up, and then the political turmoil in Europe culminating in the invasion and occupation of Belgium by a foreign army meant an end to this grad project. Nowadays there is a revival of interest in the Mundaneum [Mundaneum:Accueil] and it has a web presence under the Google Cultural Institute [Mundaneum:Google][10]. It can seen as an ancestor of both Google and the Semantic Web. The modern scholar to whom we should be grateful for unearthing the Mundaneum and reviving interest in it is W. Boyd Rayward [Rayward:1975].

Vannevar Bush [Bush:Wikipedia], as Director of the Office of Scientific Research and Development, was a very important figure in the scientific side of the American War effort during World War II. But in this context his importance is the article he published in 1945 entitled "As we may think"[Bush:1945] [NyceKahn:1991]. Bush envisaged a desk-like machine that would have the world's information stored in some accessible way, say using photographic microfilm technology, that would allow users to store trails of their explorations as they navigated the world's knowledge.

Doug Engelbart [Engelbart:Wikipedia], inventor of the computer mouse and early graphical user interfaces, as well as a developer of hypertext and networked computing, says he was much influenced by Bush's article. In addition he was both idealistic and of a philosophical inclination. Bardini and Friedewald write "After he had read the works of Benjamin Lee Whorf whose ethno-linguistic writings influenced many scientists during the late 1950s, he was convinced that technological systems were not only shaped by humans but also shaped human thinking themselves. Man and machine could not be treated separately in such a technological system. Thus Engelbart concluded that developing a tool for augmenting human intellect had to be a co-evolution of man and technology."[BardiniFriedewald:2002]

3   The World Wide Web

The early days of the World Wide Web and of the internet have been extensively discussed and reminisced over by many of the key figures in its early development. However it is the pervasiveness of its influence that I need to emphasize here. I will mention two incidents that seem to me to make the point without recourse to academically justified statistical studies.

---

[10] "Google de papier" (Le Monde); "The web time forgot" [Spiegel:2011]; "L'ancêtre génial de Google" [Spiegel:2011].

The first is the appearance of Sir Tim Berners-Lee at the opening ceremony of the London Olympic Games 27 July 2012, shown on TV putting on Twitter his message "This is for everyone" from his Next Cube computer, which flashed round the stadium on 'pixel' paddles mounted by the seats of 70,500 members of the audience, and is estimated to have reached a television audience numbering about 10^9 [Berners-Lee:Twitter]. That is a a true world-wide web presence. It affects the image of computers, technology and mathematics as well. The first Web page has just been reconstructed at CERN as a cultural icon [Berners-Lee:Web1].

Fig. 2. Tim Berners-Lee's Olympic tweet

The second example of hype that seems to me a nice illustration is what I saw recently in a magazine I picked up in Stuttgart airport[Welt der Wunder:2013]. There was a piece entitled "Will the internet be the new world power? If so, who rules it?"[11] The piece emphasized the view that, roughly speaking, the internet was a community which amounted to a culture like a country, with about 2×10^9 citizens. Of these about 50% were born in it (or with it) and 50% were immigrants (older people have to learn to navigate the internet and the world wide web). This new power has 5 declared state enemies (Bahrein, China, Iran, Syria, Vietnam) who would like very much to curb it. But it is a market, according to the Boston Consulting Group of about 4.2×10^{12}\$, so a very significant economic power. Finally, from another point of view it is approaching the size of natural brains in number of connections, so the question arises whether it is becoming some sort of new communal brain? Again we see the idea that the WWW affects everything, and certainly science and mathematics, which are the basis of its technology. But the socio-political effects on science and mathematics may be greater than the considerations arising from the technological needs.

3.1 MathML

One of the ironies early in the days of the World Wide Web seemed to be that although the birthplace of the WWW is certainly CERN, the European Organization for Nuclear Research[CERN:Website], so one might assume that

---

[11] "Wird das Internet die neue Weltmacht? Wenn ja, wer regiert sie?"

technical documentation full of equations would be of importance its early users, there was no easy way to specify the display of equations on Web pages for some time.

In May 1997 the W3C formed a Math Working Group to consider how to facilitate math on the Web[12] The Math WG contained representatives of diverse backgrounds. There were those from computer corporations and publishing, computer algebra people[13] and invited experts from organizations not members of the W3C[14].

The objectives originally considered were far-reaching: To develop an open specification for math to be used with HTML that:

1. Is suitable for teaching, and scientific communication;
2. Is easy to learn and to edit by hand for basic math notation, such as arithmetic, polynomials and rational functions, trigonometric expressions, univariate calculus, sequences and series, and simple matrices;
3. Is well suited to template and other math editing techniques;
4. Insofar as possible, allows conversion to and from other math formats, both presentational and semantic, such as TeX and computer algebra systems.
5. Output formats may include graphical displays, speech synthesizers, computer algebra systems input, other math layout languages such as TeX, plain text displays (e.g. VT100 emulators), and print media, including braille. It is recognized that conversion to and from other notational systems or media may lose information in the process;
6. Allows the passing of information intended for specific renderers;
7. Supports efficient browsing for lengthy expressions;
8. Provides for extensibility, for example through contexts, macros, new rendering schemas or new symbols. Some extensions may necessitate the use of new renderers.

The above goals were ratified by those at the initial meeting in October 1996! In essence, the participants were working up to a full-scale pasigraphy for mathematics. The connections to the computer algebra engines, which were supposed to communicate the semantics of mathematical expressions were to be part of realizing Leibniz's *calculus ratiocinator*.

Many ideas were considered early on. After all, several WG participant organizations already had markup languages that they were using to express a great deal of really useful mathematics: for instance, Scratchpad (later Axiom) from

---

[12] Dave Raggett [Raggett:Wikipedia], who produced the HTML 3.2 draft, was himself a proponent of adding some math capabilities to HTML. In fact there was some confusion over the math in HTML 3.2. books appeared with sections explaining simple extensions to HTML for math that were little more than ideas that had been suggested as to how something might be done. They were not in any draft accepted by the W3C. In fact the HTML 3.2 draft never evolved to a Recommendation of the W3C.

[13] IBM, Hewlett-Packard, Adobe, Elsevier Science, Wolfram Research, Maplesoft, SoftQuad, ...

[14] American Mathematical Society, Geometry Center, Stilo Technologies, ...

IBM, and, of course, the input languages of Maple and Mathematica. Other influences were naturally the already pervasive TeX, much used in science and spreading in scientific publishing, and the presence on the scene of Java[15].

One clear thing was that although some of the technical solutions available for mathematical markup were very clever and powerful, the W3C Math WG would have to come up with a *lingua franca* into which all could translate for no single player would be allowed by the others to dominate the field.

The WG finally decided that it would develop a markup language which accorded with XML. In fact, MathML 1.0 is an XML application, one of the first written at more than a toy level. A salient reason is that general acceptance of a math specification would only happen if it embedded well into the technology of the internet. And that was then coming to be dominated by XML and its relatives. We wanted something that really met the goal of facilitating the use of math on the Web, so we fell in line with the evolving standards of the Web. This had many implications for later work, and was not easy to take. The primary goal was to create something that would be powerful, usable and adopted. The Math WG has stuck with that intention throughout.

Many on the WG had considerable experience with TeX, and could consider that as a natural paradigm for a math language for the Web. IBM, drawing on its extensive experience with Scratchpad and then Axiom, could have itself proposed a language for math. Wolfram Research's Mathematica clearly offered a very rich language for expressing math in ASCII characters. But the disadvantage to any of these foundations would have been that we could not have been able to offer a public specification which all could agree had the expressiveness they would like to see, and none would feel they had not contributed to.

Producing an XML application meant the need for an easy input syntax for math on Web pages, would remain unmet, even for simple math. But the advantages of a machinable *lingua franca* outweighed the disadvantages. The problematic input of formulas was intended to be facilitated by interface applications tailored to the needs of their user communities — high school, research scientists, TeXies ...

The details of what was developed are naturally in the MathML specifications themselves. Other than that they are couched in XML terms and do not address input issues, the most important design decision made is probably that there are both Presentation and Content Markup sections to MathML. Presentation markup is enough to allow specification of all the layout schemata in common use today, and can be claimed to be at least as expressive as TeX for that purpose, if one does not allow the use of the aspects of TeX that make

---

[15] One then easily rejected suggestion was that each formula would be a program written in Java, which would allow different representations of it, in print or as an object for other computation, responding to others of its semantic aspects. However running many such function in a document was out of the question. It seems to me paradoxical that the current fallback for rendering of MathML, which does a remarkably good job of being universal, is MathJax [MathJax:website], written in JavaScript and using functions for each formula.

it a full programming language. Content markup is intended to be tailored to expressing the semantics of mathematical formulas as found in about the first 14 years of schooling in the US. In addition, there are arrangements made, for the purposes of both Content and Presentation that allow that allow customizing beyond the basic schemata and semantics already provided; naturally these are not very easy to do. On the Content side the MathML specification, has been, over the years, carefully harmonized with the efforts of the OpenMath Society [OpenMath:website], which has developed a very general XML notation intended to allow codifying mathematical semantics.

4   Disadvantages of the Web

What are the disadvantages of this new technology and communication medium? Too many even to enumerate here. The text below concentrates on just one aspect.

One of the problems is the interaction of a changing ethical environment and the ease that the new tools provide for repurposing material on the internet. This can be plagiarism of others' work when no references to sources are given, self-plagiarism when authors repeat their own materials whether knowingly or not, or just plain copying whether witting or not. Of course, it can be argued that there are many cases where repetition serves the reader, or where transclusion of documents is an enhancement. It has for along time been standard practice in mathematical papers to repeat notational conventions. For instance there are over 5,000 cases of "R denotes" in the reviews of Mathematical Reviews (MR) and over 388,000 on Google. We all know how troubling it is when one is apparently not privy to what an author intended some symbol to denote. For example, "Everyone knows IN means the natural numbers"—but some people think that doesn't include 0! So specifying notation is essential. Thus sometimes a whole section of one article looks like that of another simply because they are both rehearsing the same notational context required. But what if two papers seem to have pages of similar material except that the norm $\|\cdot\|_\infty$ in the one has been replaced throughout by ess. sup.·, and the authors are quite different?[16] What if one of the items is a preprint in English from a US institute with apparently Indian authors and the other item has the same formulas but a text in Hindi whose fragments seem to parallel the original very well, but there are completely different Indian authors[17]. And what if there's more coincidence, up to the extreme of everything the same except for the author (as practiced by the infamous plagiarizer Dănuţ Marcu[Marcu:wk])?

In other fields there is now increasing attention paid to plagiarism detection. It is of importance economically in, say, the medical and pharmaceutical business. But reputations have been gained through plagiarism, and destroyed

---

[16] The second paper had two authors and appeared later although the suspicion is that the first paper, with a different single author from another country, is derivative of the second on whose behalf someone complained to MR.

[17] I found two examples of this while at MR.

when it is found out. This seems to have been important in German political life recently with ministers resigning over scandals involving questioning their academic credentials. There are even questions raised about the originality of President Vladimir Putin's thesis which copies 16 pages from an earlier textbook, but this is also part of a larger political struggle [Time:2013].

Mitigation of the problem of plagiarism and repetition, with its resulting undermining of the literature's value, is partly possible using the new technologies. The main improvements will result from a renewed emphasis on professional ethics and corresponding educational attention. Some of this has started to take place in mathematics [Arnold:2009] [Arnold:2012], but slowly enough [Jackson:2002].

A striking German example of what can be done now so much material is electronically available and computing power can be deployed to implement algorithmic checking is the [VroniPlag Wiki]. It describes itself as collaborative documentation of plagiarism — critical examination of university theses on the basis of reliable locations of plagiarism. There are detailed displays, page by page, of text and unmentioned plausible earlier sources and local values for the percentage of apparently derivative material. This work is carried out by volunteers, so is an example of crowd-sourcing.

At the other end of the automation spectrum is [eTBLAST], a free web-based text comparison engine, that had its origin in [BLAST] (Basic Local Alignment Search Tool) which finds similarity in sequences, usually of nucleotides or proteins. eTBLAST's founders seem to have taken the view that documents are just sequences of letters too, and so can be treated with similar techniques. eTBLAST's corpus does include [arXiv.org], so significant mathematical material. But there has been much development since the initial start of this service 10 years ago, and they are actively involved in plagiarism detection. A byproduct of eTBLAST is [Deja Vu] that deals with MEDLINE material.

5   Final Thoughts

The tendencies today seem to be very much the opposite of the great Gauß's motto *Pauca sed matura*. Twitter's whole point [Twitter] is to encourage *CXL litterulae et immatura*.[18]

The effects of the Web have been imagined over the centuries. They will be different from much of that already imagined as well.

The medium may not be the message (Macluhan) but it does help shape it (Whorff).

References

AnaEtAl:2013. Ana, J., Koehlmoos, T., Smith, R., Yan, L.L.: Research Misconduct in Low- and Middle-Income Countries. PLoS Med 10(3), e1001315 (2013), doi:10.1371/journal.pmed.1001315 233

---

[18] "Little but ripe" contrasted with "140 lower-case letters and immature".

Arnold:2009. Arnold, D.J.: Integrity Under Attack: The State of Scholarly Publishing. SIAM News, 1–3 (December 2009), http://ima.umn.edu/~arnold/siam-columns/integrity-under-attack.pdf 241

BardiniFriedewald:2002. Bardini, T., Friedewald, M.: Chronicle of the Death of a Laboratory: Douglas Engelbart and the Failure of the Knowledge Workshop. History of Technology 23, 191–212 (2002), http://www.friedewald-family.de/Publikationen/hot2002.pdf 236

BikEtAl:2013. Bik, H.M., Goldstein, M.C.: An Introduction to Social Media for Scientists. PLoS Biol 11(4), e1001535 (2013), doi:10.1371/journal.pbio.1001535 233

Bogoshi et al.:1987. Bogoshi, J., Naidoo, K., Webb, J.: The oldest mathematical artifact. Math. Gazette 71(458), 294 (1987) (JSTOR: http://www.jstor.org/stable/3617049; with picture) 233

Bush:1945. Bush, V.: As We May Think. The Atlantic (July 1945), http://www.theatlantic.com/magazine/archive/1945/07/as-we-may-think/303881/; Reprinted in Life Magazine September 10 (1945), Wikipedia entry http://en.wikipedia.org/wiki/As_We_May_Think Reprinted Journal of Electronic Publishing 1(2) (February 1995), doi:http://dx.doi.org/10.3998/3336451.0001.101 236

Cornelissen et al.:2008. Cornelissen, E., Jadin, I., Semal, P.: Ishango, a history of discoveries in the Democratic Republic of Congo (DRC) and in Belgium. In: Huylebrouck, D. (ed.) Ishango, 22000 and 50 Years Later: The Cradle of Mathematics?, February 28-March 02, 2007, pp. 23–38. Koninklijke Vlaamse Academie van België voor Wetenschappen en Kunsten, Brussel (2008) 233

Crèvecoeur:2008. Crèvecoeur, I.: Variability of Palaeolithic modern humans in Africa. Future Prospects of the Ishango human remain (re-)study. In: Huylebrouck, D. (ed.) Ishango, 22000 and 50 Years Later: The Cradle of Mathematics?, February 28-March 02, 2007, pp. 87–97. Koninklijke Vlaamse Academie van België voor Wetenschapppen en Kunsten, Brussel (2008) 233

Friberg:2005. Unexpected Links Between Egyptian and Babylonian Mathematics. World Scientific (2005) ISBN: 9789812563286, http://books.google.com/books?id=1qQtWFHd8noC 233

Friberg:2007. Friberg, J.: A Remarkable Collection of Babylonian Mathematical Texts: Manuscripts in the Schøyen Collection. Springer (2007), doi: 10.1007/978-0-387-48977-3, ISBN:978-0-387-34543-7, ISBN:978-0-387-48977-3, http://link.springer.com/book/10.1007/978-0-387-48977-3/page/1 244

Gullberg:2007. Gulberg, J.: Mathematics: From the birth of numbers, pp. xxvii+1093. W. W. Norton & Company Inc., New York (1997) ISBN:0-393-04002-X 230

Huylebrouck:2005. Huylebrouck, D.: Afrika en wiskunde. Etnowiskunde in zwart Afrika, vanaf de koloniale tijd. VUBPress, Brussel (2005)

Huylebrouck:2008. Huylebrouck, D.: The ISShango project. Journal of Mathematics and the Arts 2(3), 145–152 (2008), http://dx.doi.org/10.1080/17513470802446319 (with pictures) 233

Jackson:2002. Jackson, A.: Theft of Words, Theft of Ideas. Notices of the AMS, 645 (June/July 2002), http://www.ams.org/notices/200206/commentary.pdf 241

Neumann:2011. Neumann, P.M.: The mathematical writings of variste Galois European Mathematical Society, Z"urich, 421 p. ISBN 978-3-03719-104-0, doi: 10.4171/104 235

NyceKahn:1991. Nyce, J.M., Kahn, P. (eds): From Memex to hypertext: Vannevar Bush and the mind's machine, 367 p. Academic Press (1991) ISBN 0125232705, 9780125232708, http://books.google.com/books/about/From_Memex_to_hypertext.html?id=oZNQAAAAMAAJ 236

Peano:1894. Peano, G.: Formulaire de mathématiques. t. I-V. Turin, Bocca frères, Ch. Clausen (1858-1932) (1894-1908) 235

ResnikEtAl:2013. Resnik, D.B., Master, Z.: Policies and Initiatives Aimed at Addressing Research Misconduct in High-Income Countries. PLoS Med 10(3), e1001406 (2013), doi:10.1371/journal.pmed.1001406 233

Schröder:1897. Schršder, E.: Über Pasigraphie, ihren gegenwärtigen Stand und die pasigraphische Bewegung in Italien. 147-162 of Verhandlungen des ersten Internationalen Mathematiker-Kongresses in Zürich vom 9. bis 11 (August 1897), http://www.mathunion.org/ICM/ICM1897/Main/icm1897.0147.0162.ocr.pdf

Rayward:1975. Rayward, W.B.: The Universe of Information: the Work of Paul Otlet for Documentation and International Organisation. All-Union Institute for Scientific and Technical Information (VINITI) for the International Federation for Documentation. Moscow, 239 p. (1975), http://lib.ugent.be/fulltxt/handle/1854/3989/otlet-universeofinformation.pdf 236

Robson:2008. Robson, E.: Mathematics in ancient Iraq: a social history, 441 p. Princeton University Press (2008) ISBN:9780691091822, LCCN:2007041758, http://books.google.com/books?id=w-e6kfvoq5gC

Standage:2007. Standage, T.: The Victorian Internet: The Remarkable Story of the Telegraph and the Nineteenth Century's On-line Pioneers (Paperback) Walker & Company, 227 p. (2007) ISBN-10: 0802716040; ISBN-13: 978-0802716040 232

Sugimori:2013. Sugimori, E., Kitagami, S.: Plagiarism as an illusional sense of authorship: The effect of predictability on source attribution of thought. Acta Psychol. 143(1), 35–39 (2013), http://dx.doi.org/10.1016/j.actpsy.2013.01.007 233

Welt der Wunder:2013. Wird das Internet die neue Weltmacht? Wenn ja, wer regiert sie? Welt der Wunder, 4/13, 34–44 (2013), http://www.bauer-plus.de/file/5704/20130417125213/welt-der-wunder.jpg 237

Web References — Valid 15 April 2013 or Thereafter

AMS:Social Media. AMS and Social Media: Connecting the Mathematics Community (Facebook, Twitter, LinkedIn, YouTube, Wordpress Blogs, podcasts, Instagram, RSS), http://www.ams.org/about-us/social 233

Archimedes:website. The Archimedes Palimpsest (Dataset), http://www.archimedespalimpsest.net/ 234

Arnold:2012. SIAM Past President Doug Arnold on mathematical literature and scholarly publishing (video) http://connect.siam.org/siam-past-president-doug-arnold-on-mathematical-literature-and-scholarly-publishing/ 241

Arnold:2007. Arnold, D., Rogness, J.: Möbius Transformations Revealed (Video), http://www.ima.umn.edu/~arnold/moebius/, http://www.youtube.com/watch?v=JX3VmDgiFnY, http://www.math.umn.edu/~rogness/ 232

arXiv.org. arXiv.org (Open access to 838,040 e-prints in Physics, Mathematics, Computer Science, Quantitative Biology, Quantitative Finance and Statistics), http://arXiv.org 241

Baez:site. Baez, J.: John Baez's Stuff, http://math.ucr.edu/home/baez/ 232

Berners-Lee:BBC. Tim Berners-Lee interviewed on the BBC, http://www.bbc.co.uk/news/technology-19492087

Berners-Lee:Twitter. Tim Berners-Lee at the London 2012 Olympics Opening Ceremony, http://www.youtube.com/watch?v=F-twBP2VCV0; http://www.zdnet.com/uk/web-inventor-tim-berners-lee-stars-in-olympics-opening-ceremony-7000001744/ 237

Berners-Lee:Web1. The First Web Page at CERN by TBL. (reconstructed) 237

BLAST. BLAST: Basic Local Alignment Search Tool, http://blast.ncbi.nlm.nih.gov/ 241

Bush:Wikipedia. Vannevar Bush on Wikipedia, http://en.wikipedia.org/wiki/Vannevar_Bush 236

CERN:Website. CERN Home Page, http://home.web.cern.ch/ 237

CERN:WWW-1. CERN's reconstruction of the First Web Page, http://info.cern.ch/hypertext/WWW/ 231

Deja Vu. Deja Vu: a Database of Highly Similar Citations, http://dejavu.vbi.vt.edu/dejavu/ 241

DPLA:website. The Digital Public Library of America, http://dp.la/

Egypt:Papyri. Egyptian Mathematics Papyri at Mathematicians of the African Diaspora, http://www.math.buffalo.edu/mad/Ancient-Africa/mad_ancient_egyptpapyrus.html 234

Engelbart:Wikipedia. Douglas Engelbart on Wikipedia, http://en.wikipedia.org/wiki/Douglas_Engelbart 236

eTBLAST. eTBLAST: a text-similarity based search engine, http://etest.vbi.vt.edu/etblast3/ 241

Friberg: page 190. [Friberg:2007] p. 190, http://link.springer.com/book/10.1007/978-0-387-48977-3/page/1#page-1 233

Gowers:blog. Gowers, T.: Gowers's Weblog: Mathematics related discussions, http://Gowers.wordpress.com 232

Interlingua:Wikipedia. Interlingua at Wikipedia, http://en.wikipedia.org/wiki/Latino_sine_flexione, https://sites.google.com/site/latinosineflexio/ 235

Ion:2003-. Ion, P.D.F.: Information and examples for N-body gravitational problems, http://www-personal.umich.edu/~pion/NBody/ 232

Ion:2011. Ion, P.D.F.: Geometry and the Discrete Fourier Transform. AMS Feature Column (November 2011), http://www.ams.org/samplings/feature-column/fcarc-geo-dft 232

Ion:UM-Home. Ion Home Page at the University of Michigan, http://www-personal.umich.edu/~pion/ 233

Ishango:website. Ishango Bone, on Mathematicians of the African Diaspora, http://www.math.buffalo.edu/mad/Ancient-Africa/ishango.html 233

Ishango:Wikipedia. http://en.wikipedia.org/wiki/Ishango_bone 233

Ishango:YouTube. "L'Os d'Ishango" Le Plus Vieil Objet Mathématique de l'Humanité (YouTube video featuring Ishango and music and illustrating the power of the current medium to allow conflation of varied agendas), http://www.youtube.com/watch?v=oUUMd5CNWbQ 233

KBI:website. Royal Belgian Institute for Natural Sciences, Brussels, Belgium, http://www.kbinirsnb.be/ 233

Lebombo:website. Lebombo, on Mathematicians of the African Diaspora, http://www.math.buffalo.edu/mad/Ancient-Africa/lebombo.html 233

Marcu:wk. Dănuţ Marcu on Wikipedia, http://en.wikipedia.org/wiki/D?nu?_Marcu 240

MathJax:website. MathJax Home Page, `http://www.mathjax.org` 239

mathoverflow. math*overflow*, `http://mathoverflow.net/` 232

MSC2010:info. MSC2010 in SKOS,
`http://msc2010.org/resources/MSC/2010/info/` 232

MSC:TiddlyWiki. MSC2010 in TiddlyWiki form,
`http://msc2010.org/MSC-2010-server.html` 232

Mundaneum:Accueil. Mundaneum Page d'Accueil,
`http://www.mundaneum.org/fr/accueil` 236

Mundaneum:Google. The Mundaneum in the Google Cultural Institute,
`http://www.google.com/culturalinstitute/`
`browse/?q.8129907598665562501=1029` 236

Nielsen:blog. Nielsen, M.: Blog, `http://michaelnielsen.org/blog/` 232

Norman:website. Jeremy Norman's From Cave Paintings to the Internet: Chronological and Thematic Studies on the History of Information and Media,
`http://www.historyofinformation.com/index.php`

NY Times:2008. The Web Time Forgot, New York Times, June 17 (2008), `http://`
`www.nytimes.com/2008/06/17/science/17mund.html?pagewanted=all&_r=0`

OpenMath:website. The OpenMath Society,
`http://www.openmath.org/society/index.html` 240

PlanetMath. PlanetMath: math for the people, by the people,
`http://planetmath.org/` 232

Polymath:Project. The polymath blog, `http://polymathprojects.org/` 232

Raggett:Wikipedia. Dave Raggett on Wikipedia,
`http://en.wikipedia.org/wiki/Dave_Raggett` 238

Smalltalk:website. `http://www.smalltalk.org/alankay.html` 232

Spiegel:2011. Internet Visionary Paul Otlet: Networked Knowledge, Decades Before Google by Meike Laaff in Der Spiegel International, July 22 (2011),
`http://www.spiegel.de/international/world/internet-visionary-paul-`
`otlet-networked-knowledge-decades-before-google-a-775951.html`,
`http://www.spiegel.de/netzwelt/web/netzvisionaer-`
`paul-otlet-googles-genialer-urahn-a-768312.html` 236

Sumer:Clay. Clay Tablets from Sumer, Babylon and Assyria at Earth's Ancient History, `http://earth-history.com/Sumer/Clay-tablets.htm` 233

Tao:blog. Tao, T.: What's new: Updates on my research and expository papers, discussion of open problems, and other maths-related topics,
`http://terrytao.wordpress.com` 232

Time:2013. Putins Ph.D.: Can a Plagiarism Probe Upend Russian Politics? By Simon Shuster/Moscow February 28 (2013), `http://world.time.com/2013/02/28/`
`putins-phd-can-a-plagiarism-probe-upend-russian-politics/` 241

Twitter. Twitter Home Page, `https://twitter.com` 241

VroniPlag Wiki. VroniPlag Wiki (kollaborative Plagiatsdokumentation; Kritische Auseinandersetzungen mit Hochschulschriften auf Basis belastbarer Plagiatsfundstellen), `http://de.vroniplag.wikia.com/wiki/Home` 241

Windley:blogsite. Phil Windley's Technometria: Alan Kay: Is Computer Science an Oxymoron? February 23 (2006),
`http://www.windley.com/archives/2006/02/alan_kay_is_com.shtml` 232

Wolfram:Alpha. Wolfram Alpha: The Computational Knowledge Engine,
`http://www.wolframalpha.com/` 232

Wolfram:mathdemos. Wolfram Demonstrations Project,
`http://demonstrations.wolfram.com/` 232

Structural Similarity Search
for Mathematics Retrieval*

Shahab Kamali and Frank Wm. Tompa

David R. Cheriton School of Computer Science
University of Waterloo
Waterloo, ON, Canada
{skamali,fwtompa}@cs.uwaterloo.ca

Abstract. Retrieving documents by querying their mathematical content directly can be useful in various domains, including education, engineering, patent research, physics, and medical sciences. As distinct from text retrieval, however, mathematical symbols in isolation do not contain much semantic information, and the structure of an expression must be considered as well. Unfortunately, considering the structure to calculate the relevance scores of documents results in ranking algorithms that are computationally more expensive than the typical ranking algorithms employed for text documents. As a result, current math retrieval systems either limit themselves to exact matches, or they ignore the structure completely; they sacrifice either recall or precision for efficiency. We propose instead an efficient end-to-end math retrieval system based on a structural similarity ranking algorithm. We describe novel optimizations techniques to reduce the index size and the query processing time, and we experimentally validate our system in terms of correctness and efficiency. Thus, with the proposed optimizations, mathematical contents can be fully exploited to rank documents in response to mathematical queries.

1 Introduction

Documents with mathematical expressions are extensively published in technical and educational web sites, digital libraries, and other document repositories such as patent collections. Retrieving such documents with respect to their math content is a challenging problem. Mathematical expressions are objects with complex structures and rather few distinct symbols and terms. The symbols and terms alone are usually inadequate to distinguish among mathematical expressions. For example, a search for documents that include the expression $\int x\sqrt{x^2 + a^2}\,dx$ is not likely satisfied by a document that includes $\sqrt{x+2}\int 2ax\,dx$. Moreover, relevant mathematical expressions might include small variations in their structures or symbols. For example, a document including $1 + \sum_{i=1}^{n} i^k$ might well be useful in

---

* Financial assistance for the research was provided by the Natural Sciences and Engineering Research Council of Canada, Mprime, and the University of Waterloo.

J. Carette et al. (Eds.): CICM 2013, LNAI 7961, pp. 246–262, 2013.

response to a query to find documents including $\sum_{j=1}^{n} j^2$. Hence exact matching of mathematical expressions is not a sufficiently powerful search strategy.

The majority of the published mathematical expressions are encoded with respect to their appearance (*presentation*), and most instances do not preserve much semantic information. Content-based mathematics retrieval systems [3,9] are limited to resources that encode the semantics of mathematical expressions, and they do not perform well with expressions encoded using presentation markup. Other systems search based on the presentation of mathematical expressions [2,5,13,18,19], but they either find exact matches only or they use a "bag of symbols" model that often returns many irrelevant results.

Because mathematical expressions are often distinguished by their structure, we should not rely merely on the symbols they include but instead consider a search paradigm that incorporates mathematical structure as well. More specifically, the similarity of two expressions, defined as a function of their structures and the symbols they share [6], can be used as an indication of the relevance of documents when a math expression is given as a query. To be useful, besides the correctness of results (i.e. their relevance to the query), the query processing time must be kept reasonably low. However, this is difficult to achieve because calculating structural similarity of expressions is computationally expensive, and many potential expressions must be considered in response to each query. Hence, efficiently processing a query is a challenging problem that we address in this paper.

The rest of this paper is organized as follows. In Sect. 2 we explain the query language and the search problem. We next describe related work. We describe a structural similarity search algorithm in Sect. 4, and we propose optimization techniques for this algorithm in Sect. 5 and 6. We finally present an evaluation of our algorithm and conclude the paper.

2   The Framework

A mathematical expression is a finite combination of symbols that is formed according to some context-dependent rules. Symbols can designate numbers (constants), variables, operators, functions, and other mathematical entities.

A text document, such as a web page, that contains a mathematical expression is a *document with mathematical content*. We assume that a query is a mathematics expression. Given a query, the search problem is to find the top-k relevant documents, where documents are ranked with respect to the similarity of their mathematical expressions to the query.

Presentation MathML is part of the W3C recommendation that is increasingly used to publish mathematics information on the web, and many web browsers support it. There are various tools to translate mathematical expressions from other languages, including LaTeX, into Presentation MathML. Thus we can assume that stored expressions are encoded in this form when they are indexed. Because forming queries directly with Presentation MathML is difficult, however, input devices such as pen-based interfaces and tablets [11,16] or more widely-known languages such as LATEX might be preferred for entering a query. Nevertheless, automatic tools can also be applied to translate queries to Presentation MathML, and therefore, regardless of the user interface, we can assume a query is also represented using this encoding. In summary, it is appropriate to assume that Presentation MathML is employed when querying mathematics information on the web.

3   Related Work

Currently, there are a few alternative approaches to math retrieval. In one approach, expressions that match the query exactly are considered as relevant. Examples include algorithms that are based on comparing images of expressions [20,21] (they calculate the similarity of images, which allows for very limited variation among the expressions returned) or using very detailed and formal query languages that enable database operations to match expressions [2,5]. We characterize such algorithms as *ExactMatch* algorithms in this paper. Some other algorithms perform some normalizations on the query and also on the expressions before exactly matching them [18]. As shown below, ExactMatch and NormalizedExactMatch perform poorly when searching for mathematical content.

As a variant, some algorithms consider retrieving expressions that share substructures with the query [3,6,9,15]. These algorithms do not consider ranking the results when many partial matches exist. We characterize all such algorithms as *SubexprExactMatch* algorithms and note that *normalized subexpression exact match* algorithms are also feasible.

Another approach to math retrieval is to transform an expression into a collection of tokens where each token represents a math symbol or a substructure [13,14,17,19,12]. Regardless of the tokenization details, some structure information is missed by transforming an expression into bags of tokens, which affects the accuracy of results as shown below.

Algorithms for retrieving general XML documents based on tree-edit distance have been proposed [10], and these could be adapted to match XML-encoded mathematical expressions. An alternative for matching based on structural similarity is to express a query in the form of a template, specifying precisely where variability is

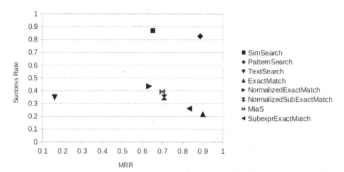

Fig. 1. Mean reciprocal rank versus success rate of each algorithm for Forum queries

permitted and where exact matching is required [7]. This *PatternSearch* approach requires more effort and skill on the part of the user who formulates queries.

Elsewhere [8] we have compared these various approaches in terms of their ability to retrieve documents that contain mathematical expressions that match a query. Some of the results are summarized in Fig. 1, which is explained in more detail in Sect. 7. For the present, we merely observe that similarity search (Sim-Search) and PatternSearch outperform the other approaches by a wide margin in terms of accuracy, and forming queries with SimSearch is much easier than with PatternSearch. The goal of this paper is to demonstrate that similarity search can also be sufficiently fast to be used in practice.

4   Structural Similarity Search

Text with XML markup such as Presentation MathML can be naturally expressed as ordered labelled trees, also called Document Object Model (DOM) trees. A DOM tree T is represented by $T = (V, E)$, where V represents the set of nodes and E represents the set of edges of T. A label $\lambda(n)$ is assigned to each node $n \in V$. In this paper we refer to a math expression and its corresponding DOM tree interchangeably.

We define similarity in terms of "tree edit distance" as follows. Consider two ordered labelled trees $T_1 = (V_1, E_1)$ and $T_2 = (V_2, E_2)$ and two nodes $N_1 \in V_1 \cup \{P_\phi\}$ and $N_2 \in V_2 \cup \{P_\phi\}$ where P_ϕ is a special node with label ϵ. An *edit operation* is a function represented by $N_1 \to N_2$ where N_1 and N_2 are not both P_ϕ. The edit operation is a deletion if N_2 is P_ϕ, it is an insertion if N_1 is P_ϕ, and a rename if N_1 and N_2 do not have the same labels. (Deleting a node N replaces the subtree rooted at N by the immediate subtrees of node N; insertion is the inverse of deletion.) A *cost* represented by the function ω is associated with every edit operation. For example, ω might reflect the design goal that renaming a variable is less costly than renaming a math operator. For ease of explanation, however, we will assume that the costs of all delete and insert operations are 1 and the cost of rename is 2. A *transformation* from T_1 to T_2 is a sequence of edit operations that transforms T_1 to T_2. The cost of a transformation is the sum of the costs of its edit operations. The *edit distance* between T_1 and T_2 is the minimum cost of all possible transformations from T_1 to T_2.

A *forest* is an ordered sequence of trees. For example deleting the root of a tree results in a forest that consists of its immediate subtrees. Note that a single tree and the empty sequence of trees are also forests. With these definitions, the following recursive formula can be used to calculate edit distance [22]:

$$
\begin{aligned}
dist(F_1, F_2) &= min \begin{cases} dist(F_1 - u, F_2) + \omega(u \to \epsilon), \\ dist(F_1, F_2 - v) + \omega(\epsilon \to v), \\ dist(F_1 - T_u, F_2 - T_v) + dist(T_u, T_v) \end{cases} \\
dist(T_u, T_v) &= min \begin{cases} dist(T_u - u, T_v) + \omega(u \to \epsilon), \\ dist(T_u, T_v - v) + \omega(\epsilon \to v), \\ dist(T_u - u, T_v - v) + \omega(u \to v) \end{cases}
\end{aligned}
\tag{1}
$$

where F_1 and F_2 are two non-empty forests such that either F_1 or F_2 contains at least two trees, T_u and T_v are the first (leftmost) trees in F_1 and F_2 respectively, u and v are the roots of T_u and T_v respectively, and $F - n$ represents the forest produced by deleting root n from the leftmost tree in forest F. The edit distance between a forest F and the empty forest is the cost of iteratively deleting (inserting) all the nodes in F. This formulation implies that a dynamic programming algorithm can efficiently find the edit distance between two trees T_1 and T_2 by building a distance matrix.

We calculate the structural similarity of two mathematical expressions E_1 and E_2 represented by trees T_1 and T_2 as follows:

$$sim(E_1, E_2) = 1 - \frac{dist(T_1, T_2)}{|T_1| + |T_2|} \qquad (2)$$

where $|T|$ is the number of nodes in tree T.

Assume document d contains mathematical expressions $E_1 \ldots E_n$. The rank of d for a query Q is calculated as the maximum similarity of expressions in d:

$$docRank(d, Q) = \max_{E_i \in d} sim(E_i, Q) \qquad (3)$$

As described, a search algorithm based on the structural similarity of math expressions would be time consuming because it requires calculating the edit distances of many pairs of trees, which is computationally expensive. A naive approach is to calculate the similarity score of every document and return the top k documents as the search result. However, this naive approach performs some unnecessary computations and can be optimized as follows:

1. Calculating the similarity of the query and an expression requires finding the edit distance between their corresponding DOM trees which is computationally expensive. On the other hand, it is not necessary to calculate the similarity of expressions that can be quickly seen to be too far from the query.
2. Many expressions are repeated in a collection of math expressions, and many share large overlapping sub-expressions. Hence, memoizing some partial results and reusing them saves us from repeatedly recalculating scores.

The next two sections address these observations.

5   Early Termination

In this section we propose a top-k selection algorithm that reduces query processing time by avoiding some unnecessary computations. More specifically, we define an upper limit on the similarity of two mathematical expressions that can be calculated efficiently, and we define a stopping condition with respect to this upper limit.

For a tree T, we designate the set of labels in T as $\tau(T) = \{\lambda(N)|N \in T\}$. For two trees, T_1 and T_2, we define τ-difference and τ-intersection as follows:

$$(T_1 -_\tau T_2) = \{N \in T_1|\lambda(N) \notin \tau(T_2)\} \qquad (4)$$

$$T_1 \cap_\tau T_2 = (\{N|N \in T_1\} - (T_1 -_\tau T_2)) \cup (\{N|N \in T_2\} - (T_2 -_\tau T_1)) \qquad (5)$$

Note that both τ-difference and τ-intersection are defined over sets of nodes, not sets of labels. As a result,

$$|T_1 \cap_\tau T_2| = |T_1| - |T_1 -_\tau T_2| + |T_2| - |T_2 -_\tau T_1| \qquad (6)$$

Consider expression E and query Q. We first calculate an upper bound on the value of $sim(E, Q)$. If the label of a node N in T_E, the DOM tree of E, does not appear in T_Q, the DOM tree of Q, their edit distance is at least equal to $1 + dist(T_E - N, T_Q)$ where $T_E - N$ is the tree that results from deleting N from T_E. A similar argument can be made for nodes in T_Q whose labels do not appear in T_E. Hence, the following lower bound on the edit distance of E and Q can be defined: $dist(T_E, T_Q) \geq |T_E -_\tau T_Q| + |T_Q -_\tau T_E|$ from which an upper bound on the similarity of the two expressions is calculated using (2) and (6):

$$sim(E, Q) \leq 1 - \frac{|T_E -_\tau T_Q| + |T_Q -_\tau T_E|}{|T_E| + |T_Q|} = \frac{|T_E \cap_\tau T_Q|}{|T_E| + |T_Q|} \qquad (7)$$

and the upper bound for the relevance of a document d to Q is calculated using (3):

$$docRank(d, Q) \leq upperRank(d, Q) = \max_{E_i \in d} \frac{|T_{E_i} \cap_\tau T_Q|}{|T_{E_i}| + |T_Q|} \qquad (8)$$

We employ a keyword search algorithm to calculate $upperRank(d, Q)$ as follows. We build an inverted index on node labels, treating each expression as a bag of words. A document is a collection of such expressions (bags of words). In general the keyword search algorithm can be modified by assigning custom weights to terms to handle arbitrary edit costs.

To find the most relevant expressions, we maintain a priority queue of length k ("the top-k list"), as presented in Algorithm 1. This algorithm produces the same results as the naive algorithm, but it reduces the query processing time by avoiding some unnecessary computations. In Sect. 7 we show that this optimization significantly reduces the query processing time.

6   Compact Index and Distance Cache

In this section we propose an indexing algorithm that i) reduces the space requirement and ii) speeds up the query processing. Our indexing algorithm is based on the observation that often many subexpressions appear repeatedly in a collection of math expressions.

Consider a collection of trees $C = \{T_1, \ldots, T_n\}$. Let $G \in_{sub} C$ denote that G is a subtree of T_i for some $T_i \in C$. The total number of subtree instances in C is equal to $|T_1| + \cdots + |T_n|$. If two subtrees G_1 and G_2 represent equivalent subexpressions, we write $G_1 \sim G_2$. This relation partitions $\{G|G \in_{sub} C\}$ into equivalence classes. Given an arbitrary tree T, its *frequency in C* is the size of the matching equivalence class in C:

$$freq(T, C) = |\{G|G \in_{sub} C \wedge G \sim T\}| \qquad (9)$$

Algorithm 1. Similarity Search with Early Termination

1: **Input:** Query Q and collection D of documents.
2: **Output:** A ranked list of top k documents.
3: Treat Q as a bag of words and perform a keyword search to rank documents with respect to $upperRank(d, Q)$.
4: Define a cursor C pointing to the top of the ranked result.
5: Define an empty priority queue $TopK$.
6: **while** true **do**
7:     $d_C \leftarrow$ the document referenced by C.
8:     **if** d_C is null **or** $upperRank(d_C, Q) < \min\limits_{d \in TopK} docRank(d, Q)$ **and** $|TopK| = k$ **then**

9:         break
10:     **end if**
11:     Calculate $docRank(d_C, Q)$.
12:     **if** $|Topk| < k$ **or** $docRank(d_C, Q) > \min\limits_{d \in TopK} docRank(d, Q)$ **then**

13:         Insert d_C in $TopK$.
14:         **if** $|TopK| > k$ **then**
15:             Remove document with smallest score from $TopK$.
16:         **end if**
17:     **end if**
18:     $C \leftarrow C.next$
19: **end while**
20: **return** $TopK$

We omit the second argument C when it is clear from context.

Given a collection of math expressions, we observe that many subtrees appear repeatedly in various expressions' DOM trees. To confirm this, we ran experiments on a collection of more than 863,000 math expressions. Details of this collection are presented in Sect. 7.1, and the experimental confirmation is included in Sect. 7.3.

The basis of our indexing algorithm is to store each subexpression once only and to allow matching subtrees to point to them. This significantly decreases the size of the index, and as we will explain later, it also effectively speeds up the retrieval algorithm. The approach can also be combined with other optimization techniques, such as the one proposed in Sect. 5, to further decrease query processing time.

We assign a signature to each subtree such that matching subtrees have the same signatures and subtrees that do not match the same expression have different signatures. Any hash function that calculates a long bit pattern from the structure and node labels and any collision resolution method can be used for this purpose.

Our index is a table, indexed by signatures, whose entries represent unique MathML subtrees (both complete trees and proper subtrees). Each entry contains the label of the root and a list of pointers to table entries corresponding to the list of the children of the root. A data structure called *exp-info* is assigned to each expression that represents a complete tree in order to store information about documents that contain it. Each entry also contains some other information, such as the frequency of the corresponding tree in the collection.

Initially, the index is empty. We add expression trees one by one to the index. To add a tree T we first calculate its signature to index into the table. If there is a match, we return a pointer to the corresponding entry in the table. We also update the *exp-info* of T if it is a complete tree. If T is not found, we add a new entry to the table for that index, storing information such as the root's label, etc. Then, we recursively insert subtrees that correspond to the children of the root of T in the index, and insert a list of the pointers to their corresponding entries in the entry of T. This algorithm guarantees that each tree is inserted once only, even if it repeats. Figure 2 shows a fragment of the index after $\frac{x^2-1}{x^2+1}$ is added.

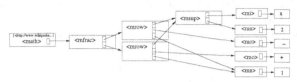

Fig. 2. The index after $\frac{x^2-1}{x^2+1}$ is added

Calculating the edit distance between two trees involves calculating the edit distance between many of their corresponding subtrees. Dynamic programming ensures that each pair of subtrees is compared no more than once within a single invocation of $sim(E_i, Q)$, but building the distance matrix involves calculating the similarity between each pair of subtrees, one from E_i and one from Q. As noted in the previous section, many subexpressions are shared among the mathematical expressions found in a typical document collection; building the distance matrix to compute the similarity of a query to each stored expression independently does not capitalize on earlier computations. We can reduce computation time significantly by memoizing some intermediate results for later reuse.

When calculating the edit distance between two trees, we store the result in an auxiliary data structure that we call a *distance cache*. More specifically, the cache stores triples of the form $[T_e, T_q, dist(T_e, T_q)]$ where T_e is a subtree of the expression, T_q is a subtree of the query, and $dist(T_e, T_q)$ is the edit distance between T_e and T_q. Effectively we are saving the distances computed by the dynamic programming algorithm (1) across similarity calls.

We implement the cache as a hash table where the key consists of the two signatures for T_e and T_q. Hence, the complexity of inserting and searching for a triple is $O(1)$. If D represents the set of all document-level expressions whose distances to Q are calculated through invocations to *docRank* in Algorithm 1, $S = \{G|G \in_{sub} D\}$, and n is the number of equivalence classes in S, the space required to store the distance cache is $O(n|Q|)$.

Each time we require the edit distance between two trees, we use the value in the cache if it is there. Otherwise we calculate the distance and store the result together with the signatures of the two subtrees in the cache.

If the available memory is limited or there are too many expressions, we may not be able to store all pairs of distances as just described. However, calculating the edit distance between small trees may be sufficiently fast that there is no benefit gained by using the cache, and storing such pairs significantly increases the size of the cache. Furthermore, storing the results for rare subtrees may not be worthwhile, as the stored results may not be reused often enough to realize the benefit of using the cache.

The benefit of memoizing the edit distance between two trees comes from the savings in processing time if the result is found in the cache instead of being calculated for the distance matrix. Following this line of reasoning, we augment the caching criteria described above to choose which distances should be stored and which should not. We calculate the benefit of storing the triple $[T_e, T_q, dist(T_e, T_q)]$ as $benefit(T_e, T_q) = calcCost(T_e, T_q) - cacheCost(T_e, T_q)$, where $calcCost(T_e, T_q)$ and $cacheCost(T_e, T_q)$ are the costs of calculating the edit distance and looking up a value in the cache respectively. We also wish to account for the number of times we will be able to realize the savings by reusing the value from the cache. Therefore, to each pair (T_e, T_q), we assign a weight $weight(T_e, T_q)$ that reflects the frequency of occurrence of that pair. We suggest how to compute the weights below.

Consider a set of tree pairs $P = \{(T_e^1, T_q^1), \ldots, (T_e^n, T_q^n)\}$ and a space constraint that allows $\mathcal{C}$ triples to be cached. Our task is then to select a set of subtree pairs $\mathcal{H}^* = \arg\max_{\mathcal{H}} \sum_{(T_e^i, T_q^i) \in \mathcal{H}} weight(T_e^i, T_q^i)\, benefit(T_e^i, T_q^i)$ such that $|\mathcal{H}| \leq \mathcal{C}$.

If we are given the set P, the problem is easily solved by choosing the $\mathcal{C}$ triples having the highest values for $weight(T_e^i, T_q^i)\, benefit(T_e^i, T_q^i)$. However, Algorithm 1 maintains a sorted list of expressions, and starting from the head of the list calculates the similarity of each expression to Q. Thus, we cannot predict exactly which pair of subtrees will be compared before the algorithm stops.

We need to assign the weight for a pair of subtrees that reflects the number of times that pair will be needed for filling a dynamic programming matrix during the remainder of the execution of Algorithm 1. Consider the following motivating example:

Example 1. Assume $freq(T_e, D) = 100$, and $freq(T_q, \{T_Q\}) = 1$. The similarity between the expressions represented by T_e and T_q will be calculated at most 100 times by Algorithm 1. While processing the query, if the edit distance function has already been called to fill 99 distance matrices for this pair, it will be called at most once more for the rest of the query processing. Caching the edit distance between T_e and T_q at this point is not likely to be as cost-effective as caching the distance for another pair of trees if those trees might still be compared 10 more times during query processing.

We want to assign a weight to each pair that reflects this declining benefit. However, we cannot afford to store frequencies for every pair of subtrees (otherwise we could store the distances instead). Therefore, we estimate the frequencies based on the frequencies for each subtree independently.

Note that T_e matches $freq(T_e, D)$ subtrees of the expressions in the collection and requires up to $|T_Q|$ entries to be made in the distance matrix during dynamic programming. We augment the index described above by adding fields $freq_D$ and $freq_{cur}$ to each node to store the frequency of that subexpression in the document collection together with a variant of that frequency, both initialized to be equal to $freq(T_e, D)$ for the node corresponding to T_e. Whenever we require a value for $dist(T_e, T_q)$, we calculate its score as the weighted benefit based on expected re-use as $score(T_e, T_q) = freq_{cur}(T_e)\, freq(T_q, \{T_Q\})\, benefit(T_e, T_q)$

where $freq(T_q, \{T_Q\})$ is the number of subtrees in the DOM tree of Q that match T_q. We also save the score in the cache along with the distance, and update $freq_{cur}(T_e)$ with the value $freq_{cur}(T_e) - \frac{1}{|T_Q|}$ to reflect the maximum number of times T_e might still be required in a distance computation.

Algorithm 2 details how the scores for each pair of trees is calculated and used to manage a limited cache. A priority queue maintains the most promising $\mathcal{M}$ pairs in the cache as similarity search progresses. Thus the cache stores quadruples $[s_e, s_q, dist(T_e, T_q), score(T_e, T_q)]$ where s_e and s_q are the signatures for T_e and T_q respectively. Because $score(x, y)$ increases monotonically with $freq_{cur}(x)$ and $freq(y)$ and because trees cannot repeat more frequently than any of their subtrees, if $dist(T_e, T_q)$ is stored in the cache for some subtree T_e stored in the document collection and some subtree T_q of the query, then $dist(T'_e, T'_q)$ is also stored for all $T'_e \in_{sub} T_e$ and $T'_q \in_{sub} T_q$, as long as $benefit(T'_e, T'_q)$ is sufficiently high.

Algorithm 2. Calculating Edit Distance with a Limited Cache

Input: Two trees T_e and T_q, $|T_Q|$ (the number of nodes in the query tree), and cache $\mathcal{M}$ storing quadruples.
Output: $dist(T_e, T_q)$ (with side-effects on $\mathcal{M}$ and $freq_{cur}(T_e)$)
Form pair $p = (s_e, s_q)$ that consists of the signatures of T_e and T_q.
$freq_{cur}(T_e) \leftarrow freq_{cur}(T_e) - \frac{1}{|Q|}$.
$v \leftarrow freq_{cur}(T_e) * freq(T_q) * benefit(T_e, T_q)$ (the score for this pair).
if p is found in $\mathcal{M}$ **then**
    $dist \leftarrow dist(T_e, T_q)$ associated with p in $\mathcal{M}$
    Replace the matched quadruple in $\mathcal{M}$ by $(s_e, s_q, dist, v)$.
else
    $dist \leftarrow$ computed $dist(T_e, T_q)$ using the distance matrix and cache for subproblems.
    $m \leftarrow min\{score(m)|m \in \mathcal{M}\}$
    if $m < v$ **then**
        if $|\mathcal{M}| = \mathcal{C}$ **then**
            Remove the entry with minimum score from $\mathcal{M}$.
        end if
        Insert $(s_e, s_q, dist, v)$ into $\mathcal{M}$.
    end if
end if
return $dist$

In the next section we show that the proposed optimization techniques significantly reduce the query processing time in practice.

7   Experiments

In this section we investigate the performance of the proposed algorithms.

7.1   Experiment Setup

Data Collection. For our experiments we use a collection of web pages with mathematical content. We collected pages from the Wikipedia and DLMF (Digital Library of Mathematics Functions) websites. Wikipedia pages contain images of expressions annotated with equivalent LATEX encodings of the expressions. We extracted the annotations and translated them into Presentation MathML using Tralics [4]. DLMF pages use Presentation MathML to represent mathematical expressions. Statistics summarizing this dataset are presented in Table 1A.

Table 1. Experimental dataset and query statistics

A	Wikipedia	DLMF	Combined
Num. pages	44,368	1,550	45,918
Num. exprs.	611,210	252,148	863,358
Avg. expr. size	28.3	17.6	25.2
Max. expr. size	578	223	578

B	Interview	Forum	Combined
Num. queries	45	53	98
Avg. query size	14.2	23.8	19.4

Query Collection. To evaluate the described algorithms we prepared two sets of queries as follows.

- *Interview:* We invited a wide range of students and researchers to participate in our study. They were asked to try our system and search for mathematical expressions of potential interest to them in practical situations. They could also provide us with their feedback about the quality of results after each search.
- *Mathematics forum:* People often use mathematics forums in order to ask a questions or discuss math-related topics. Many threads start by a user asking a question in the form of a single mathematics expression. Usually, by reading the rest of a thread and responses, the exact intention of the user is clear. This allows us to manually judge if a given expression, together with the page that contains it, can answer the information need of the user who started the thread. We manually read such discussions and gathered a collection of queries.

We only consider queries with at least one match in our dataset. Table 1B summarizes statistics about the queries, where the number of nodes in the query tree is used to represent query size.

Evaluation Measures. We evaluate the proposed algorithms using the following measures:

MRR: The rank of the first correct answer is a representative metric for the success of a mathematics search. Hence, for each search we consider the *Reciprocal Rank* (RR), that is, the inverse of the rank of the first relevant answer. The *Mean Reciprocal Rank* (MRR) is the average reciprocal rank for all queries:

$$MRR = \frac{1}{|Q|} \sum_{q \in Q} \frac{1}{\mathcal{R}(q)} \qquad (10)$$

where Q is the collection of queries, and $\mathcal{R}(q)$ is the rank of the first relevant answer for query q.

Success-Rate: If at least one of the top 20 search results returned by an algorithm for a given query is relevant, we classify the search as successful. Alternatively, if the first relevant answer is not among the top 20 results, or if no relevant result is returned at all, the search is classified as unsuccessful. The success rate is the number of successful searches divided by the total number of searches:

$$Success\ Rate = \frac{|\{q \in Q | q\ is\ successfully\ searched\}|}{|Q|} \tag{11}$$

Query Processing Time: The time in milliseconds from when a query is submitted until the results are returned. A query is encoded with Presentation MathML and if the user interface allows other formats, the time taken to translate it is ignored. Also the network delay and the time to render results are not included. For a collection of queries, we measure the query processing time of each and report the *average query processing time*.

Alternative Algorithms. We evaluate the described algorithms by comparing their performance against the following alternative algorithms:

- *TextSearch:* The query and expressions are treated as bags of words. A standard text search algorithm is used for ranking expressions according to a given query[1].
- *ExactMatch:* An expression is reported as a search result only if it matches a given query exactly. Results are ranked with respect to the alphabetic order of the name of their corresponding documents.
- *NormalizedExactMatch:* Some normalization is performed on the query and on the stored expressions: in particular, we replace numbers and variables with generic labels N and V, respectively. The normalized expressions are searched and ranked according to the ExactMatch algorithm.
- *SubexprExactMatch:* An expression is returned as a search result if one of its subexpressions exactly matches the query. Results are ranked by increasing sizes of their DOM trees and ties are broken using the alphabetic order of the name of their corresponding documents.
- *NormalizedSubExactMatch:* Normalization is performed on the query and on the stored expressions as for NormalizedExactMatch, and an expression is returned as a search result if one of its normalized subexpressions matches the normalized query..
- *MIaS:* Expressions are matched using the algorithm proposed by Sojka and Liska [17]: An expression is first tokenized, where a token is a subtree of the expression. Each token is next normalized with respect to various rules (e.g. number values are removed, or variables are removed, or both), and multiple normalized copies are preserved. The result, which is a collection of tokens, is indexed with a text search engine. Each query is similarly normalized (but not tokenized) and then matched against the index.

---

[1] We used Apache Lucene [1] in our implementation.

- *PatternSearch:* Expressions are matched against a query template as described by Kamali and Tompa [7]. Like SubexprExactMatch, results are ranked with respect to the sizes of their DOM trees.
- *SimSearch:* Expressions are matched against a query according to the algorithm described in Sect. 4.

We further refine SimSearch to cover the following algorithms that reflect the proposed optimization techniques:

- *Unoptimized:* Each expression is stored independently. The relevance score is calculated for any expression sharing at least one tag with the query.
- *ET:* The early termination algorithm described in Sect. 5. Each expression is stored independently. As described, an inverted index is used to calculate upper bounds on the scores of each document, which increases the index size.
- *Compact:* Similar to unoptimized a query is processed by comparing the relevance of each document that contains an expression with at least one node whose tag appears in the query. Each subtree is stored once only to reduce the index size as described in Sect. 6.
- *Compact-ET-NMC:* The early termination algorithm with a compact index, and no memory constraint as described in Sect. 6.
- *Compact-ET-MC:* The early termination algorithm with a compact index and a constraint on the memory that is available during the query processing (Sect. 6). The results are presented for specific amounts of available memory separately (e.g. if the memory constraint allows storing 1000 cache entries, we use the label Compact-ET-MC-1000). We consider three values for the memory constraint: 5000, 10000, and 50000 entries.
- *Compact-ET-RandMC:* Similar to Compact-ET-MC, but entries are chosen at random for being assigned space in the cache.

7.2   Correctness

Fuller descriptions of the algorithms and correctness results for the experiments are reported elsewhere [8]. For completeness, we summarize the correctness results here.

The success rate against MRR for each algorithm is plotted in Fig. 1 for the Forum queries and in Fig. 3 for the Interview queries. As both figures show, PatternSearch and SimSearch have high success rates and also high MRRs. Interestingly, PatternSearch has a higher MRR because irrelevant expressions are less likely to match a carefully formed pattern, whereas SimSeach has a slightly higher success rate because in some cases even an experienced user may not be able to guess the pattern that will yield a correct answer.

In summary, SimSearch and PatternSearch perform much better than the other approaches in terms of the correctness of results. However, because forming queries for SimSearch is easier, it is generally preferred over PatternSearch.

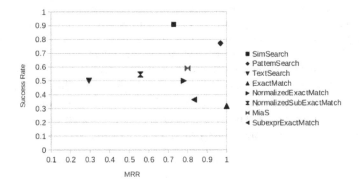

Fig. 3. MRR versus success rate of each algorithm for Interview queries

7.3 Index Size

The average number of repetitions of subtrees with sizes in specific ranges is listed in Table 2A. The average repetitions of trees whose sizes are in the range of $[1 - k]$ for various values of k is shown as a graph. As the results suggest, most subtrees repeat at least a few times. Not surprisingly, for smaller subtrees the rate of repetition is higher.

Next, we compare the compact index to an index that stores each expression independently. For our experiments, an expression's signature is computed by a conventional hash function applied to its XML string S: $S[0] * 31^{(z-1)} + S[1] * 31^{(z-2)} + \cdots + S[z-1]$ where $S[i]$ is the ith character in S and $z = |S|$. As shown in Table 2C, the size of the compact index (in terms of the number of nodes stored) is significantly smaller than that of the regular index.

7.4 Query Processing Time

Figure 4A shows that the early termination algorithm significantly reduces the query processing time — by a factor of 44. Using the compact index and memoizing partial results also reduces the query processing time by an additional

Table 2. Subtree repetitions in experimental dataset and resulting index sizes

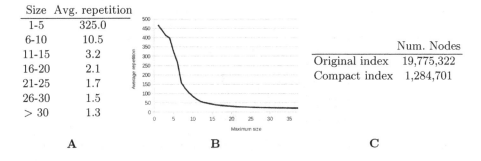

Size	Avg. repetition
1-5	325.0
6-10	10.5
11-15	3.2
16-20	2.1
21-25	1.7
26-30	1.5
> 30	1.3

	Num. Nodes
Original index	19,775,322
Compact index	1,284,701

A   B   C

factor of 1.5, to about .8 seconds per query on average. (Note that accuracy is not affected by employing any of the optimization techniques.) Figure 4B compares the proposed approach against alternative approaches. The alternative algorithms use straightforward text search or database lookup algorithms, which result in query processing times that are two to four times faster, but at the expense of very poor accuracy. To date, these approaches have been preferred to a more elaborate similarity search, largely because the latter was deemed to be too slow to be practical. However, *Compact-ET-NMC*, which applies both early termination and memoization, has practical processing speeds and far better accuracy.

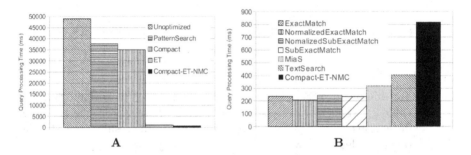

Fig. 4. The query processing time of alternative algorithms

The effect of the available memory on the query processing time is investigated in Fig. 5. For higher values of the space budget, the query processing time is very similar to that of *Compact-ET-NMC*, which assumes there is no constraint on the available memory. Even for smaller values of the constraint (e.g. when we can memoize at most 5,000 intermediate results), there is a notable improvement over the performance of *ET*.

The figure also compares the performance when the available space is managed with respect to the described algorithm and when distances for pairs of trees are chosen to be cached at random. For a small space budget, caching randomly chosen pairs has little advantage over the ET algorithm, which does not use a cache. For greater values of the space budget the performance is improved compared to ET, but not as much as when caching is applied more strategically.

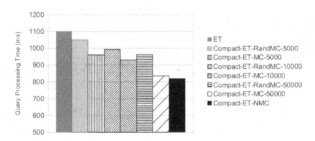

Fig. 5. The query processing time for various space budgets and cache strategies

For example, the performance of *Compact-ET-MC-50000* is very close to that of *Compact-ET-NMC*, which assumes unlimited memory is available, and *Compact-ET-MC-5000* performs similarly fast as *Compact-ET-RandMC-50000* while using only a tenth of the space budget. This validates the proposed method for choosing which pairs to cache.

8 Conclusion

Mathematics retrieval is still in an early stage of development. We have shown that in order to correctly capture the relevance of math expressions, their structures must be considered. Tree edit distance, which is a standard technique to compare structures, is computationally expensive, but optimization techniques can reduce query processing time significantly. Through extensive experiments, we showed that our algorithm significantly outperforms baseline algorithms in terms of the accuracy of results while performing comparably in terms of query processing time even when memory is limited. Additional improvements should still be explored, however, to close the remaining performance gap.

References

1. lucene.apache.org
2. Bancerek, G.: Information retrieval and rendering with MML Query. In: Borwein, J.M., Farmer, W.M. (eds.) MKM 2006. LNCS (LNAI), vol. 4108, pp. 266–279. Springer, Heidelberg (2006)
3. Einwohner, T.H., Fateman, R.J.: Searching techniques for integral tables. In: ISSAC, pp. 133–139 (1995)
4. Grimm, J.: Tralics, A LaTeX to *XML* Translator. INRIA (2008)
5. Guidi, F., Schena, I.: A query language for a metadata framework about mathematical resources. In: Asperti, A., Buchberger, B., Davenport, J.H. (eds.) MKM 2003. LNCS, vol. 2594, pp. 105–118. Springer, Heidelberg (2003)
6. Kamali, S., Tompa, F.W.: Improving mathematics retrieval. In: DML, pp. 37–48 (2009)
7. Kamali, S., Tompa, F.W.: A new mathematics retrieval system. In: CIKM, pp. 1413–1416 (2010)
8. Kamali, S., Tompa, F.W.: Retrieving documents with mathematical content. In: SIGIR (2013)
9. Kohlhase, M., Sucan, I.: A search engine for mathematical formulae. In: Calmet, J., Ida, T., Wang, D. (eds.) AISC 2006. LNCS (LNAI), vol. 4120, pp. 241–253. Springer, Heidelberg (2006)
10. Laitang, C., Boughanem, M., Pinel-Sauvagnat, K.: XML information retrieval through tree edit distance and structural summaries. In: Salem, M.V.M., Shaalan, K., Oroumchian, F., Shakery, A., Khelalfa, H. (eds.) AIRS 2011. LNCS, vol. 7097, pp. 73–83. Springer, Heidelberg (2011)
11. Maclean, S., Labahn, G.: A new approach for recognizing handwritten mathematics using relational grammars and fuzzy sets. In: IJDAR, pp. 1–25 (2012)

12. Mišutka, J., Galamboš, L.: System description: Egomath2 as a tool for mathematical searching on wikipedia.org. In: Davenport, J.H., Farmer, W.M., Urban, J., Rabe, F. (eds.) Calculemus/MKM 2011. LNCS, vol. 6824, pp. 307–309. Springer, Heidelberg (2011)

13. Munavalli, R., Miner, R.: Mathfind: a math-aware search engine. In: SIGIR, pp. 735–735 (2006)

14. Nguyen, T.T., Chang, K., Hui, S.C.: A math-aware search engine for math question answering system. In: CIKM, pp. 724–733 (2012)

15. Schellenberg, T., Yuan, B., Zanibbi, R.: Layout-based substitution tree indexing and retrieval for mathematical expressions. In: DRR (2012)

16. Smirnova, E.S., Watt, S.M.: Communicating mathematics via pen-based interfaces. In: SYNASC, pp. 9–18 (2008)

17. Sojka, P., Líska, M.: The art of mathematics retrieval. In: ACM Symposium on Document Engineering, pp. 57–60 (2011)

18. Youssef, A.: Search of mathematical contents: Issues and methods. In: IASSE, pp. 100–105 (2005)

19. Youssef, A.S.: Methods of relevance ranking and hit-content generation in math search. In: Kauers, M., Kerber, M., Miner, R., Windsteiger, W. (eds.) MKM/CALCULEMUS 2007. LNCS (LNAI), vol. 4573, pp. 393–406. Springer, Heidelberg (2007)

20. Zanibbi, R., Yu, L.: Math spotting: Retrieving math in technical documents using handwritten query images. In: ICDAR, pp. 446–451 (2011)

21. Zanibbi, R., Yuan, B.: Keyword and image-based retrieval of mathematical expressions. In: DRR, pp. 1–10 (2011)

22. Zhang, K., Shasha, D.: Simple fast algorithms for the editing distance between trees and related problems. SIAM J. Comput. 18(6), 1245–1262 (1989)

Towards Machine-Actionable Modules
of a Digital Mathematics Library
The Example of DML-CZ

Michal Růžička[1,2], Petr Sojka[1], and Vlastimil Krejčíř[1,2]

[1] Masaryk University, Faculty of Informatics
Botanická 68a, 602 00 Brno, Czech Republic
`mruzicka@mail.muni.cz`, `sojka@fi.muni.cz`
[2] Masaryk University, Institute of Computer Science
Botanická 68a, 602 00 Brno, Czech Republic
`krejcir@ics.muni.cz`

Abstract. Publishing and archiving mathematical literature presents its own sets of problems. Reaching the goal of building global digital mathematics library (DML), smaller DMLs play an inevitable role in collecting, validating, digitizing and checking data from smaller publishers.

In this paper, we overview the technical challenges of building a machine-actionable set of modules we have developed over almost a decade of evolution of the Czech Digital Mathematics Library (DML-CZ). Firstly, we survey methods of effective automated data acquisition from the content providers. Then we show OCR processing of mathematical documents and automated segmentation of plain text references for metadata enhancement and effective DOI look up. Finally we describe connection to the European Digital Mathematics Library (EuDML) project and public interfaces of DML-CZ for the best visibility and accessibility.

Keywords: DML-CZ, EuDML, DOI, ParsCit, references, validation, DSpace, OAI-PMH, TeX, LaTeX, Tralics, Infty, machine-actionable digital library, library automation, Google Scholar, webometrics.

1   Introduction

Publishing and archiving mathematical literature is a unique and challenging task in many respects. It revolves around handling of mathematical formulae in papers, dealing with the number and diversity of math publishers' size and approaches, as well as the existence of reference databases as Mathematical Reviews and Zentralblatt Math community services. There are specific citation patterns across a great diversity of topic areas throughout their long evolution. Handling all these specifics in a local digital mathematical library (DML) possesses variety of challenges.

One possible approach to run and sustain a project of a small digital library is to try to automate as many processes as possible, while still maintaining high-quality, checked data in the repository. The running costs of such a system can

J. Carette et al. (Eds.): CICM 2013, LNAI 7961, pp. 263–277, 2013.

be too high to allow it to survive in the digital publishing ecosystem unless machine-actionable modules are developed for a DML.

This paper provides an overview of the technical challenges we have coped with in the design, development and adoption of technologies and technical solutions during almost a decade of evolution of the Czech Digital Mathematics Library (DML-CZ, http://dml.cz/).

The structure of the paper is as follows: The following section reports on the interfaces and formats we have settled on in the DML-CZ with our data providers — a machine-actionable input module that validates the data from them. In Section 3 we discuss the tools and the conversions we perform on the data collected from data providers: getting the full text including math formulae, enhancements of bibliographies, reference and DOI matching modules. Section 4 deals with the interfaces we use to export the enhanced and checked data to the wider public: it is available to EuDML, Google Scholar and review databases.

The schema of the modules described in this paper can be seen in Figure 1. Ellipses refers to external entities. The automatic modules, depicted as rectangles, are integrated in our two main subsystems, Metadata Editor and DSpace, shown here as circles.

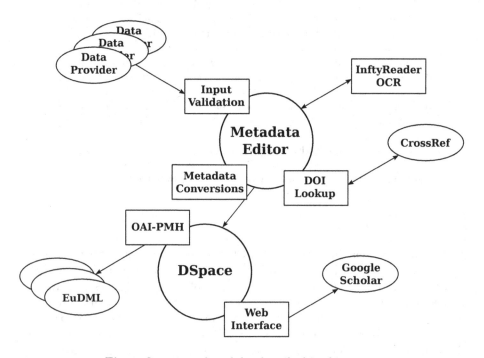

Fig. 1. Overview of modules described in this paper

2   Inputs

The aim of the DML-CZ project is the digital preservation of the content of the bulk of the mathematical literature that has ever been published in the Czech lands. Since the start of the project several years ago, most of the old publications have been processed. With fewer papers yes to be retro-digitized, it is increasingly important to cooperate with editors of the active Czech mathematical journals on the continuous inclusion of new publications. Journal papers are the core contents that are regularly added to the library. DML-CZ holds ten journals that constitute the content, which amounts to several hundreds new papers per year, on a regular basis:

1. Acta Universitatis Carolinae. Mathematica et Physica
2. Acta Universitatis Palackianae Olomucensis. Facultas Rerum Naturalium. Mathematica
3. Applications of Mathematics
4. Archivum Mathematicum
5. Commentationes Mathematicae Universitatis Carolinae
6. Communications in Mathematics
7. Czechoslovak Mathematical Journal
8. Kybernetika
9. Mathematica Bohemica
10. Pokroky matematiky, fyziky a astronomie

For the long term sustainability of the DML-CZ project it was vital to reduce the costly manual labour in the routine processing of new publications while maintaining the good quality of metadata. To achieve this, we cooperate closely with the publishers who prepare the DML-CZ data as an integral part of their publishing process. Data providers prepare the DML-CZ data according to the DML-CZ input format specification. This simple data format is related to the internal data format of the DML-CZ Metadata Editor tool (see Section 3) and consists of these parts:

1. XML metadata file describing the publication (title, authors, abstract, keywords, ...),
2. XML metadata file containing a semantically marked up list of references of the paper, i.e. each of the references has properly marked author names, title, year of publication etc.,
3. full text of the paper in the PDF format, and,
4. optional but highly recommended bunch of source files[1] suitable for generation of the paper.

An XML metadata file containing a semantically marked-up list of references is important for the proper presentation of metadata via the web interface of the

---

[1] Being a *mathematical* digital library, the DML-CZ content providers use almost exclusively the TeX typographic system for the preparation of their publications.

digital library and especially for further internal processing (e.g. DOI lookup, see Section 3.2) and for use by third parties (see Section 4.2).

The inclusion of sources in the data package is optional, but we strongly recommend it. The availability of the original source codes enables us to instantly correct some sorts of errors that are occasionally found in the metadata provided. One example of such corrections is the substitution of the authors' custom TeX macros for their LaTeX equivalents in the metadata. We support fixed set of macros, including the ones used in LaTeX packages developed by the $\mathcal{AMS}$.

There are three basic options for DML-CZ data providers to prepare the data for DML:

1. Develop and use their own tools for DML-CZ data generation,
2. adopt a 'complex' DML-CZ LaTeX-based processing system,
3. integrate a 'lightweight' TeX extension for DML-CZ.

These options were documented in our earlier publications. [Růž08, RS10, RS11]

No matter which option the data provider uses, the result is a data archive ready to be delivered to the DML-CZ Metadata Editor for further processing and subsequent publication. To be automatically processable, the data archives have to follow the above-mentioned rules. As there is a variety of different ways our publishers prepare the data the *input validation* module of the Metadata Editor checks compliance of the provided data with the DML-CZ requirements.

The validation process includes

- data integrity tests,
- tests of the completeness of the data set,
- the validation of XML metadata (title of the publication and the list of references),
- validity of the LaTeX code included in the metadata.

Mathematical publications collected in the DML-CZ contain a lot of mathematical expressions, and these expressions often appear in the metadata. Thus, DML-CZ allow the use of mathematical expressions in the metadata encoded in the LaTeX notation. The use of LaTeX markup is allowed only for the mathematical statements. The rest of the metadata is plain text without any special markup. Moreover, the set of allowed LaTeX macros is fixed, restricted to the LaTeX and subset of $\mathcal{AMS}$ packages. No new macros may be used in titles, abstracts or the bibliography. Compliance with these restrictions is automatically checked, and the metadata validated. These tests save us from further manual corrections as the DML-CZ workflow requires the automatic conversion of the mathematical expressions for various purposes, e.g. conversion of mathematical formulae to MathML [Aus+10] for indexing, exporting as well as further processing like conversion of MathML formulae to text for better accessibility.

Fatal errors detected by the input validation module — such as invalid XML metadata — prohibits data upload to the DML-CZ system. In addition, the validation module produces a variety of warnings. These cover optional parts of the data package such as source codes and possible errors that it cannot be fully to checked automatically, e.g. absence of the list of references can be an error in the case of regular article but might be completely acceptable for an editorial.

3   Processing

At the beginning of the project, the Metadata Editor [BKŠ08] was developed to enable DML-CZ team to organize a large amount of old scanned publications and label them with metadata descriptions manually. In the later stages of the project, the Metadata Editor was used mostly as an entry point for new born-digital publications provided mainly by the editorial staff of the journals included in the DML-CZ. Incoming documents are then checked and assigned to appropriate collections. The web interface of the Metadata Editor allows quick fixes and provides operators with tools for publishing final version of the documents to the public DML-CZ repository which is available to all readers at `http://dml.cz/`.

Now that new data comes in high quality directly from the publishers, the Metadata Editor is now used mainly as a control interface for the processes and as an interface to run enhancement modules on the data and metadata. During the publishing process

- articles and items of their references are checked against review databases — if a match is found, the identifier is attached to the item and presented as a direct link in the DML-CZ public repository,
- PDF full text is equipped with TEX-typeset cover page containing document metadata and providing users with the persistent DML-CZ link to the publication landing page,
- new PDFs are optimized and re-compressed for faster download and viewing in the browser,
- finally, PDF documents are marked with a digital signature and together with metadata are published in the public DML-CZ repository.

The similarity of documents is periodically recomputed over the DML-CZ content enabling the DML-CZ public repository to provide its users with an indication of the similarity of the given documents across the DML-CZ repository.

3.1   Maths Optical Character Recognition

DML-CZ participates in the European Digital Mathematics Library (EuDML) project (see Section 4). With a large proportion of the old publications scanned the necessity to make accessible versions of DML-CZ documents available to the EuDML have lead us among other reasons[2] to reprocess the DML-CZ content with the optical character recognition software (OCR). Previous version of texts extracted from the scanned images by FineReader did not contain mathematical formulae, which often bear the main semantic message in mathematical literature.

The tool used for OCR processing is InftyReader.[3] [Suz+03] This software incorporates the unique ability to recognize mathematical expressions. InftyReader

---

[2] Such as preparation of data for development and testing of improved mathematical documents similarity computations and maths-aware search engines.

[3] `http://www.inftyproject.org/`

accepts various bitmap image formats on the input (TIFF, BMP, GIF, PNG, PDF) and saves recognized objects in its own XML format. This rich internal representation of the document can be consequently transformed to various formats including LaTeX, XHTML+MathML and other XML formats. This transformation is a challenging task, as conversion ideally goes from presentation to content markup and disambiguation is needed.

Even though the initial Infty internal representation is common for all the different conversion export drivers (to MathML, LaTeX), the drivers seems to be of varying quality. Our main goal was the use of InftyReader generated LaTeX output that could be consequently processed similarly to the LaTeX code contained in the DML-CZ metadata, i.e. converted to the MathML by Tralics [Gri10]. However, InftyReader 2.9.5 generated LaTeX source code proved to contain several types of systematic errors that make its direct use difficult. For example, some math mode commands, such as `\ddagger`, are generated outside the math mode or the math mode itself is occasionally not opened/closed properly (missing a dollar sign). This leads to a substantial amount of subsequent errors during the processing of the rest of the LaTeX code. There is also the use of non-existent commands such as `\napos`, `\uu` instead of `\u{u}`, etc. Thus, use of the InftyReader generated TeX files results in at least one error during their processing by Tralics for more than 60% of the TeX inputs.

On the other hand, the development team of the InftyReader is willing to help us. We managed to correct several kinds of errors tweaking internal configuration tables[4] and other fixes were developed by the InftyReader team. We believe the TeX output will be significantly improved in future releases of the InftyReader.

Luckily, use of InftyReader generated XHTML+MathML seems more reliable with the current version of the transformation module. XHTML+MathML driver outputs less than 5% of invalid output files. Not being directly presented to the users these outputs seem to be good enough for internal use: indexing for similarity search, MathML to text conversion for document similarity computations, document classification and clustering.

Being available for MS Windows only, the InftyReader processing runs on separate server on a virtual machine. Batch processing was further complicated by random crashes of the InftyReader on certain input files. As it required constant monitoring, we used AutoIt software to automate attention handling required for InftyReader; we log all peculiarities of the OCR process to be reviewed only after the whole recognition batch has been processed. Running in four parallel threads on a server with today's standard hardware configuration[5], the content of the DML-CZ with more than 33,000 papers on more than 300,000 pages can be reprocessed by InftyReader-based workflow in approximately two weeks.

3.2 DOI References Parsing

An important part of scientific publications are their lists of references. The usefulness of the digital repository increase if references to other documents

---

[4] These tweaks were then integrated to newer releases of the InftyReader.

[5] Quad-core CPU at 2.8 GHz, 4 GiB RAM.

contains widely used unambiguous persistent identifiers, such as DOI (Digital Object Identifier), linking directly to the target of the ID.

However, not all authors use DOI or other identifiers as part of the reference and the markup used is not uniform. Moreover, DOI can be assigned to a publication arbitrarily long after the document is published, i.e. author can cite the publication long before the DOI was assigned.

Thus, a DOI look up is often the responsibility of the digital library maintainer and it is necessary to periodically look up for the existence of DOIs for documents that do not have assigned this identifier so far.

CrossRef provides tools for DOI lookup.[6] To achieve the best results, high quality markup for references is necessary, i.e. important elements such as authors, title, year of publication, journal name, publisher, pages etc. have to be properly indicated in the data.

Unfortunately, this is not the usual case. New issues of DML-CZ journals with metadata provided directly by the publishers are of reasonable quality according to the detail of markup references. However, even here we are provided with just basic 'authors — title — the rest' segmentation of the reference string by some publishers as the preparation of richly marked up metadata is a time consuming and costly operation.

A large proportion of the DML-CZ contains old, scanned publications with the only available texts obtained from OCR processing. For these papers, semiautomatic basic 'authors — title — the rest' segmentation required a vast investments in time, money and human resources during the development of the project. Even then, the quality of the metadata cannot be guaranteed. Thus, we have a great interest in the automatic segmentation of unstructured reference strings.

Our first attempt was the use of the Perl module Biblio::Citation::Parser.[7] This tool has proved to be too simple to successfully cope with various citation formats that are in common use in the DML-CZ. A very promising solution to this challenging problem seems to be the ParsCit tool.[8] [CGK08, LNK10]

For example, the plain text reference string

```
[5] Lambe, L., Stasheff, J.: Applications of perturbation theory
to iterated fibrations. Manuscripta Math. 58 (1987), 363-376.
```

is segmented by Biblio::Citation::Parser (version 1.10) as follows:

```
<authors>Lambe, L., Stasheff, J.</authors>: <title>Applications
of perturbation theory to iterated fibrations. Manuscripta Math.
58 (1987), 363-376</title>
```

It should be noted that the citation string was written without a line break. If the line break is part of the reference string, the tool fails completely. In contrast, ParsCit (version 110505) segmented the reference string as shown in Figure 2.

---

[6] http://www.crossref.org/guestquery/; http://help.crossref.org/#ID5824

[7] http://search.cpan.org/ mjewell/Biblio-Citation-Parser-1.10/
    lib/Biblio/Citation/Parser.pm

[8] http://wing.comp.nus.edu.sg/parsCit/

```xml
<?xml version="1.0" encoding="UTF-8"?>
<algorithms version="110505">
  <algorithm name="ParsCit" version="110505">
    <citationList>
      <citation valid="true">
        <authors>
          <author>L Lambe</author>
          <author>J Stasheff</author>
        </authors>
        <title>Applications of perturbation theory to iterated
               fibrations.</title>
        <date>1987</date>
        <journal>Manuscripta Math.</journal>
        <volume>58</volume>
        <pages>363--376</pages>
        <marker>[5]</marker>
        <rawString>Lambe, L., Stasheff, J.: Applications of
                   perturbation theory to iterated fibrations.
                   Manuscripta Math. 58 (1987), 363-376.</rawString>
      </citation>
    </citationList>
  </algorithm>
</algorithms>
```

Fig. 2. Reference segmentation done by ParsCit

```xml
<?xml version="1.0" encoding="UTF-8"?>
<references xmlns:str="http://exslt.org/strings">
  ...
  <reference id="5">
    <prefix>[5]</prefix>
    <title>Applications of perturbation theory to iterated
           fibrations</title>
    <authors>
      <author>Lambe, L.</author>
      <author>Stasheff, J.</author>
    </authors>
    <journal>Manuscripta Math.</journal>
    <volume>58</volume>
    <year>1987</year>
    <pages>363-376</pages>
    <suffix>Manuscripta Math. 58 (1987), 363-376.</suffix>
  </reference>
  ...
</references>
```

Fig. 3. Example of the hand made metadata of references from the publisher

The conversion was equally successful regardless of how many line breaks were included in the reference string. In fact, ParsCit should even be able to recognize individual parts of the full text such as title, abstract, list of references etc. For the purpose of segmenting references we use a plain text file with the string 'References' at the very beginning which is followed by the list of references. Each of the reference strings is one line with no line breaks. One blank line is used as the separator of the references.

We are currently considering using this tool for DOI lookups. Our first tests involving ParsCit as a preprocessor for CrossRef DOI look up by HTTP XML Query[9] suggest reasonable accuracy.

The CrossRef XML query schema defines a large set of elements for the structural description of various parts of the reference string and special element `<unstructured_citation>` that can contain the raw citation string. Structural elements from the ParsCit output XML can be easily transformed into the CrossRef XML query schema. It can be mostly done by renaming ParsCit XML output elements to their CrossRef XML query counterparts. Instead of a full list of authors, it proved better to use just the first author in the 'Lastname, Firstname' notation, i.e. ParsCit output

```
<authors>
  <author>L Lambe</author>
  <author>J Stasheff</author>
</authors>
```

becomes

```
<author search-all-authors="false">Lambe, L</author>
```

in the CrossRef XML query.

To provide a DOI lookup service with as much information as possible the CrossRef query is constructed not only from the structural elements ParsCit has identified in the input, but the raw citation string from the ParsCit `<rawString>` element is added to the query as the `<unstructured_citation>` element.

Hand made metadata record of our sample reference from the publisher is shown in Figure 3. Its transformation to the CrossRef XML query format is similar to the ParsCit output transformation and the resulting XML query varies in small details such as punctuation only.

As a basic test we have used articles of volume 48, issue 5 of the *Archivum Mathematicum* journal (`http://dml.cz/handle/10338.dmlcz/143106`) that was published at the end of 2012. As we have high quality metadata for references from the publisher of the journal we compared the results of the DOI lookup using these hand made metadata and automatically segmented plain text reference strings harvested from the landing pages of the articles.

As can be seen in the summary in Table 1 and Figure 4 both hand made metadata and ParsCit generated metadata led to a surprisingly similar result of

---

[9] `http://help.crossref.org/#ID5829`

Table 1. Comparison of DOI lookup results using hand made and ParsCit automatically segmented references from the articles of *Archivum Mathematicum*, volume 48, issue 5

	number of refs.	ParsCit		hand made	
		resolved	not resolved	resolved	not resolved
article #2	10	1 (10.00%)	9 (90.00%)	2 (20.00%)	8 (80.00%)
article #3	6	3 (50.00%)	3 (50.00%)	3 (50.00%)	3 (50.00%)
article #4	19	4 (21.05%)	15 (78.95%)	4 (21.05%)	15 (78.95%)
article #5	12	3 (25.00%)	9 (75.00%)	3 (25.00%)	9 (75.00%)
article #6	17	6 (35.29%)	11 (64.71%)	6 (35.29%)	11 (64.71%)
article #7	16	6 (37.50%)	10 (62.50%)	4 (25.00%)	12 (75.00%)
article #8	9	3 (33.33%)	6 (66.67%)	3 (33.33%)	6 (66.67%)
article #9	10	1 (10.00%)	9 (90.00%)	2 (20.00%)	8 (80.00%)
article #10	13	5 (38.46%)	8 (61.54%)	5 (38.46%)	8 (61.54%)
article #11	13	2 (15.38%)	11 (84.62%)	1 (7.69%)	12 (92.31%)

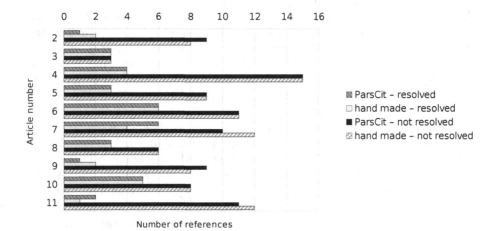

Fig. 4. Visualization of DOI lookup results from Table 1

CrossRef DOI lookup.[10] ParsCit always correctly identified all the references and segmented them well enough to achieve results comparable to the hand made metadata. Further investigation of the possibilities of deployment of this tool will be of great interest to us in the near future.

4   Output Modules

The content of the DML-CZ digital library is collected to be read and used. To achieve this, it has to be visible and usable. To achieve high visibility, it has to be indexed by search engines, especially by Google, given that it is used for 85% of searches today.

DML-CZ is available to the outside world via the DSpace repository software. This includes the end user interface (classic web based on HTML/JavaScript) and the OAI-PMH server providing the DML-CZ metadata together with links to the DML-CZ data in various XML formats. [Kre08] This section describes the complete workflow from the DML-CZ internal metadata to the EuDML specific NLM formats exported via OAI-PMH. The last subsection discusses the cooperation with Google Scholar.

4.1   Metadata Transformation

The project of the European Digital Mathematics Library EuDML [Syl+10], http://eudml.org/, is based on metadata and data of smaller regional DML projects. It was realized that almost every EuDML content provider uses a *different* internal format for their holdings. For example, the DML-CZ internal metadata format was established during the development of DML-CZ several years before the EuDML project was started. To adopt the EuDML format to be used by the DML-CZ internal tools would be quite difficult and time consuming, causing troubles in the well established DML-CZ workflow and would create a lot more work on the publishers' side. Now that the DSpace OAI-PMH is able to provide reliable metadata and their transformations into various formats, we took advantage of it.

Thus, a great deal of efforts went into mapping the metadata into the unique metadata format that is required for central processing and global enhancement methods.

The basic schema stands on the OAI-PMH and assumes that local repositories make their metadata available via an OAI server and are harvested by the EuDML central repository. There are now two ways of exposing the local metadata to the EuDML.

The first possibility is to expose metadata in an internal format which local DML repositories use natively. Metadata is harvested in this format and on the harvester side transformed into the specific EuDML metadata format. This approach puts almost no demands on the local repositories and most of the work is done on the EuDML side.

---

[10] Articles #1 and #12 are not present as they do not contain any references.

The second option, used in DML-CZ, is to expose metadata in the EuDML specific formats. For this purpose two formats have been set up — one for journals (*eudml-article2*[11]) and one for books (*eudml-book2*[12]). This assumes most of the work has to be done on the local repository side which can however bring some benefits — metadata has to be validated, errors have to be identified and corrected. This process leads to improved local repository metadata. For the DML-CZ this approach has been adopted.

Both *eudml-article2* and *eudml-book2* formats are based on the NLM Journal Archiving and Interchange Tag Suite format (version 3.0) [Dig08]. The NLM format is suitable for describing journal articles and is used without any changes in EuDML. To describe books, the NLM format has been extended and several new tags have been added. In the rest of the text both formats are referred to just NLM. [NIS12]

As the metadata for most of the DML-CZ content is very static and hardly ever changes, the transformations from DML-CZ internal XML metadata [BKŠ08] into the NLM are done in a batch with the help of XSL transformation. The result is a not a fully compatible NLM (pre-NLM) file stored directly next to journal/book metadata file in the internal structures. The file is not in the final NLM format because there are certain kinds of dynamic information (e.g. links to fulltexts) that have to be added on-the-fly by OAI-PMH at the moment the metadata are requested.

The XSL 2.0 transformations are used to obtain the NLM format from the internal format, including EuDML specific XSL functions for handling metadata like language codes. As a transformer the Saxon is used. The transformation process is integrated into the internal tools via set of bash scripts.

4.2   OAI-PMH

The pre-NLM file is then stored (on demand or automatically during import/ /update operations) in the DSpace repository. A DSpace digital object (called *Item*) schema allows various kinds of files to be stored. These are logically separated into so called *Bundles*. The pre-NLM file is stored in such a specific *Bundle* and used later by the DML-CZ OAI-PMH server.

DSpace OAI-PMH server is fully and easily configurable and provides various methods (XSL crosswalks, plugins in Java) how to add a custom format. A special Java plugin for exposing NLM formats has been developed. The plugin is called when metadata for an *Item* (or *Record* in terms of OAI-PMH) are requested on the OAI-PMH server. The plugin loads stored pre-NLM XML and adds links to fulltexts. These links cannot be added during the first transformation phase (pre-NLM) because at that time the information about the publisher's moving wall is not known (articles behind the publisher's moving wall can be used only for indexing and not for exposing fulltexts). The resulting XML is in the final NLM format and is served in an OAI-PMH <record> element.

---

[11] Namespace `http://jats.nlm.nih.gov`.

[12] Namespace `http://eudml.org/schema/2.0/eudml-book`

However, the way DSpace works with OAI-PMH *Sets* in the default configuration is not meaningful in DML-CZ, because the DSpace core structures represented by *Community* and *Collection* objects are handled in a different semantic way in the DML-CZ. The *Sets* are treated as the *Collections* which represent journal issues in the DML-CZ. The DSpace has been patched to change the *Sets* to be top level *Communities*, thus the *Sets* represent whole journals, proceedings series and monographs collections (as can be seen at the DML-CZ homepage). The patch includes a new database index table of articles and chapters for the top communities in DSpace and necessary code changes to work with it.

4.3   Google Scholar

The connection to Google Scholar is made in the way they recommend — the HTML header `<meta>` tags are used to fill up necessary article/chapter metadata. While there is no precise specification of the `citation_` format, the example provided by Google is followed. Every HTML page in DML-CZ is generated on-the-fly via XSL so the `<meta name="citation\_(spec)" content="(value)"/>` tags are processed the same way. Indexing these `<meta>` tags allows Google Scholar to link directly to the fulltexts in DML-CZ without the necessity of parsing paper metadata from landing HTML pages and PDFs. We believe that agreement on this interface might contribute slightly to the Page ranking of papers, as the metadata are contributed from the verified DML-CZ source.

Looking at the Google Analytics statistics over the last five years, the ratio of DML-CZ traffic generated by Google searches continually increases, reaching more than 85% at the time of writing. This shows the importance of this export interface, together with the sitemap updates we regenerate regularly and thus point Google to the newly published items automatically. DML-CZ is now ranked among the top ten repositories in Central and Eastern Europe, and best repository in the Czech Republic, measured by `http://repositories.webometrics.info/`.

5   Conclusions

We have reported on the workflows, interfaces and modules we have developed for the low-cost running of DML-CZ. When agreeing on formal interfaces that could be enforced by validation, considerable savings of manual work have been achieved, while in parallel increasing data quality and services of the library. After introducing the modules described, there is almost no manual intervention necessary. We perform manual checks of uploaded data before sending them to the public library but this is not strictly necessary.

Exporting data via the agreed interfaces to Google and EuDML skyrocketed the visibility of DML-CZ repository content, as measured by `webometrics.info` or similar metrics. We believe that our DML-CZ example demonstrates that by maintaining solid information technologies, Computer Science methodologies and web standards, even a small digital library can be run at a moderate cost.

Our future plans include adding further machine-actionable modules and functionalities into DML-CZ. Having full texts with math formulae by math OCR now, it is natural to add formulae searching with our Math Indexer and Searcher system MIaS [SL11], which already works well in EuDML. One of the most challenging remaining issues is the improvement of the process of the automated OCR of mathematics and its tighter integration into the rest of the system.

Also, new EuDML external APIs will be employed, namely for article similarity — similar articles will be acquired from the EuDML set instead of from a local set of articles in DML-CZ only. We also expect extensive development of the module for automatic parsing of the references. We hope we could significantly improve the quality of references metadata and eliminate the necessity of their costly and inefficient manual corrections.

Acknowledgements. This work was partially supported by the European Union through its Competitiveness and Innovation Programme (Information and Communication Technologies Policy Support Programme, 'Open access to scientific information', Grant Agreement No. 250503, a project of the European Digital Mathematics Library, EuDML).

References

[Aus+10] Ausbrooks, R., et al.: Mathematical Markup Language (MathML). Version 3.0. W3C Recommendation. World Wide Web Consortium (W3C) (October 21, 2010), Carlisle, D., Ion, P., Miner, R. (eds.),
`http://www.w3.org/TR/2010/REC-MathML3-20101021/`
(visited on January 06, 2013)

[BKŠ08] Bartošek, M., Kovář, P., Šárfy, M.: DML-CZ Metadata Editor: Content Creation System for Digital Libraries. In: Sojka, P. (ed.) Towards a Digital Mathematics Library, pp. 139–151. Masaryk University, Birmingham (2008) ISBN: 978-80-210-4658-0, `http://dml.cz/dmlcz/702537` (visited on January 09, 2013)

[CGK08] Councill, I.G., Lee Giles, C., Kan, M.-Y.: ParsCit: An open-source CRF reference string parsing package. In: Language Resources and Evaluation Conference (LREC 2008), Marrakesh, Morocco (May 2008),
`http://www.comp.nus.edu.sg/~kanmy/papers/lrec08b.pdf` (visited on March 13, 2013)

[Dig08] Digital Archive of Journal Articles National Center for Biotechnology Information (NCBI) and National Library of Medicine (NLM). NCBI Book Tag Library version 3.0 (November 2008), `http://dtd.nlm.nih.gov/book/`

[Gri10] Grimm, J.: Producing MathML with Tralics. In: Sojka, P. (ed.) Towards a Digital Mathematics Library, pp. 105–117. Masaryk University, Paris (2010) ISBN: 978-80-210-5242-0, `http://dml.cz/dmlcz/702579` (visited on January 09, 2013)

[Kre08] Krejčíř, V.: Building Czech Digital Mathematics Library upon DSpace System. In: Sojka, P. (ed.) Towards a Digital Mathematics Library, pp. 117–126. Masaryk University, Birmingham (2008) ISBN: 978-80-210-4658-0, `http://dml.cz/dmlcz/702539` (visited on January 09, 2013)

[LNK10] Luong, M.-T., Nguyen, T.D., Kan, M.-Y.: Logical Structure Recovery in Scholarly Articles with Rich Document Features. International Journal of Digital Library Systems 4, 1–23 (2010), http://www.comp.nus.edu.sg/~kanmy/papers/ijdls-SectLabel.pdf, doi: 10.4018/jdls.2010100101 (visited on March 13, 2013)

[NIS12] National Information Standards Organization NISO. JATS: Journal Article Tag Suite, ANSI/NISO Z39.96-2012 (August 2012), http://jats.niso.org/

[RS10] Růžička, M., Sojka, P.: Data Enhancements in a Digital Mathematics Library. In: Sojka, P. (ed.) Towards a Digital Mathematics Library, pp. 69–76. Masaryk University, Paris (2010) ISBN: 978-80-210-5242-0, http://dml.cz/dmlcz/702575 (visited on January 13, 2013)

[RS11] Růžička, M., Sojka, P.: Redakční systém odborného časopisu s podporou exportu do digitální knihovny v MathML. In: Zpravodaj CSTUG, pp. 4–20 (January 2011), doi: 10.5300/2011-1/4

[Růž08] Růžička, M.: Automated Processing of TeX-typeset Articles for a Digital Library. In: Sojka, P. (ed.): Towards a Digital Mathematics Library, pp. 167–176. Masaryk University, Birmingham (2008) ISBN: 978-80-210-4658-0, http://dml.cz/dmlcz/702533 (visited on January 13, 2013)

[SL11] Sojka, P., Líška, M.: The Art of Mathematics Retrieval. In: Proceedings of the ACM Conference on Document Engineering, DocEng 2011, pp. 57–60. ACM, Mountain View (2011) ISBN: 978-1-4503-0863-2, doi: 10.1145/2034691.2034703

[Soj08] Sojka, P. (ed.): Towards a Digital Mathematics Library. Masaryk University, Birmingham (2008) ISBN: 978-80-210-4658-0, http://dml.cz/dmlcz/702564 (visited on January 13, 2013)

[Soj10] Sojka, P. (ed.): Towards a Digital Mathematics Library. Masaryk University, Paris (2010) ISBN: 978-80-210-5242-0, http://dml.cz/dmlcz/702567 (visited on January 13, 2013)

[Suz+03] Suzuki, M., Tamari, F., Fukuda, R., Uchida, S., Kanahori, T.: INFTY–An integrated OCR system for mathematical documents. In: Vanoirbeek, C., Roisin, C., Munson, E. (eds.) Proceedings of ACM Symposium on Document Engineering, pp. 95–104. ACM, Grenoble (2003)

[Syl+10] Sylwestrzak, W., Borbinha, J., Bouche, T., Nowiński, A., Sojka, P.: EuDML– Towards the European Digital Mathematics Library. In: Sojka, P. (ed.) Towards a Digital Mathematics Library, pp. 11–24. Masaryk University, Paris (2010) ISBN: 978-80-210-5242-0, http://dml.cz/dmlcz/702569 (visited on January 13, 2013)

A Hybrid Approach for Semantic Enrichment of MathML Mathematical Expressions

Minh-Quoc Nghiem[1], Giovanni Yoko Kristianto[2],
Goran Topić[3], and Akiko Aizawa[2,3]

[1] The Graduate University for Advanced Studies
[2] The University of Tokyo
[3] National Institute of Informatics,
Tokyo, Japan
{nqminh,giovanni,goran_topic,aizawa}@nii.ac.jp

Abstract. In this paper, we present a new approach to the semantic enrichment of mathematical expression problem. Our approach is a
combination of statistical machine translation and disambiguation which
makes use of surrounding text of the mathematical expressions. We first
use Support Vector Machine classifier to disambiguate mathematical
terms using both their presentation form and surrounding text. We then
use the disambiguation result to enhance the semantic enrichment of a
statistical-machine-translation-based system. Experimental results show
that our system archives improvements over prior systems.

Keywords: MathML, Semantic Enrichment, Disambiguation, Statistical Machine Translation.

1 Introduction

The semantic enrichment of mathematical documents is among the most significant areas of math-aware technologies. It is the process of associating semantic
tags, usually concepts, with mathematical expressions. We use MathML [10]
Presentation and Content Markup to represent mathematical expressions and
their meaning. The semantic enrichment task then becomes the task of generating Content MathML outputs from Presentation MathML expressions. It is an
important technology towards fulfilling the dream of global digital mathematical
library (DML).

The semantic enrichment of mathematical expression is a challenging task.
Mathematical notations are ambiguous, context-dependent, and vary from community to community. Given a Presentation MathML element, there are many
potential mappings to its Content MathML element. For example, the token δ
can be mapped to $Kronecker Delta$, $DiracDelta$, $DiscreteDelta$, or δ. By correctly disambiguating these token elements, we can get a more accurate semantic
enrichment system.

Disambiguation of mathematical elements is an important component in the
semantic enrichment system. Basic methods for dealing with ambiguities so

J. Carette et al. (Eds.): CICM 2013, LNAI 7961, pp. 278–287, 2013.

far were either rule-based [1] or statistics-based [7]. The rule-based approach is of course generally not able to derive meaning from arbitrary Presentation MathML expressions. The statistics-based approach resolves ambiguities based on the probabilities, and thus gets better results than the rule-based system. In this paper, we enhance the statistics-based approach by combining it with a disambiguation component.

So far, there has been limited discussion about the contribution of surrounding text to mathematical element disambiguation problem. It is becoming increasingly difficult to ignore the surrounding text of mathematical expressions. For example, the token δ can be mapped to $Kronecker Delta$ if its surrounding text contains the word 'Kronecker delta'. It is difficult to disambiguate using only the presentation of mathematical expression. The combination of mathematical expression itself and its surrounding text can lead to improvements in disambiguation process.

The aim of this paper is to examine and solve the ambiguity when mapping Presentation MathML elements to their Content elements. This paper also attempts to find the contribution of surrounding text to mathematical element disambiguation problem. We use a Support Vector Machine (SVM) learning model for MathML Presentation token element (mi) disambiguation. Both presentation of mathematical expression and its surrounding text are encoded in a feature vector used in SVM. We evaluate the efficacy of the system by incorporating it into an SMT-based semantic enrichment system.

We formulate the problem as follows: given a Presentation MathML expression and its surrounding text, can we interpret its Content MathML expression? This paper provides contributions in three main areas of mathematical semantic enrichment problem. First, we show that combination of a disambiguation component and the SMT-based system improves the system's performance. Second, we show that the text surrounding the mathematical expressions contributes to the disambiguation process. Third, we show that the name of the category that a mathematical expression belongs to is the most important text feature for disambiguation.

The remainder of this paper is organized as follows. Sections 2 provides a brief overview of the background and related work on semantic enrichment of mathematical expressions. Section 3 presents our method. Section 4 describes the experimental setup and results. Section 5 concludes the paper and points to avenues for future work.

2   Related Work

MathML [10] is the best-known open markup format for representing mathematical formulas. It is recommended by the W3C Math Working Group as a standard to represent mathematical expressions. MathML is an application of XML for describing mathematical notations and encoding mathematical content within a text format. MathML has two types of encoding: Content MathML addresses the meaning of formulas; and Presentation MathML addresses the display of formulas. We use MathML Presentation Markup to display mathematical expressions and MathML Content Markup to convey mathematical meaning.

Most major computer algebra systems, such as Mathematica [5] and Maple [6], are capable of importing and exporting MathML of both formats. These importing and exporting functions enable the conversion from Presentation to Content MathML. Importing, of course, depends on the interpretation of each computer algebra systems engine.

There is a project called SnuggleTeX [1], which addresses the semantic interpretation of mathematical expressions. The project provides a direct way to generate Content MathML from Presentation MathML based on manually encoded rules. The current version at the time of writing this paper supports operators that are the same as ASCIIMathML [2]. For example, it uses the ASCII string "\in" instead of the symbol "∈". One major drawback of this approach is that it always makes the same interpretation for the same Presentation MathML element.

A recent study by Nghiem et al. [7] also addressed the semantic interpretation of mathematical expressions. This study applied a method based on statistical machine translation to extract translation rules automatically. This approach contrasted with previous research, which tended to rely on manually encoded rules. This study also introduced segmentation rules used to segment mathematical expressions. Combining segmentation rules and translation rules strengthened the translation system and the best system achieved 20.89% error rate. The shortcoming of this approach is that it did not make use of text information surrounding mathematical expressions.

Wolska et al. [8,9] presented a knowledge-poor method of finding a denotation of simple symbolic expressions in mathematical discourse. The system used statistical co-occurrence measures to classify a simple symbolic expression into one of seven predefined concepts. They showed that the lexical information from the linguistic context immediately surrounding the expression improved the results. The lexical information from the larger document context also contributed to the best interpretation results. This approach had been evaluated on a gold standard manually annotated by experts, achieving 66% precision.

3   Our Approach

The system has two phases, a training phase and a running phase, and consists of three main modules.

- Statistical-based rule extraction: Extracts rules for translation, given the training data. We establish two types of rules: segmentation rules and translation rules. Each rule is associated with its probability.
- SVM-based disambiguation: An SVM training algorithm builds a model that assigns to identifiers (mi) their correct content. Features are extracted from both the presentation of mathematical expressions and their surrounding text.

– Translation: The input of this module includes a Presentation MathML expression, a set of rules for translation, and the output from the disambiguation module. This module translates Presentation into Content MathML expression.

Figure 1 shows the system framework.

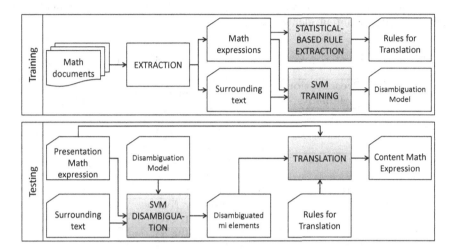

Fig. 1. System Framework

3.1 Statistical-Based Rule Extraction

The rules for translation were extracted according to the procedure used by Nghiem et al. [7]. Given a set of training mathematical expressions in MathML parallel markup, we extracted two types of rules: segmentation rules and translation rules. Translation rules are used to translate (sub)trees of Presentation MathML markup to (sub)trees of Content MathML markup. Segmentation rules are used to combine and reorder the (sub)trees to form a complete tree. The output of this module is a set of segmentation and translation rules, each rule is associated with its probability.

3.2 SVM Disambiguation

An mi token element in MathML presentation markup can be translated into many different elements in MathML content markup. In this paper, we assumed that one mi element can be translated into one of a limited predefined set of Content elements. Given an mi element, we use an SVM training algorithm to build a model that assigns to its correct Content element. When translating, each of the Presentation mi elements will be disambiguated before generating Content MathML expressions. The accuracy of the SVM disambiguation is a crucial preprocessing step for a high-quality MathML Presentation to Content translation.

We used the alignment output of GIZA++[1] [11] to generate training and testing data for the disambiguation problem. Given a training data consists of several parallel markup expressions, we used GIZA++ to align the Presentation terms to the Content terms. From this alignment results, we extract pairs of Presentation *mi* elements and their associated Content elements. Only *mi* elements that have ambiguities in their translation are kept to generate training and testing data. Table shows 1 the examples of Presentation *mi* elements and their associated Content elements.

Table 1. Presentation *mi* elements and their associated Content elements

Presentation elements	Content elements
\<mi\> σ \</mi\>	\<ci\>Weierstrass Sigma\</ci\>
	\<ci\>Divisor Sigma\</ci\>
	\<ci\> σ \</ci\>
\<mi\> μ \</mi\>	\<ci\>MoebiusMu\</ci\>
	\<ci\> μ \</ci\>
\<mi\>H\</mi\>	\<ci\>StruveH\</ci\>
	\<ci\>Harmonic Number\</ci\>
	\<ci\>Hankel H1\</ci\>
	\<ci\>Hankel H2\</ci\>
	\<ci\>Hermite H2\</ci\>
	\<ci\>H\</ci\>
\<mi\>y\</mi\>	\<ci\>Bessel Y Zero\</ci\>
	\<ci\>Spherical Bessel Y\</ci\>
	\<ci\>y\</ci\>

For each mathematical expression, an *mi* element has only one correct translation. In other mathematical expressions, the same *mi* element might have another correct translation. Assume that an *mi* element e has n ways of translating from Presentation into Content MathML. For each mathematical expression, we create one positive instance by combining e and its correct translation. We also create $n - 1$ negative instances by combining e and its incorrect translations.

The features used in the SVM disambiguation may be divided into two main groups: Presentation MathML features and surrounding text features. Presentation MathML features are extracted from the Presentation MathML markup of the mathematical expression. Surrounding text features are extracted from the text surrounding the mathematical expression. The category which the mathematical expression belongs to is also used. Table 2 shows the features we used for classification.

There were six Presentation MathML features in our experiment. The first one determines whether the *mi* element is the only child of its parent. The relation

---

[1] https://code.google.com/p/giza-pp/

Table 2. Features used for classification

Feature		Description
Presentation MathML feature	Only child	Is it the only child of its parent node
	Preceded by mo	Is it preceded by an <mo> node
	Followed by mo	Is it followed by an <mo> node
	&#8289;	Is it followed by a Function Application
	Parent's name	The name of its parent node
	Name	The name of the identifier
Text feature	Category	Relation between category name and candidate translation
	Unigram	Vector represents unigram feature
	Bigram	Vector represents bigram feature
	Trigram	Vector represents trigram feature
Candidate translation		One of n candidate translations of the mi element

between the mi element and its surrounding mo elements is encoded in the following three features. The last two features represent the name of the mi element and its parent. Among these features, the name of the mi element is the most important feature.

Among the text features, the first one is the category that mathematical expression belongs to. In mathematical resource websites, such as the Wolfram Functions Site, mathematical expressions belong to different categories. But usually we do not have the text surrounding these mathematical expressions. We then can calculate the relation between the category name and the Content translation of each mi element. The relation has one of three values: the same as the Content translation, contains the Content translation, or does not contain the Content translation.

In case we have the text surrounding or the description of the mathematical expressions, we can use n-gram features [12]. In this paper, we use unigram, bigram and trigram features. These features are implemented as the vectors containing the n-grams which appear in the training data. We will assign each instance into one of two classes, depending on the candidate translation. The class is 'true' if the candidate translation is the correct Content translation of the mi element, and 'false' otherwise.

3.3   Translation

After disambiguation, we use the result to enhance the semantic enrichment of a statistical-machine-translation-based system. The input of this module includes a Presentation MathML expression, a set of rules for translation, and the output from the disambiguation module. The output of this module is the Content MathML expression which represents the meaning of the Presentation MathML expression. If there is only one mapping from a Presentation element, that Content element is chosen. If the disambiguation module accepts more than two mappings from a Presentation element, the Content element with higher probability is chosen.

4   Evaluation

The first dataset for the experiments is the Wolfram Functions site [3]. This site was created as a resource for educational, mathematical, and scientific communities. All formulas on this site are available in both Presentation MathML and Content MathML format. The only text information on this dataset is the function category of each mathematical expression. In our experiments, we used 136,685 mathematical expressions divided into seven categories.

The second dataset for the experiments is the Archives of the Association for Computational Linguistics Corpus [4] (ACL-ARC). It contains mathematical expressions extracted from scientific papers in the area of Computational Linguistics and Language Technology. Currently, we use mathematical expressions drawn from 20 papers which were selected from this dataset. We have manually annotated all mathematical expressions with MathML parallel Markup and their textual descriptions. Out of 2,065 mathematical expressions in the dataset, only 648 expressions have their own description. Table 3 shows examples of mathematical expressions and their description in ACL-ARC dataset.

The evaluation was done using two metrics: accuracy score for disambiguation and tree edit distance rate score for semantic enrichment. The accuracy score of disambiguation is the ratio of correctly classified instances to total instances. The tree edit distance rate (TEDR) score [7] is defined as the ratio of (1) the minimal cost of transforming the generated into the reference Content MathML tree using edit operations and (2) the maximum number of nodes of the generated and the reference Content MathML tree. We also compare our semantic enrichment results to the results of Nghiem et al.

First, we set up an experiment to examine the disambiguation result on each Presentation MathML mi element. In this experiment, we compare three systems. The first system uses both Presentation MathML and text features. The second system uses only Presentation MathML features. The last system chooses the interpretation with highest probability.

Training and testing were performed using ten-fold cross-validation. For each category, we partitioned the original corpus into ten subsets. Of the ten subsets, we retained a single subset as validation data for testing the model, remaining subsets are used as training data. The cross-validation process was repeated ten times, and the ten results from the folds then averaged to produce a single estimate. Table 4 shows the results of the disambiguation component.

The results in Table 4 show that disambiguation result using SVM outperformed the 'most frequent' method. The reason 'most frequent' method got high scores is because mathematical elements often have a preferred meaning. The systems that used only Presentation MathML features achieved even better

Table 3. Examples of mathematical expressions and their description in ACL-ARC dataset

Textual description	MathML Presentation expression	MathML Content expressions
a word to be translated	<mrow> <mi>w</mi> </mrow>	<ci>w</ci>
a word in a dependency relationship	<mrow> <mi>w</mi> </mrow>	<ci>w</ci>
a matrix	<mrow> <mi>t</mi> </mrow>	<ci>t</ci>
a similarity matrix which specifies the similarity between individual elements	<mrow> <mi>sim</mi> </mrow>	<ci>sim</ci>
argument	<mrow> <msub> <mi>S</mi> <msub> <mi>j</mi> <mi>i</mi> </msub> </msub> </mrow>	<apply> <selector /> <ci>S</ci> <apply> <selector /> <ci>j</ci> <ci>i</ci> </apply> </apply>
The LM probabilities	<mrow> <mi>P</mi> <mo></mo> <mrow> <mo>(</mo> <mrow> <mi>v</mi> <mo>	</mo> <mrow> <mi>Parent</mi> <mo></mo> <mrow> <mo>(</mo> <mi>v</mi> <mo>)</mo> </mrow> </mrow> </mrow> <mo>)</mo> </mrow> </mrow>	<apply> <ci>P</ci> <apply> <ci>	</ci> <ci>v</ci> <apply> <ci>Parent</ci> <ci>v</ci> </apply> </apply> </apply>

scores, because they use surrounding mathematical elements. It is interesting to note that on the ACL-ARC data, the 'most frequent' system get higher score than the system with text features. Overall, on WFS data, we gained 5 to 16 percent accuracy improvements.

The systems that also used text features outperform the systems that used only Presentation MathML features in most of WFS categories. This result may be explained by the fact that the category of a mathematical expression is closely related to that expression. Contrary to expectations, this study did not find any improvement in ACL-ARC data. It seems possible that these results are due to the lack of training data and the sparseness of n-gram features. This finding was unexpected and suggests that in order to use n-gram text features, we need more data.

Second, we set up an experiment to examine the semantic enrichment result. The results from disambiguation component are used in the semantic enrichment

Table 4. Disambiguation accuracy

Category	Number of instances	With text features	Without text features	Most frequent
ACL-ARC	2,996	92.9573	**93.7583**	93.4246
Bessel-TypeFunctions	1,352	**92.8254**	92.3077	86.0947
Constants	714	**91.1765**	90.3361	83.7535
ElementaryFunctions	6,073	96.1963	**96.3774**	89.6427
GammaBetaErf	3,816	**95.2830**	94.4706	78.0136
HypergeometricFunctions	72,006	**97.5571**	97.0697	88.0746
IntegerFunctions	11,955	**95.8009**	95.1652	90.0711
Polynomials	5,905	**98.2388**	95.3091	87.3328
All WFS Data	320,726	**98.9243**	98.4398	92.7025

system. We compare three systems: with text feature, without text feature, and the system of Nghiem et al. which used 'most frequent' method. In this experiment, we use 90 percent of expressions for training both SVM-based disambiguation and translation components. We use the other 10 percent of expressions for testing. Table 5 shows the translation result.

Table 5. Semantic enrichment TEDR

Category	Number of expression	With text feature	Without text feature	Most frequent
Bessel-TypeFunctions	701	**18.0604**	**18.0604**	18.4118
Constants	555	**33.9016**	34.0328	34.6230
ElementaryFunctions	9,537	7.4879	**7.4809**	7.7343
GammaBetaErf	1,558	**17.2308**	17.2851	18.4796
HypergeometricFunctions	9,347	**49.4678**	49.4797	49.6902
IntegerFunctions	1,175	**20.5292**	20.5874	20.9945
Polynomials	727	**19.6309**	19.7987	20.2685
All WFS Data	23,600	**29.0707**	29.0869	29.2769

The results in Table 5 show that combining disambiguation and statistical machine translation improved the system. Expressions in 'Gamma Beta Erf' category benefit from the disambiguation module the most with 1.2 percent error rate reduction. Less ambiguity in elementary functions might lead to lower performance in 'Elementary Functions' category. We did not evaluate on ACL-ARC data because the disambiguation result was almost the same as the 'most frequent' method. Overall, on WFS data, we achieved 0.2 to 1.2 percent error rate reduction.

5   Conclusion

In this paper, we have presented a new approach to the semantic enrichment for mathematical expression problem. Our approach, which combines statistical machine translation and disambiguation component, shows promise. This study has shown that the disambiguation component using presentation features improved the system performance. The use of text features, especially the category of each expression, also played an important role in the disambiguation of mathematical elements. Experimental results of this study showed that our system achieves improvements over prior systems.

This research has raised many questions in need of further investigation. One question is finding and combining new features, such as the style of the font, for the disambiguation task. Another possible improvement is making use of co-occurrence of mathematical elements in the same document. In the scope of this paper, we only disambiguated lexical ambiguities of mathematical expressions. Structural ambiguities should also be considered to achieve better results. The evidence from this study suggests that in a small dataset, descriptions of mathematical expressions did not improve the system performance. Further work needs to be done to establish whether descriptions of mathematical expressions contribute to the the task in a larger dataset.

References

1. McKain, D.: SnuggleTeX version 1.2.2, `http://www2.ph.ed.ac.uk/snuggletex/`
2. ASCII MathML, `http://www1.chapman.edu/jipsen/mathml/asciimath.html`
3. The Wolfram Functions Site, `http://functions.wolfram.com/`
4. The archives of the Association for Computational Linguistics, `http://acl-arc.comp.nus.edu.sg/`
5. Wolfram Research, Inc.: Mathematica Edition: Version 8.0. Wolfram Research, Inc. (2010)
6. Maplesoft: Maple. Maplesoft, a division of Waterloo Maple Inc., Waterloo, Ontario (2013)
7. Nghiem, M.Q., Yoko, G.K., Matsubayashi, Y., Aizawa, A.: Automatic Approach to Understanding Mathematical Expressions Using MathML Parallel Markup Corpora. In: The 26th Annual Conference of the Japanese Society for Artificial Intelligence (2012)
8. Wolska, M., Grigore, M., Kohlhase, M.: Using Discourse Context to Interpret Object-Denoting Mathematical Expressions. In: Towards a Digital Mathematics Library, pp. 85–101 (2011)
9. Wolska, M., Grigore, M.: Symbol Declarations in Mathematical Writing. In: Towards a Digital Mathematics Library, pp. 119–127 (2010)
10. Ausbrooks, R., et al.: Mathematical Markup Language (MathML) version 3.0. W3C Recommendation, World Wide Web Consortium (2010)
11. Och, F.J., Ney, H.: A systematic comparison of various statistical alignment models. Computational Linguistics 29(1), 19–51 (2003)
12. Cavnar, W.B., Trenkle, J.M.: N-Gram-Based Text Categorization. In: Proceedings of SDAIR 1994, 3rd Annual Symposium on Document Analysis and Information Retrieval, pp. 161–175 (1994)

Three Years of DLMF: Web, Math and Search*

Bruce R. Miller

Information Technology Laboratory,
National Institute of Standards and Technology, Gaithersburg, MD
bruce.miller@nist.gov

Abstract. DLMF was released to the public in May 2010 and is now completing its 3rd year online. As a somewhat early adopter of large-scale MathML content online, and exposing a math-aware search engine to the public, the project encountered situations distinct from those with our previous web sites. In the hopes that our experiences may inform developers of current and future Digital Library projects, we describe some of our observations delivering MathML content and trends in both web usage and browser evolution. We will also look at the the ways our readers have used math search, attempting to assess whether they found what they sought, and ways the engine might be improved.

1 Introduction

Three years ago, after a considerable gestation, the Digital Library of Mathematical Functions (DLMF) [5] was released as a free resource to the public. As it is the successor to the *Handbook of Mathematical Functions* by Abramowitz & Stegun [1], we also served the traditional audience with the commercial publication of a companion Handbook [9]. We faced certain challenges [7]: to use what were (when we started) cutting edge technologies like MathML [3] to enhance reuse and accessibility; given the heavy mathematical content, math-aware search seemed essential; we needed to develop tools to assist in authoring XML and MathML content.

In our role as proponents of Mathematical Knowledge Management (MKM), we are enthusiastic about MathML and work to develop enabling technologies such as math-aware search; we'll happily promote these technologies in venues such as the current one. On the other hand, the goal of the DLMF project itself is to provide and make useful the mathematical knowledge it contains. It uses the technologies, even quietly encourages their use, but doesn't loudly force or announce them. The effect of this is that our users are not, for the most part, coming to our site with expectations of MathML or math search; when they submit a search query they are more likely to simply use what comes to mind, than following explicit instructions for 'how to search for math'.

Our DLMF is just one of several different kinds of 'Digital Library' and so not all of our experiences will be relevant to all other library developers. Nonetheless, we anticipate that many of them will be helpful to other developers. After looking at general usage of the DLMF, we will focus on the delivery of MathML and use patterns of math-aware search.

---

* The rights of this work are transferred to the extent transferable according to title 17 U.S.C. 105.

J. Carette et al. (Eds.): CICM 2013, LNAI 7961, pp. 288–295, 2013.

2   Response

The reviews and feedback we have had on the DLMF and Handbook, at least to our faces, has been almost completely positive. Most complaints that we have received question the accuracy of formulae, occasionally with justification. In a few other cases, we have apparently overlooked listing someones favorite software package. Another handful of comments concerned technical problems with MathML, typically missing fonts, or other browser issues. In response, we have made 5 minor updates of the DLMF to include 20 corrections in errata along with various technical and conversion improvements.

An internal study of the citation indices indicates that citations of the Handbook and DLMF are gradually displacing citations of Abramowitz & Stegun. Citations specifically of the online DLMF seem to be a small portion (17%) of the total. Although the continued increase of citations of Abramowitz & Stegun had originally been an important motivation for the DLMF project, the total citations over the last 3 years, ironically, seem to be leveling off. A speculation is that use of the DLMF is indeed displacing both the new *and* old printed handbooks, but that since citation of online materials is unfamiliar to most authors, it is either not being cited as consistently, or its citations are harder to recognize in the citation indices. [We've added a 'How to Cite' page, to try to improve the outlook.] It is difficult to confirm this theory, however.

3   General Log Analysis

Web server logs are notoriously difficult to interpret, or are simply unreliable [4]. A surprising amount of traffic, *half the bandwidth*, seems to be web indexing robots (or worse). We seemingly expend as many resources preparing to find material as we do using it once we've found it! Many of those robots routinely masquerade as familiar web browsers. Normally, various caches shield the server from many page requests and thus skew the statistics. However, since we use content negotiation based on agent identification strings, we mark the pages as uncacheable and thus may be less affected by this inaccuracy. Nevertheless, the server logs are what we have and they will have to serve.

The traffic to our site has been gradually increasing since its unveiling (Figure 1). Initially around 200 visits per day increasing to over 600 — we've arbitrarily defined a 'visit' to end when a visitor has not requested a new page for more than 30 minutes. A visit appears to average around 4–5 pages. We're restricting our attention to what we believe are humans in the following discussions.

Sketchy information encourages speculation. Do the trends of average visit duration, shown in Figure 2, indicate a gradually decreasing attention span? Apparently people spend an average of 5 minutes at the site, but near 10% stay for more than a half-hour.

We spent some effort making snippets of our formula available as TeX or Presentation MathML(pMML). There seems to be a pay-off, as it appears that on average, each visitor (385/day) downloads at least a pMML (202/day) or TeX (196/day) or BibTeX (25/day) snippet.

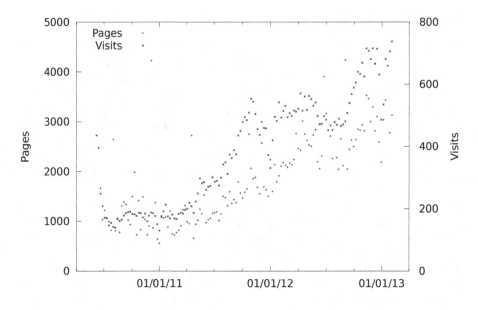

Fig. 1. Daily visits to DLMF

4   MathML Specific Issues

Setting up a website to use MathML may encourage us to be a bit too clever for our own good. We want to serve MathML whenever possible, but otherwise fall back to images for the math. Portal pages, or forcing users to understand and choose the appropriate format, are awkward and intrusive. So we set up our server to determine, by user-agent sniffing, which format the browser supports and send that automatically. It actually works fine, although we have to occasionally update the agent rule base.

One complication is due to breaking the assumptions of stock web analyzers. The analyzer may no longer correctly classify requests as being requests for pages, for example. Moreover, it can yield a wildly misleading picture of browser share which is typically based on the number of successful requests originating from each kind of browser.

The DLMF has a total 1,613 pages, with some 38K math expressions. Thus, to view a typical page, a MathML supporting browser will fetch the single page, with embedded MathML and is ready to display. A non-MathML agent will load the page without MathML, and make an average of 24 extra requests to fetch the images of the math before it can view the page. This severely biases the apparent market share. Interestingly, the total amount of data downloaded in both cases is comparable.

It seems more interesting for our purposes to divide up browsers into 3 categories. DLMF makes fairly heavy demands on the rendering agent, and so only the most complete implementations, whether native or via plugin, are served

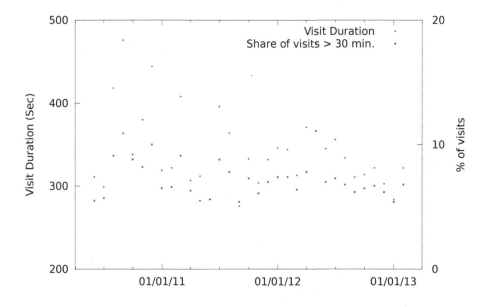

Fig. 2. Lengths of visits to DLMF

MathML by default. A second category has partial (or even sporadic) support for MathML, but not quite good enough to cover DLMF's material; for example, supporting only the CSS profile, or missing crucial elements like prescripts and multiscripts; we'll call these 'almost MathML'. The final category is without MathML support. Figure 3 shows the trend in page views between these three categories of browser. (We'll avoid 'naming names', since support is evolving, and our main interest here is the (positive) outlook.)

One should be careful over-interpreting Figure 3, as these figures seem rather sensitive to the patterns used for robots, and as the robots vary their choice of browser to mimic. Nevertheless, it seems encouraging that MathML supporting browsers, and particularly browsers that could support it, with a bit more effort[1], sum up to such a large and growing share.

In the meantime, there have been two other encouraging developments affecting MathML support. One is the inclusion of MathML in HTML5 [2], along with the support of most browsers. Although many browsers claim to support HTML5, few in fact implement MathML (yet). This lack is partially ameliorated by the second development, the advancement of MathJax [6] which implements MathML rendering using JavaScript and CSS.

One 'take away' lesson is the following. Even if XML and namespaces seems unloved by the HTML5 community, it is only through the use of XML infrastructure that DLMF can almost trivially track this change. We will be adding HTML5 with MathML as a formatting option in the near future; perhaps using

---

[1] Not that we are offering business advice.

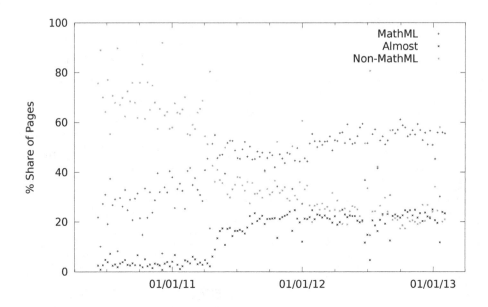

Fig. 3. Browser trends

MathJax as the fallback rendering engine. Additionally, although the agent-sniffing machinery may seem less necessary with the advent of MathJax, it may still be useful for handling other contingencies, such as mobile agents and tablets.

5   Search Issues

DLMF provides a search engine which supports mathematics-specific search, as well as conventional text search. Taking guidance from modern search engines we let the user type whatever query they expect to work, and attempt to make the best of it. We infer, based on the tokens in the query, whether it was intended to find math or text, what kind of notation they're using, and so on. In the following, we'll try to assess whether or not we have succeeded, but it is also interesting to see what queries these untrained users did in fact submit.

Our search engine converts the math to a serialized text equivalent, both in the document during indexing and in the query during search, and then leverages a text search engine to perform the actual search look-up [8]. This type of search engine is oriented towards the more informal usage that we envision. However, it is certainly not the only approach to math-aware search in all contexts.

From our web server analysis, we see that there were an average of 24 searches a day; roughly 1 for every 16 visits. About 16% of them appear to be intended to be math searches, although perhaps a third of these are simple terms, like exp, that are easily interpreted as either math or text.

Figure 4 shows the number of 'tokens' (basically sequences of contiguous letters or numbers or individual noise characters) used in math and text queries.

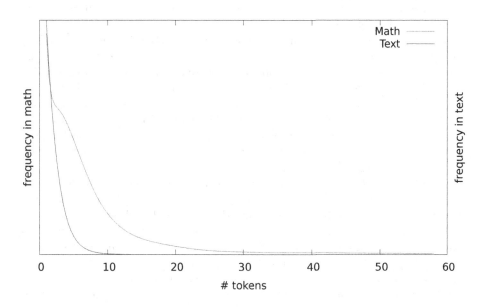

Fig. 4. Distribution of number of search terms per query

The shapes indicates that math queries are most commonly expressions of several terms, but queries up to 58 were seen. Text queries tend to be at most a few words, and a superficial scan of the lists suggest they are almost always phrases.

Out of the math searches, common patterns include

- $ wildcard (10%: suggesting they *did* read the help file after all!);
- LaTeX markup (5%);
- various identities (e.g. $c^2 = a^2 + b^2$);
- pairs of math symbols presumably expected in the same formula, but not a math expresssion, as such;
- simple formula fragments: `sin 2x`, `sinh cos`.

Some surprising math queries include

- cut & pasted long formula;
- examples from Help page;
- `sinx+cosy` or `sinacosb`;
- `\hbox` and `\vbox`;
- `x_sub{0}`;
- + apparently used as query meta-operator (or url encoding?).

But does it work? At a commercial website, a sale is easily recognized as a success, but for a Digital Library, success is your readers discovering the information they desire. Did they leave the site because they found the information, and are now going on to do some productive work, or did they leave out of frustration?

Short of a survey, with its own set of problems, how can we tell if the search engine works for the users?

We therefore turn back to our server logs to attempt to infer success or failure from the users' sequences of actions. What do they do *after* they have executed a search? If they follow the link to one of the search results (we'll call that 'Click Thru' in the following), what do they do after that? One might idealize a a successful outcome as when a user performs a search, inspects one of the results and then, having found the desired item, will visit random other pages within the site. Less successful outcomes would have the user floundering, checking the next page of hit results ('Next 10'), trying other search results or formulating alternative searches ('New Search').

Table 1 collects the tracks derived from our web logs, showing how the behavior of users searching for math differed from those searching for text. It would seem that searchers for text often follow that idealized path suggested above. While it isn't clear that searchers for math are unsuccessful, they certainly appear to need more fishing around to find what they wanted (assuming they did); they were more likely to check the next page of hits or try a different query. Moreover, even after they've clicked on one search result, they were more likely to come back to the search results and try another result or another search. Whether this is somehow due to the different nature of 'searching for a math expression', or is a measure of poor search results is hard to tell.

Table 1. What users did after a search, or after clicking on a search result

	Total	Click Thru	Next 10	New Search	Other Page	Left DLMF
			Next page request			
Searches 23190						
Math	3644 (16%)	37%	12%	30%	14%	7%
Text	19308 (83%)	43%	5%	26%	17%	8%
ClickThru 20888						
Math	2963 (14%)	18%	13%	33%	30%	6%
Text	17751 (85%)	24%	5%	18%	44%	9%

6   Conclusions

The DLMF is online and appears to be appreciated and used after 3 years. Of course, mathematics, let alone, special functions, is a niche, not mainstream, interest; we don't expect web traffic to rival Google or Amazon. Nor do we expect browser implementers to pay as much attention to MathML as to video. Nevertheless, there are reasons for optimism about delivering math on the web; solutions sometimes appear where you least expect them.

We find math search to be used modestly, but this is not surprising given that users don't expect it and we have ruled out being confrontational to promote it. Our log analysis suggests that math searches require a bit more work to find results than do text searches, but nevertheless appear to serve the users. A more convincing analysis of search behaviors, and indications of search success, would likely require instrumenting the search engine to generate search-specific logs.

Disclaimer: Certain products, commercial or otherwise, are mentioned for informational purposes only, and do not imply recommendation or endorsement by NIST.

References

1. Abramowitz, M., Stegun, I. (eds.): Handbook of Mathematical Functions with Formulas, Graphs, and Mathematical Tables. National Bureau of Standards Applied Mathematics Series 55, U.S. Government Printing Office, Washington, D.C. (1964)
2. Berjon, R., Leithead, T., Navara, E.D., O'Connor, E., Pfeiffer, S.: HTML 5.1, http://www.w3.org/TR/html51/
3. Carlisle, D., Ion, P., Miner, R., Poppelier, N.: Mathematical Markup Language (MathML), W3C http://www.w3.org/TR/MathML/
4. Kathuria, P.: I, robot? Don't believe your web stats (March 10, 2013), http://www.limov.com/library/do-not-believe-your-web-stats.lml
5. NIST Digital Library of Mathematical Functions, http://dlmf.nist.gov/, Release 1.0.5 of 2012-10-01. Online companion to [9]
6. MathJax, http://www.mathjax.org
7. Miller, B., Youssef, A.: Technical Aspects of the Digital Library of Mathematical Functions. Annals of Mathematics and Artificial Intelligence 38, 121–136 (2003)
8. Miller, B.R., Youssef, A.: Augmenting Presentation MathML for Search. In: Autexier, S., Campbell, J., Rubio, J., Sorge, V., Suzuki, M., Wiedijk, F. (eds.) AISC/Calculemus/MKM 2008. LNCS (LNAI), vol. 5144, pp. 536–542. Springer, Heidelberg (2008)
9. Olver, F.W.J., Lozier, D.W., Boisvert, R.F., Clark, C.W. (eds.): NIST Handbook of Mathematical Functions. Cambridge University Press, New York (2010), Print companion to [5]

Escaping the Trap of Too Precise Topic Queries

Paul Libbrecht

Center for Educational Research in Mathematics And Technology (CERMAT),
Martin Luther University Halle-Wittenberg, Germany
http://www.cermat.org/

Abstract. At the very center of digital mathematics libraries lie controlled vocabularies which qualify the *topic* of the documents. These topics are used when submitting a document to a digital mathematics library and to perform searches in a library. The latter are refined by the use of these topics as they allow a precise classification of the mathematics area this document addresses. However, there is a major risk that users employ too precise topics to specify their queries: they may be employing a topic that is only "close-by" but missing to match the right resource. We call this the *topic trap*. Indeed, since 2009, this issue has appeared frequently on the i2geo.net platform. Other mathematics portals experience the same phenomenon. An approach to solve this issue is to introduce tolerance in the way queries are understood by the user. In particular, the approach of including fuzzy matches but this introduces noise which may prevent the user of understanding the function of the search engine.

In this paper, we propose a way to escape the topic trap by employing the navigation between related topics and the count of search results for each topic. This supports the user in that search for close-by topics is a click away from a previous search. This approach was realized with the i2geo search engine and is described in detail where the relation of being *related* is computed by employing textual analysis of the definitions of the concepts fetched from the Wikipedia encyclopedia.

Keywords: mathematical documents search, topics search, web mathematics library, search user interface, learning resources, mathematics classifications, mathematics subjects.

1 Searching by Mathematical Topics

The problem to retrieve mathematical documents in large collections becomes an everyday challenge with the ever-growing collection of digital texts that mathematics professionals and learners have access to.

Word based search in mathematical documents, while still prevailing, suffers from broad issues: formulæ are not covered and, more importantly, the same concepts may be expressed in different terms or with the same terms but in a different sentence organization. For example, the concept of *right angle* is often expressed in sentences such as *the angle in A is right*, or as $\alpha = 90°$.

J. Carette et al. (Eds.): CICM 2013, LNAI 7961, pp. 296–309, 2013.
© Springer-Verlag Berlin Heidelberg 2013

To alleviate this issue, classifications have been established and mathematical documents annotated by them. The document is *indexed* by them. They can be called topics or subjects, even though philosophy of science would treat these concepts as fundamentally different, we consider these names to be equivalent in this paper. Nowadays, most research papers in mathematics are annotated using the *Mathematics Subjects Classification (MSC)*[1] and platforms to offer search with them are available. In education, several repositories have been created to share learning resources in communities of practice: the contributed resources are annotated with the topics being covered, as well as many other properties (rights, typical age, instructionnal function, ...). All offer users to search and contribute using topics.

Topic based engines employ a classification of topics. They allow users to search not only by words (and maybe by formulæ) but by the concepts or scientific domains that the resource treats. Examples of such classifications include the MSC, the EUN LRE Thesaurus,[2] and the GeoSkills ontology (described in Section 3).

The usage of such classifications effectively diminishes the ambiguity of the search by words in the text as it diminishes the choices of possible expressions. Moreover, the classifications can, generally, be expressed in multiple languages and thereby allow to find search results in multiple languages, even if the user does not understand them. We shall survey below different methods of choosing the topic.

Even though topic based search can be an effective way to drill down the number of documents matching a word, it also strongly depends on the chosen topic and users may find themselves very quickly *trapped in the niche* of a topic that is too precise. Innovative ways to relax the queries when facing an insufficient search result are the focus of this paper.

In the experience of the author, including reports of multiple users using the i2geo platform, searching by topic is permanently compared to searching by words: users will often attempt and mistrust a topic that has been entered if they realize that some documents may be about that topic but have not matched the search. Doing so, they quickly realize that word based search suffers from multiple issues (sketched, among others, in [LKM09]). Reports such as the logbooks analyzed in [LK13] indicate this hesitation.

With word searches, at least three solutions exist to offer tolerant or fuzzy matching:

---

[1] The Mathematics Subjects Classification is the most commonly used classification for mathematical documents at the research level. It can be browsed through its main catalog at `http://msc2010.org/`. A description of recent developments is given in [IS12].

[2] The EUN LRE Thesaurus is a classification of learning material topics aimed at comprehensiveness across Europe but not at a very deep precision. It is available in 15 languages and contains, for mathematics, less than 20 topics. See `http://lreforschools.eun.org/web/guest/lre-thesaurus` for its list.

Partial Results. In almost all search engines as well as i2geo.net the search method assumes that, by default, most of the ingredients of a query (for example the different words) are not queried as a conjunction but as a weighted disjunction: show first the results that match all words, but include also those that only contain single words. This common practice often fails at informing the user about the quality of a series of match and thus introduces *noise*.

Latent Semantics. Another way, which has been the basis of [LD97] is to employ a vector space model of the word/document occurrence and deduce a distance between documents from distance in these spaces. It has been caled latent semantic analysis (LSA). Given a relatively homogenous corpus of texts, this provides effective ways to detect relationships between words and between documents and thus allows search results to include documents that match terms *nearby*. Little research has investigated this approach for mathematics (we know of [Car04] alone).

Suggested Terms. Another way, commonly practiced in contemporary web search engines, is the suggestion of terms that are likely to match the user's query while the original one would lead to no search result as described in [Hea09, §4.3]. A version of this feature is the widespread Google search engine's "did you mean" as well as its suggested queries. Thus far, little query suggestion has been used for mathematics search engines.

1.1 Outline

This paper first draws a panorama of the user interfaces that search by topics. It then describes the search mechanism of the i2geo.net platform, followed by the core contribution: a method to suggest *related* topics having searched for a topic and its implementation. A discussion sets the contribution in perspective. As conclusion, we sketch future works.

2 Selecting Topics and Searching

The choice of the right topic in a search process is a key step. While seasoned mathematicians will often only search the topics of the communities they are used to, commonly calling them by number in the case of the MSC, there are multiple usage types where the user is not necessarily aware of the complete classification. They include searches for literature by a mathematician outside his or her domain of expertise, searches for literature by non-researchers, and searches for learning resources by newcomers in the learning resources platform.

An example is the concept of complexity of an algorithm. It can be found in 7 domains of the MSC, including Mathematical logic and foundations and Computer science. What would a student enter as classification key? The process of choosing the right topic is often a preparation step that is not done in conjunction with the search engine.

Several user interfaces paradigms exist to select a topic:

One way to do so is to employ a multi-stage hierarchy where the user first selects a broad topic then finer grained ones. This approach suffers from one major drawback: the hierarchy must be somewhat natural or be learned. As an example, [Meg12] reports incomprehension of most pupils when they are told that the history branch is part of humanities when using the EduTube-Plus educational video repository[3] browsing by the topics of the EUN-LRE thesaurus. Similarly, the Cosmos portal[4] has a fine grained classification of physical topics which avoids repetitions;

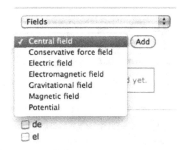

Fig. 1. Selecting a topic in two steps in the Cosmos portal

a normal user, thus, needs to search multiple times to find the classical topic of optics which has been put under the main topic of waves while it could have been under the main topic of light or fields.

An enriched version of these approaches is the facetted navigation approach described in [Hea09, §8.6]. This approach is richer because it indicates to the user the amount of matching documents before refining or generalizing a query along a hierarchy. In this research we leverage this witness of the query total as an important support to inform the user before choosing a query. However, we claim that navigating up (generalizing) or down (specializing) along a hierarchy may be too restricted and that the user may need alternative ways to navigate to related topics.

Another method of displaying topics to be chosen is by offering *tag clouds* which lay out the topics so as to fill the plane by attributing a size to a topic dependent on its frequency in the collection being searched. This method works only well for a small number of topics which are not too inhomogeneous.

Finally, another method to access the topics is that adopted by both major mathematical search engines Zentralblatt[5] and MathSciNet[6]: they suppose that the user will employ *subject codes* of the MSC to denote the topics. Both of these search engines do not offer a way to search for the topic, they suppose the codes to be known and let users use other tools like the MSC's main catalog to identify the codes. The euDML library varies this by combining browsing for content and browsing for topics.[7] This approach makes it natural for users to switch between topics of the same parent.

---

[3] The EduTubePlus repository ran during the EU project of the same name. See `http://www.edutubeplus.info/`. Successor repositories are being built.

[4] The Cosmos portal is a learning objects repository to share resources pertaining to astronomy in classroom. `http://portal.discoverthecosmos.eu/`

[5] Zentralblatt Math is a service giving access to abstracts of most of the current and past mathematics research. It is accessible at `http://www.zentralblatt-math.org`.

[6] MathSciNet, also called Math Reviews, is a also service to crawl through abstracts of most of the current and past mathematics research. It is accessible at `http://www.ams.org/mathscinet/`.

[7] The euDML library gives access to a considerable amount of free access mathematics research papers. It can be browsed by topics at `https://eudml.org/subjects/MSC`.

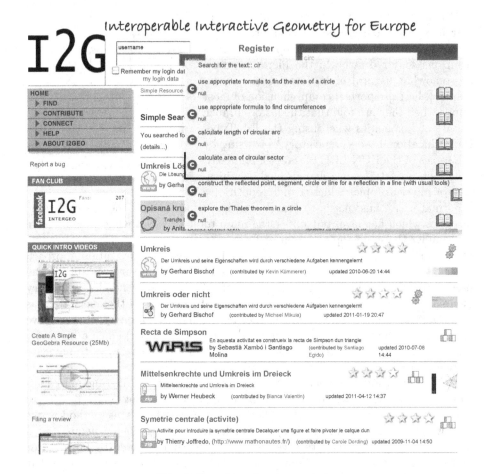

Fig. 2. Searching for circumcircle

The approach employed by the i2geo.net platform is a mix of the approaches above. It employs the search paradigm to allow the user to find the right topic (this is similar to what a newbie can achieve by browsing MSC's catalog), but also allows other forms of navigation by displaying the topic behind a hyperlink everytime the topic is mentioned. This allows users to meet the topic in other places and search for it.

The approach of i2geo.net aims at being easy to use by teachers of varying proficiency in mathematical science and in the usage of the computer tools. It has, thus, been kept on a single platform with a somewhat consistent user interface.

The auto-completion phase, where the user searches for a topic, competency, or educational level, is depicted in the Ficture 2 where the user searches for the concept of circumcircle, choosing it from the possible choices and find it in several languages below.

In all these approaches, the niche trap of a user stuck in searching for a topic that is too precise is wide open. In the MSC based search above, it is very likely that a user would not attempt to query all the topics pertaining to the complexity; if also including a keyword such as *graph* to get informed about complexity of graph algorithms, he or she will either get too many results or too little. Similarly in i2geo, it has been easy to choose a topic that yields zero results and thus fall into an annoying trap.

3   The Cross-Curriculum Search of i2geo

The i2geo.net platform, described in [LKM09] and available at `http://i2geo.net/`, is a learning object repository where teachers of Europe come to share learning resources to learn using dynamic geometry tools. The resources include simple documents of the dynamic geometry systems as well as learning material to support such a learning (exercise sheets, teacher advice, ...). The platform has been built during the Inter2geo project[8] gathering mathematics education experts from France, Spain, Germany, the Netherlands, and the Czech Republic until 2010. It is now maintained by the University of Halle.

The i2geo search supports search by word and by topic. The topics represent concepts, capacities (which are called competencies), and educational levels of the compulsory mathematics education, they are encoded in an ontology called GeoSkills, each being expressed in the languages of the countries above. The multilinguality of this ontology and the topic based search offered in i2geo supports the mainly graphical nature of many learning resources using dynamic geometry: it is easy to translate a resource of dynamic geometry from one resource to another. The i2geo search has been called a *cross-curriculum search*.

The main method of searching is by the choice of a topic, competency, or educational level, and obtaining the matching search results which obeys the following rules which form the ontology based query-expansion mechanism of i2geo search:

- If a *topic* is chosen, i.e. an abstract mathematical concept, resources that have been annotated with this topic are returned and, less preferred, resources that have been annotated by more specific topics.
- If a *competency* is chosen, i.e. a capacity including a verb and topics (see [LD09]), resources that have been annoated with this competency are returned first and, if not, resources annotated with one of the ingredient of this topic.
- If a *level* is chosen, only resources annotated with that level are returned.

The resulting search is effective to find learning resources as soon as the topic is correctly identified but its usage in the last three years has been frustrating for users which often fall in the niche trap of a topic that is too fine as reported, for example, in the log-books of teacher's usage found in [LK13].

---

[8] More information about the Inter2Geo eContentPlus project can be read in `http://i2geo.net/Main/About`.

One of the particularly annoying situation has been that of the search competencies which are expected to describe precise capacities that a learner is epxected to acquire. These are typically narrowly defined so that they correspond to a rather isolated place in the curriculum standard. Thus searching for a competency is a very precise action, so much that it returned zero results in most of the cases. Thus, we have made it deliberately tolerant: it should not only return the resources annotated with that competency but resources with are matching any of its ingredients; for example, the comptency use the intercept theorem to calculate magnitudes[9] has as ingredients, the competency process use in calculating magnitudes and the topics intercept-theorem, rational number, measure, and proportional. Even if a *highlighting* method would be used (as in [Hea09, §5.5]), the user would still only understand the relationship to the query if he would completely understand the ingredients of the competency.

One of the actions that users can make, if they realize that they are lost in a too precise topic, is to click on the topic which is displayed and see its hierarchy, then click *add this topic* so that resources matching this parent or this child are displayed. However this operation has been repeatedly evaluated as too heavy, with users switching to words queries. So as to realize a simpler switch, we have developed a query suggestion mechanism which suggests to the user *related* queries, including parent and children nodes (for concepts), referenced concepts and competency process (for competencies), and related concepts which we describe in the following section.

Moreover, the search index of the terms (topics, competencies, levels) has been enriched with counts of matching resources. This allows the count to be displayed everytime the term is displayed. This inclusion allows the input of topics to be restricted to those that would not yield empty search results. The query suggestions are depicted in figure 3.

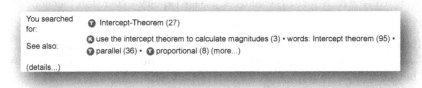

Fig. 3. Suggested queries related to the concept of intercept theorem

4 Computing Related Topics through LSA and Popular Definitions

In order for related topics to be computed, an ontology was created which bases on the GeoSkills' ontology and adds the property of *being related* as a reified

---

[9] This competency can be browsed at http://i2geo.net/comped/showCompetency.
html?uri=use_intercepttheorem_to_obtain_magnitudes .

relationship between the nodes of the GeoSkills ontology. Such a relationship is enriched with a *similarity* between 0 and 1.

- As a first step, a part of the relationships are authored manually using Protégé OWL 4[10] to state the important relationships between concepts that are likely not to be achieved by the text similarity methods below. Such a relationship exists, for example, to connect the topic of circular diagram and of pie chart which are two concepts that are very common in statistics education. They almost have the same semantics but they have not been clearly identified as equivalent with certainty (several experts have hypothesized that circular diagrams allow chords that are not diagrams to be allowed as limits of the domains). These different topics have been contributed by mathematics education experts, mostly teachers, so we contend that such a differentiation of topics is useful even though it is subtle. Coping with such a differentiation without disorienting is, thus, part of the mission of the i2geo search engine and the query suggestion mechanism is a step forward: these two concepts are flagged as close relatives (with the biggest similarity).
- As a second step, relationships are simply built by policy: parent concepts, competency ingredients, educational levels, age, or educational regions.
- As a third step, relationships are built by analogy between definitions. To this effect, concepts have been annotated with the URLs to the sections of a Wikipedia page stating a definition of it, for each of the language. Currently this annotation has been done for the languages the author masters, English, French, and German. In each language, a text analysis process, similar to the Latent Semantic Analysis [LD97], is performed on these definition texts which can compute similarity distance between the concepts. The SemanticVectors library is used for this [WC10].

Because it is built on different corpora for different languages, this suggestion mechanism takes in account the semantic fields of each language. For example, the fact that a disc is the surface inside of a circle in English and French makes it close to other regions of the plane while the closest German translation of the mathematical disc, "eine Scheibe", is a cylinder, with a thickness. While a strict organization might have decided that "Scheibe" is not the appropriate translation, other namings that attempt to denote the interior of a circle ("Kreisfläche", "Kreisscheibe") in German seem to have never been widespread.

5   i2geo Implementation

The search engine of i2geo is based on its sharing platform repository, which stores learning resources and displays them, supporting a complete learning objects sharing mechanism described in [LKM09] which derives from the XWiki

---

[10] Protégé is a widespread ontology editor. See `http://protege.stanford.edu/`. It has been used for most ontology engineering tasks around GeoSkills in versions 3.

Collaborative Learning Asset Managements System (XCLAMS[11]). The resources search index extends Apache Lucene with i2geo and XWiki information. This user interface extends the basic text search mechansim with the search weighting described in section 3 and with the chooser for topics, competencies, and levels. This chooser is a JavaScript component which fetches results from the auto-completion index as the user types.

The auto-completion index, also based on Lucene, is populated by reading the ontologies: the GeoSkills ontology [LD09], the ontology of relationships (called `GeoSkills-Relatives`), and the ontology of subjects. The ontology is read using the library OWLapi and the reasoner Pellet.

Similarly as the GeoSkills ontology and its displays such as the CompEd web-application [DL09], which are designed to be updated on a daily basis to take advantage of the changes by the curriculum experts, the GeoSkills-Relatives ontology is designed to be updated when changes are made either in the definitions' texts or in the manual relationships. While GeoSkills would receive updates from curriculum experts, the GeoSkillsRelatives ontology would receive updates from developers receiving feedback from users and encoding them using Protégé, and from the definition texts fetched from their web-sources, currently Wikipedia. These ontologies are then published and they enter the rebuild of the auto-completion index which, also, queries the numbers of matching resources for each term.

The code of the software and the ontologies are open-source, under the Apache Public License, and available from `http://github.com/i2geo/`. It has not yet entered the production site of i2geo.net, as it is being polished.

6   Discussion

Having described the technical realization of this approach, we now discuss its relevance to research areas which are closely related to it.

6.1   Navigation Between Topics

The approach of suggesting queries we have developed supports the user in exploring the topics' organization, inviting them to look at topics which are laterally related to the queried topics. This form of navigation shares the practice of facetted navigation of [Hea09, §8.6] in that it suggests queries, but it does so without being restricted to the hierarchy of topics. Moreover, users can explore the structure of topics: they can click on the queried topic and see a *topic page*. The topic pages of i2geo, served by the CompEd server, perform a similar role as the MSC catalog: they allow users that wish to inquire about the classifications' structure to navigate through it. Zentralblatt Math and euDML present any occurrence of an MSC subject as a hyperlink to the query for this topic but do not provide a link to the context, that is, its hierarchy. In contrast, the i2geo

---

[11] Technical details on the open-source XCLAMS project can be found `http://xclams.xwiki.org`

search links them to *topic pages* (except for the suggested queries which display the number of matching resources and link to the search result). It is probable that the approach of presenting this as a pop-up window, as i2geo does it, is inappropriate for contemporary web habits. However the topic pages provide information to users. The MSC Catalog and euDML search engine also render the topic in its environment wrt other topics; however no further information is provided about each of the subjects.

Should such links be such as euDML? As a drill-down of the search? As a topic page? Topic pages could include illustrations, or links to a Wikipedia or MathWorld, which would draw on users visual memory, so that they are supported in recognizing topics and thus better remember them.

Alternatively, a few attempts have been made to navigate through mathematical topics by the display of a graph that represents a map of the topics' relationships. Examples include the server `http://thesaurus.maths.org/` (now closed), the knowledge map of the Khan Academy's Knowledge Map (`https://www.khanacademy.org/exercisedashboard`), or the MSC map at `http://map.mathweb.org`. However, all these approaches consume a large screen space and are thus difficult to combine with the search activity.

It remains open if additional navigation mechanisms, be them topic pages or graphical knowledge maps, are effective in providing the users of the search tool an awareness of the structure and nature of the topics so that they can be chosen effectively.

6.2   Applicability for MSC Subjects

What are the challenges to port the approach we described to the subjects of the Mathematics Subjects Classification? It relies on a few ingredients: the ontological structure between topics, the multiple names of each so as to allow search and identifiable display, and the existence of Wikipedia pages to describe them so that additional relatedness connections are computed.

One of the challenges lies in the user interface: when unknown to the user, subjects are best understood when displayed within the hierarchy as this provides context information. For example, the subject 14Q20 Effectivity, complexity if suggested, has much chances to confuse users, since its belonging to algebraic geometry is not explained. Similarly the subject 97H60 Linear algebra when displayed would miss its belonging to the subject 97-XX Mathematics education.

Another challenge lies in finding pages that form a source of descriptions of subjects to perform the computation of relatedness between the subjects. While pages in Wikipedia exist for main subjects, they are largely missing for more refined subjects.

6.3   Mix and Match on the World Wide Web

More and more markup formats are available to allow web-page authors to indicate inside web-pages such properties as *being of a given topic*. The anchoring of topics within a structure such as a taxonomy or ontology makes them a particularly good source for identifiable keys to describe the topics.

Among the main markups that allow this indication is the family of microdata annotations at `http://schema.org/`, in particular `CreativeWork`'s about property. This property can carry both a text and a URI so that such initiatives as the standardized thesaurus encoding of MSC [IS12] in the SKOS language or the ontological nature of GeoSkills in the OWL language can be easily encoded there.

It appears, from schema.org's intro pages, that such an inclusion in web-pages of digital mathematics libraries would allow the main web search engine to give a significant weight to the topics annotations and be able to exploit the classification structure. Thus they could suggest generalizations to the domain of **category theory** when mentioning **topoi**, or to **perpendicular bisector** when mentioning **circumscribed triangle**. This could complement the suggested queries offered by these systems which are mostly based on the of earlier users' queries and are, thus, almost always absent when inputting refined mathematical topics.

We have observed that regular users commonly use main web search engine in parallel to the search engine of the digital libraries. We expect that the search engines' complementary features are likely to enrich each others, where users are able to depict features of one to describe desired features of others.

7   Conclusion

In this paper, we have described a way to enrich the search engines that employ topics so as to avoid the trap of a too precise topic. This trap seems to be common to all search engines that offer this function, including the exemplary facetted search system of [Hea09, §8.6].

The solution we propose is to enable the user in choosing topics that are closely related to the query by the presentation of suggested topics decorated by the number of matches. For example, it allows a user who has search results about **ellipse** to go in one click to the search results for **conic sections**, to **cone**, **disc**, or **meridian**.

The mathematics learning resources' sharing platform `http://i2geo.net` has integrated this suggestion mechanism. This integration supports the user to explore other search queries that may satisfy better their expectactions. Anchoring this choice in the display of the available data appears to be an important step to guide the user while still avoiding the pages with empty search results.

Users typically lack the knowledge of the topics classifications employed by search engines. The suggestion mechanism allow them to explore related topics. This lack of knowledge is stated quite clearly by the teachers trying to find the history domain but needing to navigate through social studies first (see Section 2). It is also echoed by multiple i2geo users who prefer to switch to text search.

It could be suggested that such a lack of knowledge could be alleviated by designing a more natural hierarchy or by educating the users in the usages of widespread classifications. This would be quite artificial: many hierarchy decisions are natural in one culture and not in others; this is the case for the history domain which is a part of social studies in the American education systems but not in most European ones. Thus, offering a tool to stimulate the users in exploring the related topics is an important way to support the user into gaining a better knowledge of how topics are navigated and to let them come out of the trap of a too precise topic.

7.1 Future Work

Studies currently planned ahead of this research include the following:

A stabilization of the server code is the closest objective. The current implementation, requiring delicate versions of each of the software libraries, has prevented a full deployment to the server. This implementation will reach the users of the `http://i2geo.net` platform which is used regularly by users of Europe and beyond (since last year, a mean of 90 search queries per day has been observed).

Another facet is a broader coverage of definition URLs of topics. On the one hand, a fairly modest count of topics has been enriched with the URL of Wikipedia definitions, while a broad part has been automatically guessed without validation. On the other hand, only three languages have been taken up, French, English, and German. We intend to employ the Curriculum-encoders' voluntaries community to organize such contributions (`http://i2geo.net/xwiki/bin/view/Group_Curriculum-Encoders/`).

These two development works are the basis to get a usable implementation. They are likely to enhance productivity of users that employ topic based search. Early and recent experiments in which pre-service teachers were discovering i2geo.net have confirmed the need for more stimulations to attempt search by other topics. Such discoveries, generally coupled with an introduction in a course, form an important field-trial-like evaluation of the search platform: the little time they allocate to discover the platform and the relative neutrality of students make them good candidates to judge the quality of a prime-time experiment.

Finally, as indicated in [IS12], it is likely that linking across several taxonomies will emerge as a common practice. One of the attempts to do so has been done on `http://i2geo.net/xwiki/bin/view/Subjects/` which takes a handful of relevant mathematical domains as search subjects and maps them to GeoSkills nodes allowing users to search for dynamic geometry constructions in probability and statistics, for example. The mapping there is created by ontological axiom statements making equivalent the subjects to the union of topics and competencies. Such a mapping is currently being leveraged by a team of the Open University of Cyprus for the subjects of the MathTax taxonomy[12]. It will allow fine-grained subjects search on `http://i2geo.net` and on the upcoming `http://opendiscoveryspace.eu` and thus become more compatible with the MSC.

Linking across several classifications, either as done in the i2geo subjects or in other ways described in Section 6.3, is likely to shed a new light on the

---

[12] The MathTax taxonomy is in use in parts of the USA's National Sciencce Digital Library. It can be browsed at `http://people.uncw.edu/hermanr/MathTax/`.

suggestions of related terms since the display of them would be less homogenous. At the same time, such a mix is likely to distill more relevant information from the user's culture and context and thus create more relevant search experiences. It has the potential of exploiting the user's locality or earlier search to influence the weight of suggested queries and thus, for example, prefer suggested queries that are in the user's region's curriculum standard, closer to the user's known concepts, or to the user's currently research topics.

Acknowledgements. This research work has been partially funded by the European Commission under the Policy Support Programme. The opinions expressed in this paper are that of the author. The author wishes to thank Yannis Harlambous for seminal discussions, Cyrille Desmoulins for contributing large parts of the ontological engineering approach of the GeoSkills ontology, and the workers of the Inter2geo eContentPlus project for having provided a realistic environment to deploy modern technology. Moreover, he wishes to thank the shepherd of the programme comittee for the detailed and dedicated review work.

References

Car04.   Cairns, P.: Informalising formal mathematics: searching the mizar library. In: Asperti, A., Bancerek, G., Trybulec, A. (eds.) MKM 2004. LNCS, vol. 3119, pp. 58–72. Springer, Heidelberg (2004),
http://www-users.cs.york.ac.uk/~pcairns/papers/index.html

DL09.   Desmoulins, C., Libbrecht, P.: CompEd, a web-based competency ontology editor for dynamic geometry. In: Dicheva, D., Mizoguchi, R., Pinkwart, N. (eds.) Proceedings of SWEL Workshop 2009 (2009),
http://hcis.in.tu-clausthal.de/pubs/2009/aied/ontologies_and_social_semantic_web_for_intelligent_educational_systems.pdf

Hea09.   Hearst, M.A.: Search User Interfaces. Cambridge University Press (2009),
http://searchuserinterfaces.com/

IS12.   Ion, P., Sperber, W.: MSC 2010 in SKOS – the Transition of the MSC to the Semantic Web. EMS Newsletter 84, 55–57 (2012) ISSN 1027-488X,
http://www.ems-ph.org/journals/newsletter/pdf/2012-06-84.pdf

LD97.   Landauer, T.K., Dumais, S.T.: A solution to plato's problem: The latent semantic analysis theory of acquisition, induction, and representation of knowledge. Psychological Review 104(2), 211 (1997) ISBN 978-1-60750-062-9,
http://lsa.colorado.edu/papers/plato/plato.annote.html

LD09.   Libbrecht, P., Desmoulins, C.: A cross-curriculum representation for handling and searching dynamic geometry competencies. In: Dicheva, D., Mizoguchi, R., Greer, J. (eds.) Semantic Web Technologies for e-Learning. The future of learning, vol. 4. IOS Press (2009) ISBN 978-1-60750-062-9

LK13.   Libbrecht, P., Kortenkamp, U.: The role of metadata in the design of educational activities. In: Mariotti, M.A., et al. (eds.) To appear in Proceedings of the Conference on European Research in Mathematics Education (2013), Pre-conference papers are available at
http://cerme8.metu.edu.tr/wgpapers/wg15_papers.html

LKM09. Libbrecht, P., Kortenkamp, U., Mercat, C.: I2Geo: a Web-Library of Interactive Geometry. In: Proceedings of DML 2009 (July 2009),
`http://i2geo.net/xwiki/bin/view/Main/Proceedings`

Meg12. Megalou, E.: Re: Multilang-search-lors-overview.rtf. Received per email with message id `000301cd2a1a\$e4c11db0\$ae435910\$@gr` (May 2012)

WC10.   Widdows, D., Cohen, T.: The semantic vectors package: New algorithms and public tools for distributional semantics. In: Fourth IEEE International Conference on Semantic Computing (IEEE ICSC 2010). IEEE Computer Society (2010),
`https://code.google.com/p/semanticvectors/wiki/RelatedResearch`

Using MathML to Represent Units of Measurement for Improved Ontology Alignment

Chau Do and Eric J. Pauwels

Centrum Wiskunde & Informatica, 1098 XG Amsterdam, The Netherlands
{do,eric.pauwels}@cwi.nl

Abstract. Ontologies provide a formal description of concepts and their relationships in a knowledge domain. The goal of ontology alignment is to identify semantically matching concepts and relationships across independently developed ontologies that purport to describe the same knowledge. In order to handle the widest possible class of ontologies, many alignment algorithms rely on terminological and structural methods, but the often fuzzy nature of concepts complicates the matching process. However, one area that should provide clear matching solutions due to its mathematical nature, is units of measurement. Several ontologies for units of measurement are available, but there has been no attempt to align them, notwithstanding the obvious importance for technical interoperability. We propose a general strategy to map these (and similar) ontologies by introducing MathML to accurately capture the semantic description of concepts specified therein. We provide mapping results for three ontologies, and show that our approach improves on lexical comparisons.

Keywords: MathML, ontology matching, ontology alignment, units of measurement.

1 Introduction and Motivation

An increasing number of scientific and technological areas, including multi-agents, bioinformatics and the semantic web, are making use of ontologies to better represent their knowledge domain. An ontology describes a domain of interest by presenting a vocabulary as well as definitions of the terms used in the vocabulary [1]. With independent individuals and groups developing their own ontologies, we are faced with the problem of heterogeneous representation across ontologies. This is quite problematic when it becomes necessary to amalgamate or link data between various sources. Over the years, several solutions have been proposed for matching ontologies (i.e. identify corresponding or matching terms in different ontologies). Most take a generic approach in order to deal with the widest possible variety of ontologies from various domains. Consequently, these matchers do not take advantage of domain specific attributes which could lead to better matches.

One area of application where a domain-agnostic approach might be suboptimal concerns *units of measurement*. Units are particularly interesting since,

J. Carette et al. (Eds.): CICM 2013, LNAI 7961, pp. 310–325, 2013.
© Springer-Verlag Berlin Heidelberg 2013

unlike more common concepts which often carry multiple meanings, they have a clear definition due to an inherent mathematical structure. For example, *person* can be interpreted as either equivalent to *human* or as a subclass of *human*, both alignments are acceptable depending on the application. Contrast this to units of measurement, where there are well established rules, such that a unit defined in one ontology should only be matched to equivalent units in another ontology. Aligning units of measurement ontologies is of particular importance to areas requiring data sharing or conversion between units, just to name a few. For example, independent sensor networks may use different ontologies to represent their measurements and a mapping is required when their data is consolidated.

The aim of this paper is to propose a semi-automatic solution for the problem of aligning units of measurement ontologies. The solution we propose hinges on the use of MathML to extend the semantic description of the units that already exist in the ontology. To understand the underlying idea of our approach, consider two ontologies, Ω_1 and Ω_2, containing definitions for units of measurement. Assuming that these definitions contain both the dimensions and conversion values of the units, it cannot be assumed that the way this information is represented and encoded is similar. For example, assume in ontology Ω_1 the unit *degree Celsius* is denoted as *degreeCelsius* respectively, whereas in ontology Ω_2 it is known as *ThermoUnit_C*. More often than not, these two ontologies will have been developed by groups working independently. To circumvent this problem, we propose to insert for each unit in an ontology a MathML-encoded description using the information available in the ontology. In the *degree Celsius* example, a straightforward search based on a generic lexical comparison would find it difficult to spot this match. But if both ontologies contain the MathML-encoded relationship between Celsius and the base unit Kelvin, i.e. $T_c = T_k - 273.15$, then matching of these terms across the ontologies becomes trivial.

Our choice for using MathML in these encodings is motivated by the fact that this is already a widely accepted language for describing mathematical equations. Furthermore, due to its standardization, it is possible to write a generic program that can process different equations. Although in general, matching ontologies is difficult, the problem is made more manageable through the creation of a richer set of structures and relationships, which encode the precise mathematical relationships that exist between measurement units. This allows for more exact matchings as well as non-obvious ones (e.g. *NewtonPerMeter* and *joule_per_square_metre*).

This paper focuses on units defined in RDFS/OWL ontologies. For ontologies, even the ones within the confines of RDFS and OWL, there is no explicit requirement to represent the mathematical structure of units. The definitions vary from extremely minimal (for example only the names of the units) to more complex (some ontologies define dimensions, conversions, alternative symbols and so on). Even in the latter case, there is no clear and consistent manner for representing the mathematical structure. For example, to denote division between two units, one ontology defines the properties *numerator* and *denominator*, while another merely defines the property *hasOperand* and indicates the division by the inversion of the unit (e.g. *perKilogram*). To make matters worse, the labeling of concepts in ontologies are different (some examples encountered were: *cubic_metre* vs. *meterCubed*, *Vector_L1* vs. *SI_length_dimension_exponent* for the length dimension). Additionally, the structure and organization of concepts within ontologies can vary. Due to these variations, there is no logical link between units in different ontologies.

The semantic web is composed of layers, each building upon the previous one. The ontology vocabulary level defines the terms and relationships for concepts. The layer above this, the logic level, builds upon this foundation using reasoners to provide inferences. Reasoners lack the arithmetic skills to spot the correspondence between a statement such as "1 week *hasDuration* 7 days" in one ontology and the concatenation of statements "1 day *hasDuration* 86400 secs" and "1 week *hasDuration* 604800 secs" in another. Providing mathematical descriptions of these facts (encoded in MathML which is amenable to manipulations by software such as Mathematica) creates new opportunities for more effective identification.

The rest of the paper is structured as follows: first, in section 2 we describe some background information and related work. This is followed by a description of our proposed solution in section 3. Section 4, outlines the application of our approach to real life ontologies. Finally, section 5 outlines the results, followed by the conclusions and future work in section 6.

2   Background and Related Work

While this paper focuses on units represented within RDFS/OWL ontologies, units have been considered and represented in other related areas. OpenMath for example, deals with units using content dictionaries (CD). In [2] and [3], the representation of units in CDs is proposed and discussed. A question of whether or not RDF is a more suitable means of representation is also raised, but not definitively dealt with by the authors. [4] builds upon these CDs and suggests changes for better conformance to the SI standard. As will be explained later on, our method of inserting MathML into existing ontologies utilizes the information available within them. The information is extracted and MathML is automatically generated without any recognition of the unit that is being processed. Therefore we do not attempt to match the units to ones available in CDs.

An alignment between ontologies is described as a set of correspondences. Correspondences represent a relationship (equivalence, subclass, disjointness, etc.) between the entities of the two ontologies being aligned. Entities here can refer to classes, properties, individuals and so on. Consider two ontologies, Ω_1 and Ω_2, to be aligned, where Ω_1 has the class *dog*, denoted here as Ω_1:dog and Ω_2 has the classes *animal* and *canine*, denoted as Ω_2:animal and Ω_2:canine. The correspondences that make up the alignment would be:

1. Ω_1:dog *is a subclass of* Ω_2:animal
2. Ω_1:dog *is equivalent to* Ω_2:canine

The matching process can have additional inputs, such as a partial initial alignment, weights and thresholds (varies between matching algorithms) and sources for common knowledge (e.g. WordNet [5], UMLS [6]). Depending on the matching algorithm, the correspondence may have a level of confidence (normally between 0 and 1) associated with it [7].

Over the years, many ontology matching systems have been proposed, some of which are summarized in [1], [8], [9]. Although the approach taken by each system is different, most are based on terminological and structural methods. Terminological methods refers to the use of lexical comparisons of the labels, comments and/or other annotations of each entity. Structural comparisons look at for example similarities in the hierarchy of the ontology structure or the corresponding neighbors of matched entities. Semantic methods can also be applied for verification of matches or building on initial matches. These methods include looking at the range of values, cardinality, the transitivity and symmetry of the entities [7].

In an effort to find a common basis on which to compare ontology matching systems, the Ontology Alignment Evaluation Initiative (OAEI [10]) was formed. The initiative is composed of several tracks dealing with ontologies from different areas such as biomedical, conferences and anatomy. In particular the benchmark tests (see [11] for more information) have generated quantitative results, allowing for the comparison between different matching systems and tracking of advancements in these systems. It is clear that many of the systems are generic matchers, while some have more inclination towards specific areas (e.g. ASMOV towards the biomedical area). This is further indicated in the test case ontologies, which deal mainly with concepts from various domains.

Our proposed matching allows for an n to m cardinality (n entities in one ontology can align to m entities in the second ontology), which is an important improvement over a simple lexical comparison. Matching systems commonly only produce a one to one alignment [7]. Ones that provide an n to m alignment are AgreementMaker, COMA++ and ASMOV.

AgreementMaker comprises of a first layer, which produces similarity matrices based on concepts between the two provided ontologies. The features of the concepts (e.g. label, comments, annotations) are compared using syntactic and lexical comparisons. The results are fed into the second layer, which uses conceptual or structural methods to improve the results. Descendant's Similarity Inheritance (DSI) and Sibling's Similarity Contribution (SSC) are examples of the algorithms used for this stage. The last layer outputs a final matching or alignment by combining two or more matchers from the previous layers. For the first two layers, several matchers are available for comparison [12].

Similarly, COMA++ is based on an iterative process constructed of three main steps. The first is component identification, where relevant components for matching are determined. The second step is the matcher execution which applies multiple matchers in order to compute component similarities. The final step is similarity combination, where the correspondences between components is found from the calculated similarities [13].

ASMOV initially applies a pre-processing step to the two input ontologies. This step is terminological based and uses either an external thesaurus or string comparison method. The next step comprises of the structural methods, the first being a calculation of relation or hierarchical similarity. The second part comprises of internal or restriction similarity based on the established restrictions between classes and properties. Finally, an extensional similarity is found using data instances in the ontology [14].

Clearly these matchers are generically designed to deal with a wide variety of ontologies. Our approach focuses on the area of units of measurement and applies MathML to better represent the semantics of the units in order to increase correct equivalence alignments.

The incorporation of MathML into ontologies has been done before. For example, the Systems Biology Ontology (SBO) from the European Bioinformatics Institute [15] incorporates subject related equations using MathML. However, other than representing equations, the MathML is not being used further. More interesting usages of MathML can be seen in the Systems Biology of Microorganisms initiative, which has the aim of producing computerized mathematical models representing the dynamic molecular process of a micro-organism [16]. Within this initiative, SysMO Seek is an "assets catalogue" representing information such as models, experiments, and data. MathML is used to represent the mathematical models [17].

Another notable area where MathML and ontologies merge, is the OntoModel tool. Utilized for pharmaceutical product development, OntoModel allows for model creation, manipulation, querying and searching. It uses a combination of Content MathML and OWL. The former is used to represent the mathematical equations and the latter is used for the ontologies that represent the mathematical models and other related information [18].

While SysMo Seek and OntoModel use MathML to represent mathematical equations/models, the MathML is not used to align ontologies as we propose here.

3   Proposed Alignment Approach

The main contention of this paper is that MathML could play a pivotal role in this effort of efficient ontology alignment. MathML comes in two distinct flavors: *Presentation MathML* simply specifies what formulas should look like, while the aim of *Content MathML* is to encode the exact semantics of mathematical expressions. With the introduction of version 3.0, MathML has come closer in-line with OpenMath, particularly with respect to content dictionaries [19]. Obviously, we are interested in *Content MathML* and in the remainder of this paper we will use *MathML* as shorthand for *Content MathML version 3.0*.

Ontologies describing units of measurement routinely provide information on their fundamental physical dimensions (e.g. *length, mass, time*, etc.) and their conversion value (e.g. *Celsius* = 5/9(*Fahrenheit* - 32)). As can be seen, the conversion value includes both a multiplier and an offset. A special case is dimensionless units, which are sometimes represented by an additional "dimension", and must be handled in a slightly different manner (more on this later). Usually, it is understood that the conversions convert back to the SI base and/or derived units. For example *watt* can be described as either *joule per second* ($W = J/s$) or *kilogram-meter-squared per second-cubed* ($W = kg \cdot m^2/s^3$). This wording in itself illustrates another problem with lexical representation. Does the term "squared" correspond to the meter only or to kilogram-meter? Although to a person, this is clear, to a machine different interpretations are possible, unless a convention has already been established. The introduction of MathML resolves this ambiguity. To determine if two units are equivalent, their dimensions and conversion values must match.

The basic idea underpinning our approach is very straightforward. Suppose we have two ontologies, say Ω_1 which relates concepts $\alpha, \beta, \gamma, \ldots$ (we will denote this as $\Omega_1 = \{\alpha, \beta, \gamma, \ldots\}$) and a second ontology $\Omega_2 = \{\xi, \eta, \zeta, \ldots\}$. In addition, let us assume that we are given as prior knowledge that concept α in Ω_1 is equivalent to concept ξ (denoted here as $\Omega_1 : \alpha \leftrightarrow \Omega_2 : \xi$), as well as $\Omega_1 : \beta \leftrightarrow \Omega_2 : \eta$. If we now are able to determine (e.g. using MathML) that $\gamma = \alpha/\beta$ but also that $\zeta = \xi/\eta$ then we can confidently infer the previously unknown match $\Omega_1 : \gamma \leftrightarrow \Omega_2 : \zeta$.

Specifically, our matcher requires three inputs: Ω_1', Ω_2' and initial matchings (prior knowledge). Ω_1' and Ω_2' are ontologies, which have been modified by inserting MathML into them as an alternate representation of their units. In order to align the units in the two ontologies, the MathML representation is compared, with the initial matchings acting as a common reference point. Our matcher will provide correspondences with only equivalence relations. A more detailed explanation is given in the following sections.

3.1 Minimum Prior Knowledge

As pointed out earlier, all units can be described both in *SI base units* and in *derived units* which, in turn, can be re-expressed in base units. Therefore, it can be concluded that all units can be described using the seven SI base units: *meter, kilogram, kelvin, second, candela, ampere* and *mole*. In view of this, in order to match two unit ontologies, only the SI base units need to be matched as a starting point. This is used as the minimum prior knowledge that is required to process the MathML labels. An assumption that our general approach makes is that the unit conversion values are always relative to the SI base units and this is reflected in the MathML comparison. When two units of measurement ontologies are to be matched, the user must supply an initial matching between the base units found in each ontology.

3.2 Generation and Insertion of MathML

The difficulty in generating MathML and inserting it into an existing ontology depends on the structure of the ontology and what information is available in it. For instance, different ontologies will use different properties to indicate that one unit is the quotient of two other units. However, given a well structured ontology that is consistent in how it represents the units, a repeatable pattern will emerge as to where the necessary information for the MathML encoding can be found. A program can be written to automatically process these patterns and recast them in MathML code. We inserted MathML into three exisiting ontologies, one of which will be looked at in more detail later on. But in the end, the effort of inserting MathML will vary from ontology to ontology.

For the purposes of our approach, the MathML need only be inserted such that it is accessible by our matcher. Consequently, there is no need for integrating the MathML with the existing ontology such that any external semantic reasoner (e.g. Fact++, Pellet) can process it. With this in mind, we take a similar approach as OntoModel, which generates the MathML for an equation and incorporates it as a string into a *hasML* property [18]. While a specific data property could be developed, it did not make sense to create a new ontology just for one data property for the MathML code. Consequently, it was decided to incorporate the MathML into the ontologies as an `rdfs:comment` with an `rdf:parseType="Literal"` to indicate that markup language is being used (see [20] and [21]).

3.3   Processing of MathML

Once the ontologies have MathML inserted to represent their units, the matching process can begin to determine equivalent units. Before a comparison of the MathML code can be done, it must first be extracted from each ontology for every unit. By this, we mean that a search is done in each ontology for an `rdfs:comment` containing MathML code. The assumption by the matcher is that each unit that will be considered and aligned already has the corresponding MathML inserted. Expanding upon this initial matching of units (i.e. aligning entities without corresponding MathML) is a topic of future work (section 6). As noted previously, some units describe their conversions not in SI base units, but derived units. Both approaches are commonly used. Therefore, when comparing the MathML code, it must be checked to see if the units can be broken down further if they are not expressed in terms of base units.

As an example, consider we are given by the user the initial base units matchings of:

$$\Omega_1\text{:meter} \leftrightarrow \Omega_2\text{:metre}$$
$$\Omega_1\text{:kilogram} \leftrightarrow \Omega_2\text{:kilogram}$$
$$\Omega_1\text{:second} \leftrightarrow \Omega_2\text{:second\_time}$$
$$\Omega_1\text{:kelvin} \leftrightarrow \Omega_2\text{:kelvin}$$
$$\Omega_1\text{:candela} \leftrightarrow \Omega_2\text{:candela}$$
$$\Omega_1\text{:ampere} \leftrightarrow \Omega_2\text{:ampere}$$
$$\Omega_1\text{:mole} \leftrightarrow \Omega_2\text{:mole}$$

Where Ω_1 is the first ontology and Ω_2 is the second ontology. We encounter the units Ω_1:joule and Ω_2:newton_metre during the matching process. They are given by the following equations (represented in MathML):

$$\Omega_1 : joule = (\Omega_1 : newton) \times (\Omega_1 : meter) \tag{1}$$

$$\Omega_2 : newton\_metre = \frac{(\Omega_2 : metre)^2 \times (\Omega_2 : kilogram)}{(\Omega_2 : second\_time)^2} \tag{2}$$

Here eq. (1) is expressed in the derived unit of newton, while eq. (2) is expressed completely in base units. Having only the base units as initial matchings, in order to compare these two units, the unit of Ω_1:newton needs to be processed first. Therefore the following equation has to be first determined by the matcher:

$$\Omega_1 : newton = \frac{(\Omega_1 : meter) \times (\Omega_1 : kilogram)}{(\Omega_1 : second)^2} \tag{3}$$

Knowing eq. (3), when the matcher encounters eq. (1), it searches for Ω_1:newton and upon finding it, reconstructs eq. (1) in its base units. Now the two units can be compared, with reference to the initial matchings. Once the dimensions and conversion values match, it can be concluded that the units are equivalent. This does not apply however to dimensionless units. For example, the units radian and steradian are both dimensionless and have a conversion multiplier of 1 and 0 offset. In this case, a lexical comparison (i.e. using different distance measurements) is used. When the comparison is completed, equivalence rules representing the alignment can be created and the results outputted to a file for later processing.

4   Application of Approach

As a proof of concept, the approach outlined in the previous section was applied to three ontologies. The implementation is divided into two phases.

Phase I. involves the following pre-processing steps for each individual ontology:

1. Find dimension and conversion data for the units
2. Generate MathML based on information of previous step and insert as `rdfs:comment`
3. Output modified file of ontology with the MathML code

Phase II. compares two modified ontologies given initial matchings of the base units:

1. Read in the initial matchings file and two ontology files. Extract the MathML.
2. Compare the units (specifically their dimensions and conversion value) and determine which are equivalent
3. Output a file containing equivalence rules between the units of the two ontologies

The approach is broken down into two phases in order to make the implementation more modular . A general program can be written for Phase II, since the MathML is standardized. This program can be reused for comparison between any two units ontologies. The onus of inserting the MathML into the ontology can be placed on either its creator or a third party.

4.1   Phase I

Inserting MathML into Existing Ontologies. The three ontologies looked at in this work are: 1) Quantities, Units, Dimensions and Types (QUDT) [22], 2) Ontologies of units of measure 1.8 (OM) [23] and 3) Semantic Web for Earth and Environmental Terminology (SWEET) version 2.2 [24]. QUDT was originally developed by NASA for the NASA Exploration Initiatives Ontology Models project. It is currently being developed by TopQuadrant (see [25]) and NASA. The OM ontology was developed at Wageningen UR - Food & Biobased Research by the Intelligent Systems group. OM was designed to improve upon the deficiencies found in other units ontologies and was based on standards found in the field of units of measure. Additionally, the ontology is made more accessible by providing web services for things such as listing units by application area and unit conversion [26], [27]. SWEET is another ontology developed by NASA, but this time from the Jet Propulsion Laboratory. The focus of this ontology is on the Earth sciences and it bases its terms on the keywords found in the NASA Global Change Master Dictionary [28]. The three ontologies are supported by prominent organizations, while OM purports to be an improved ontology, designed in light of the flaws of previous units ontologies.

All three ontologies are fairly different in their structures and labeling. As a result different programs were written to insert the MathML into each ontology. However the general approach is similar in that patterns in the structure of the ontologies were first identified. A program was then written, using the Apache Jena library (http://jena.apache.org/) for Java and SPARQL queries, to utilize these patterns in order to extract the necessary information for the resulting MathML equation. Due to space restrictions, we provide a description of only the insertion of MathML into the OM ontology.

The OM ontology is well structured with units broken down into groups based on whether they are single, a multiple, an exponent or comprising of a division and so on. The units' mathematical relationship to other units is further expressed by object and data properties. To make this discussion more concrete, figure 1 shows an example of how unit division (in this case *millimetre per day*) is structured in the OM ontology. Here the unit division breaks down into the numerator and denominator object properties. In this example, they point to *millimetre* and *day* respectively. The two terms and their position in the division operation are clearly indicated through these properties. By following the numerator arm, it is seen that millimetre is comprised of the singular unit metre, which also happens to be a base unit. Millimetre also has a prefix *milli* with a value of 1e-3. This value comprises part of the overall conversion necessary to convert *millimetre per day* to *metre per second*. The denominator path breaks down the

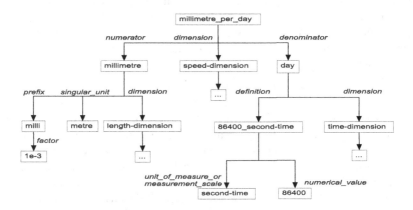

Fig. 1. Breakdown of a unit based on the division of two other units in the OM ontology. Length-dimension and time-dimension further break down into the basic seven dimensions: time, length, mass, amount of substance, temperature, electric-current and luminosity.

unit day into a numerical value of 86400 (the number of seconds in a day) and a unit of measure or measurement scale, second-time. The conversion value is determined as 1e-3/86400. The dimensions can be extracted in several ways. Either the speed-dimension can be directly processed or the length-dimension and time-dimension can be processed with the knowledge that one is the numerator and the other the denominator. Not shown in the figure, is that these dimensions break down into the seven dimensions: time, length, mass, amount of substance, temperature, electric-current and luminosity.

Other units consisting of a division, are represented similarly. The organization is different for unit multiplication, exponentiation and so on. What this example shows is that there are patterns in the OM structure which, once recognized, can be used to automatically determine the conversion value and dimensions of a unit. For example, numerical values found in the numerator section should be divided by the values found in the denominator section. In the case of multiplication where the unit breaks down into term 1 and term 2, values found after processing terms 1 and 2 should be multiplied. The dimensions are either processed directly if the unit has dimension data or constructed from the dimensions of the base units that comprise it.

After studying the OM ontology, we found that only a handful of these patterns exist. Recognizing this made it possible to write a program that searched the ontology, extracted the dimension data and calculated the conversion data.

The approach for the QUDT and SWEET ontologies was the same. In all three, we were able to identify patterns that covered the majority of units. Some units which did not fall within these patterns had to be handled manually. Reasons for this non-conformity vary from unusual units to errors in the ontologies.

Generating MathML. Each unit is represented by an equation which incorporates its dimension and conversion data. In other words the unit is described in terms of its SI equivalent units and the conversion values necessary to convert to these SI units. The general structure of such a conversion equation is shown in eq. 4 below:

$$unit = a \times \frac{[n_1^{x1}][n_2^{x2}][n_3^{x3}]\cdots}{[d_1^{y1}][d_2^{y2}][d_3^{y3}]\cdots} + b \tag{4}$$

Here a represents the conversion multiplier and b the conversion offset of the unit. The variables n_i and d_j represent the different units this unit is comprised of. So as noted before, the latter can be base SI units or derived units. Basically, for our approach to work they can be any other unit as long as it is possible to trace them back to a base SI unit. At least one of n_i and d_j should be present, but there is no limitation on the combination of these variables, this depends on the unit. The structure of the general conversion eq. 4 is fairly straightforward, simplifying the generation of the MathML encoding. An example of MathML code generated and inserted as a label is given in figure 2 for the unit newton.

As can be seen in the figure, the references to the other units in the ontology are given by the `id` attribute. The variables $n1$, $n2$, $d3$ are equivalent to the n_i and d_j in eq. 4. In this manner, the variables show the relationship of one unit to other ones in the ontology, which can eventually be traced back to the SI base units. After the MathML is inserted into the `rdfs:comment` of each unit, the modified model of the ontology is outputted to a file.

4.2   Phase II

The implementation of this phase can be a standalone program. It will process ontology files containing MathML in their `rdfs:comment`. In addition, an initial alignment containing equivalences between the seven SI base units is provided to the program. Below is a detailed description of the steps.

Extract MathML. A search through the ontologies for all individuals containing MathML code is initially done. This is done by conducting a SPARQL query for all `rdfs:comment` and a filter is applied for only comments containing MathML. The results of this query are assumed to be all the processable units. In other words, anything without MathML is ignored (see Future Work, section 6). The MathML is then parsed to extract the dimension and conversion data.

Compare Units. Once all the units in each ontology have the necessary information extracted, a comparison can be made using the initial matching data. Since no further information is known about the ontologies, a very general approach was taken. In the first pass, each unit in one ontology is compared to all the ones in the other. To compare the dimension data, the initial matching units and units that have already been found to be the same, are referred to. The reason for this is, as mentioned before in section 3.3, some units may be described in terms of other ones. Hence, a second pass is necessary to catch all the

```
<math xmlns="http://www.w3.org/1998/Math/MathML">
    <bind><csymbol cd="fns1">lambda</csymbol>

        <bvar><ci id="myOntology:Meter">n1</ci></bvar>
        <bvar><ci id="myOntology:Kilogram">n2</ci></bvar>
        <bvar><ci id="myOntology:Second-Time">d3</ci></bvar>

        <apply><csymbol cd="arith1">divide</csymbol>

            <apply><csymbol cd="arith1">times</csymbol>
                <apply><csymbol cd="arith1">power</csymbol>
                    <ci xref="myOntology:Meter">n1</ci><cn>1</cn>
                </apply>
                <apply><csymbol cd="arith1">power</csymbol>
                    <ci xref="myOntology:Kilogram">n2</ci><cn>1</cn>
                </apply>
            </apply>

            <apply><csymbol cd="arith1">power</csymbol>
                <ci xref="myOntology:Second-Time">d3</ci><cn>2</cn>
            </apply>

        </apply>
    </bind>
</math>
```

Fig. 2. Sample MathML for unit *newton* (N). It encodes the fact that $N = (m \cdot kg)/s^2$ Notice how the SI base units are identified using the `id` attribute. The `xref` to the same unit in the $< ci >$ tag makes the relationship more explicit.

units which were not matched due to this reason. The steps of the comparison are summarized in figure 3.

- **Step 1**: First the simplest comparison is made by checking if the offsets of the conversion value are the same. If they are not, the units are not equivalent and a false is returned by the function.
- **Step 2**: Second, the multiplier of the conversion value is compared to see if they are the same.
- **Step 3**: Once the conversion value is confirmed to be the same, the dimensions are looked at next. If the units are expressed in units other than base SI units, these must first be broken down or matched. For example, *tesla* can be given as $T = N/(A \times m)$. If *tesla* is described in terms of *newton (N)* in both ontologies and *newton* has already been matched, then no breakdown is required. Otherwise a search is done for *newton* (already checked units are stored in memory) and if found, T will be modified to $T = (kg \times m)/(A \times m \times s^2)$.
- **Step 4**: The next step, reduce dimensions, checks if there are the same units in the denominator and numerator and reduces them, resulting in $T = kg/(A \times s^2)$.
- **Step 5**: Once the units have been reduced as necessary, they can be compared with reference to the initial mappings and already matched units, to see if they are the same. If they are, the units match. Output the matched units.

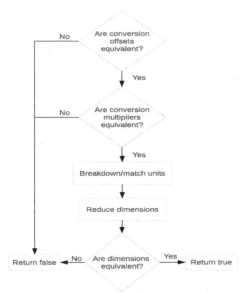

Fig. 3. The steps taken when comparing two units. First the conversion value is compared. If these are equal, the units (representing the dimensions) are matched or broken down into their base SI units if possible. Next, the reduce dimensions step checks if there are the same units in the denominator and numerator and reduces them. Finally, they are compared with reference to already matched units to determine if the dimensions are the same.

5   Results

To evaluate our matching approach, we manually aligned the ontologies for comparison (referred to as reference alignments). Following suit with the OAEI comparisons, we calculate the precision, recall and F-measure. The measurements of precision and recall are well known in information retrieval, but have been modified to take into consideration the semantics of alignments for the purposes of evaluating ontology alignments [29]. For this reason, we use the Alignment API version 4.4 to compare the generated alignments from our method with the reference alignments. The results are given in table 1.

The F-measures, being a combination of the precision and recall values, are fairly good. As a point of reference, the highest F-measure produced by the matchers participating in the OAEI competition from 2007-2010 was around 0.86 [1]. As can be seen the recall values are very good, indicating that most of the alignments in the reference are covered by the generated ones. The precision values are lower, indicating there are a number of false positives (i.e. units that were incorrectly identified as equivalent by the MathML approach). Looking closer at the results, the false positives fall into the following categories:

Table 1. Precision, recall and F-measure values

Alignment	Precision	Recall	F-Measure
OM-QUDT	0.81	0.95	0.87
SWEET-QUDT	0.77	0.97	0.86
SWEET-OM	0.82	0.99	0.90

1. Mathematically equivalent but conceptually different units: There are two sub-types within this category. The first covers matches such as *hertz = becquerel*. While they are mathematically equivalent (both being equal to *1/s*), conceptually they are different, with the former representing frequency and the latter representing radioactive decay. The second sub-type encompasses matches such as *(square meter · steradian) = square meter*. When reduced completely, *steradian* becomes dimensionless and the equation is once again mathematically equivalent. This problem could be dealt with by modifying the *Reduce dimensions* step in the comparison. Both problems could also be handled by adding additional checks (e.g. lexical comparison of the labels).
2. Incorrect information in the ontologies: The insertion of the MathML is dependent on the information in the ontologies and if this information is incorrect, the resulting MathML and therefore the comparison is affected. Several problems were found in each of the ontologies. For example in QUDT, there are incorrect conversion values for the units of *teaspoon, tablespoon* and *centistokes*. Also there are no conversion values for the units of *degree Celsius per minute* and *year tropical*, to name a few. In the OM ontology the dimensions were wrong for the current density dimension and the permittivity dimension. In the SWEET ontology, some of the units were incorrectly composed. For example, the unit joule is only composed of *perSecondSquared* and *kilogram*, missing the *meter squared*.

These issues can be improved upon in future work, which will increase the precision values. Supporting documents for the results can be found at [30].

6   Conclusion and Future Work

Ontology alignment is a difficult problem, but by harnessing domain specific attributes, this problem can be simplified. We have shown that in the area of units of ontologies, MathML can be used to better represent the semantics of the units in order to compare them between ontologies. The generated alignments provide good precision and recall values when compared to manually created reference alignments.

For future work, we intend to improve upon the results by using further checks to ensure that the matched units are conceptually correct as well as mathematically. Furthermore, it will be interesting to look at combining this approach with other methods of ontology alignment. For example, the MathML matching can be used as an initial match in combination with lexical comparisons for non-mathematical concepts. This initial mapping is then fed into an algorithm which considers structural similarities between the two ontologies to build upon the initial matching. Another advantage of inserting MathML is that the information for conversion between units is more explicit. Instead of having to find the dimension information (to see if the units are compatible) and the conversion information within the ontologies, the MathML can be referred to. We intend to explore this area in the future for different applications, such as automatic unit conversion of sensor data between different networks.

Acknowledgements. The authors gratefully acknowledge financial support by CWI's Computational Energy Systems project (sponsored by the Dutch National Science Foundation NWO) and wish to thank Dr. Christoph Lange for his valuable comments.

References

1. Shvaiko, P., Euzenat, J.: Ontology matching: state of the art and future challenges. IEEE Transactions on Knowledge and Data Engineering (2012)
2. Davenport, J.H., Naylor, W.A.: Units and dimensions in OpenMath (2003)
3. Stratford, J., Davenport, J.H.: Unit knowledge management. In: Autexier, S., Campbell, J., Rubio, J., Sorge, V., Suzuki, M., Wiedijk, F. (eds.) AISC/Calculemus/MKM 2008. LNCS (LNAI), vol. 5144, pp. 382–397. Springer, Heidelberg (2008)
4. Collins, J.B.: OpenMath content dictionaries for SI quantities and units. In: Carette, J., Dixon, L., Coen, C.S., Watt, S.M. (eds.) Calculemus/MKM 2009. LNCS, vol. 5625, pp. 247–262. Springer, Heidelberg (2009)
5. Princeton University: WordNet, http://wordnet.princeton.edu/ (accessed on March 2, 2013)
6. U.S. National Library of Medicine: Unified Medical Language System, http://www.nlm.nih.gov/research/umls/ (accessed on March 2, 2013)
7. Euzenat, J., Meilicke, C., Stuckenschmidt, H., Shvaiko, P., Trojahn, C.: Ontology Alignment Evaluation Initiative: six years of experience. In: Spaccapietra, S. (ed.) Journal on Data Semantics XV. LNCS, vol. 6720, pp. 158–192. Springer, Heidelberg (2011)
8. Noy, N.F.: Semantic integration: a survey of ontology-based approaches. SIGMOD Record 33(4), 65–70 (2004)
9. Choi, N., Song, I.Y., Han, H.: A survey on ontology mapping. ACM Sigmod Record 35(3), 34–41 (2006)
10. Ontology Alignment Evaluation Initiative: 2012 campaign, http://oaei.ontologymatching.org/2012/ (accessed on January 9, 2013)
11. Rosoiu, M., dos Santos, C.T., Euzenat, J., et al.: Ontology matching benchmarks: generation and evaluation. In: Proc. 6th ISWC Workshop on Ontology Matching (OM) (2011)
12. Cruz, I.F., Antonelli, F.P., Stroe, C.: AgreementMaker: efficient matching for large real-world schemas and ontologies. Proceedings of the VLDB Endowment 2(2), 1586–1589 (2009)

13. Aumueller, D., Do, H.H., Massmann, S., Rahm, E.: Schema and ontology matching with COMA++. In: Proceedings of the 2005 ACM SIGMOD International Conference on Management of Data, pp. 906–908. ACM (2005)

14. Jean-Mary, Y.R., Shironoshita, E.P., Kabuka, M.R.: Ontology matching with semantic verification. Web Semantics: Science, Services and Agents on the World Wide Web 7(3), 235–251 (2009)

15. The European Bioinformatics Institute: Systems Biology Ontology, http://www.ebi.ac.uk/sbo/main/ (accessed on December 10, 2012)

16. Bechhofer, S., Buchan, I., De Roure, D., Missier, P., Ainsworth, J., Bhagat, J., Couch, P., Cruickshank, D., Delderfield, M., Dunlop, I., et al.: Why linked data is not enough for scientists. Future Generation Computer Systems (2011)

17. Lange, C.: Ontologies and Languages for Representing Mathematical Knowledge on the Semantic Web. Semantic Web Journal 4(2) (2013)

18. Suresh, P., Hsu, S.H., Akkisetty, P., Reklaitis, G.V., Venkatasubramanian, V.: OntoMODEL: ontological mathematical modeling knowledge management in pharmaceutical product development, 1: conceptual framework. Industrial & Engineering Chemistry Research 49(17) (2010)

19. W3C: Mathematical Markup Language (MathML) Version 3.0, http://www.w3.org/TR/MathML3/ (accessed on February 11, 2013)

20. W3C: RDF Vocabulary Description Language 1.0: RDF Schema, http://www.w3.org/TR/rdf-schema/ (accessed on November 20, 2012)

21. W3C: RDF Primer, http://www.w3.org/TR/2004/REC-rdf-primer-20040210/#xmlliterals (accessed on March 2, 2013)

22. Hodgson, R., Keller, P.: QUDT - Quantities, Units, Dimensions and Data Types in OWL and XML, http://www.qudt.org/ (accessed on November 14, 2012)

23. Rijgersberg, H., Wigham, M., Broekstra, J., van Assem, M., Top, J.: Ontology of units of Measure (OM), http://www.wurvoc.org/vocabularies/om-1.8 (accessed on November 14, 2012)

24. NASA: Jet Propulsion Laboratory: Semantic Web for Earth and Environmental Terminology (SWEET), http://sweet.jpl.nasa.gov/ (accessed on November 14, 2012)

25. TopQuadrant, http://www.topquadrant.com/ (accessed on November 15, 2012)

26. Rijgersberg, H., Wigham, M., Top, J.: How semantics can improve engineering processes: A case of units of measure and quantities. Advanced Engineering Informatics 25(2) (2011)

27. Rijgersberg, H., van Assem, M., Top, J.: Ontology of units of measure and related concepts. Semantic Web 4(1) (2013)

28. Raskin, R.: Semantic Web for Earth and Environmental Terminology (SWEET), esto.ndc.nasa.gov/conferences/estc2003/papers/A7P2(Raskin).pdf (accessed on November 14, 2012)

29. Euzenat, J.: Semantic precision and recall for ontology alignment evaluation. In: Proc. 20th International Joint Conference on Artificial Intelligence (IJCAI), pp. 348–353 (2007)

30. Do, C.: Personal website, http://homepages.cwi.nl/~do/home.html (created on December 4, 2012)

A Web Interface for Isabelle: The Next Generation

Christoph Lüth and Martin Ring*

Deutsches Forschungszentrum für Künstliche Intelligenz
Bremen, Germany

Abstract. We present Clide, a web interface for the interactive theorem prover Isabelle. Clide uses latest web technology and the Isabelle/PIDE framework to implement a web-based interface for asynchronous proof document management that competes with, and in some aspects even surpasses, conventional user interfaces for Isabelle such as Proof General or Isabelle/jEdit.

1 Introduction

Recent advances in web technology, which can succinctly if not quite accurately be summarised under the 'HTML5' headline, let us develop interfaces of near-desktop quality, leveraging the advantages of the web without diminishing the user experience. Web interfaces do not need much resources on the user side, are portable and mobile, and easy to set up and use, as all the user needs is a recent web browser (in particular, there can be no missing fonts or packages). The question arises how far we can exploit this technology for a contemporary interactive theorem prover.

This paper reports on such an attempt: a modern, next-generation web interface for the Isabelle theorem prover. Isabelle is a particularly good candidate for this, because it has an interface technology centered around an asynchronous document model. As demonstrated by the system presented here, Clide, we can answer the motivating question affirmatively, modulo some caveats. Readers are invited to try the public test version of the system at `http://clide.informatik. uni-bremen.de` .

2 Basic System Architecture

The basic system architecture is clear: we need a web server to connect with Isabelle on one side, and with web browsers on the other side. Hence, the questions to address are, firstly, how to connect Isabelle with a web server, and secondly, how to use a browser to edit Isabelle theory files?

---

* Research supported by BMBF grant 01IW10002 (SHIP).

J. Carette et al. (Eds.): CICM 2013, LNAI 7961, pp. 326–329, 2013.

Isabelle on the Web. Isabelle poses some specific challenges when implementing a web interface, most of which are common to most interactive theorem provers (or at least those of the LCF family). Firstly, Isabelle's syntax is extremely flexible and impossible to parse outside Isabelle. Thus, the interface needs to interact closely with the prover during the syntactic analysis. Moreover, the provided notation is quite rich, and requires mathematical symbols, super- and subscript, and flexible-width fonts to be displayed adequately.

Secondly, Isabelle's document model is asynchronous [1], meaning that at any time changes of the document can be made by the user (editing the text) or by the prover (parsing the text, or annotating it with results of it being processed). Further, the prover may be slow to respond, or may even diverge. Hence, the communication between the web server and the browser needs to be asynchronous too — the browser needs to be able to react to user input at any given time, and simultaneously needs to be able to process document updates from the prover communicated via the web server.

Finally, there is also something of a 'cultural' gap [1], with sequential theorem provers written in higher-order functional languages on one side, and asynchronous web applications written in imperative languages like Java on the other side. Fortunately, the Scala programming language provides the foundation to unify these worlds, and the Isabelle/PIDE framework [2] crosses that chasm to a large extent. It provides access to Isabelle's document model and interaction from a Scala API, encapsulating Isabelle's ML LCF core; all that remains is to connect it to a web server, and thence a browser. In our application, we use Scala together with the Akka library and Play, a fast, reliable and fully concurrent state-of-the-art framework for web development — and because of Isabelle/PIDE, it seamlessly integrates with Isabelle.

Editing Theory Files. HTTP does not allow *server pushes* where the server initiates messages to the browser, which is essential for an asynchronous model as needed here. There are workarounds such as AJAX and Comet, but for an application like this, where up to a couple of thousand small messages need to be sent per minute, the resulting message overhead is prohibitively expensive. The solution here are *WebSockets*, as introduced in HTML5. They allow for a full-duplex TCP connection with just one single byte of overhead, and are supported by all major browsers in their recent incarnations.

Secondly, Javascript (JS) is still the only viable choice for cross-browser client-side logic. Its significant weaknesses can be ameliorated by libraries such as *BackboneJS* for an MVC architecture on the client, *RequireJS* for modularisation and dependency management, and *CoffeeScript*, a language that compiles to JS but exhibits a clean syntax and includes helpful features like classes.

3   Implementation Considerations

The single most important design constraint was the asynchronous communication between server and browser; Fig. 1 shows the system architecture.

Server   User

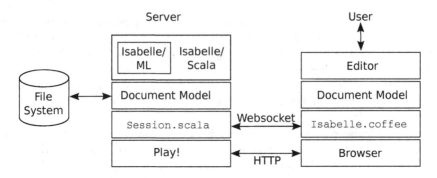

Fig. 1. The Clide system architecture

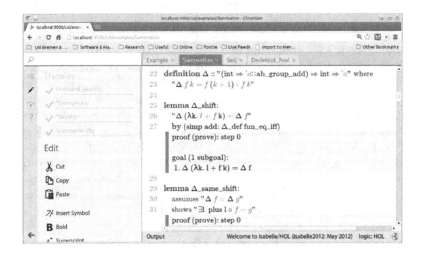

Fig. 2. The Clide interface, with the editor component on the right

System Architecture. The asynchronous document model is implemented by two modules, `Session.scala` and `isabelle.coffee` in Scala and JS, which run on the server and in the browser respectively, and synchronise their document models. The two modules communicate via WebSockets, using a self-developed thin communication layer which maps the relevant Scala types to JS and back, using Scala's dynamic types and JSON serialisation; this way, we can call JS functions from Scala nearly as if they were native, and vice versa.

Interface design. The visual design of the interface is influenced by the Microsoft Design Language [3]. It eschews superfluous graphics in favour of typography (Fig. 2), reducing the interface to the basics such that it does not distract from the center of attention, the proof script. The prover states can be shown inline in the proof script (useful with smaller proof states, or when trying to understand a proof script), or in a dedicated window (useful for large proof states).

The Editor. The interface itself is implemented in JS, using jQuery and other libraries. Its key component is the editor. It needs to be able to display mathematical symbols, Greek letters and preferably everything Unicode; perform on-the-fly replacements (as symbols are entered as control sequences); use flexible-width fonts; allow super- and subscripts; and allow tooltips and hyperlinks for text spans. No available JS editor component provided all of these requirements, so we decided to extend the CodeMirror editor to suit our needs. For mathematical fonts, we use the publicly available MathJax fonts. This results in an editing environment allowing seamless editing of mathematical notation on the web.

4   Conclusions

Clide provides a much richer user experience than previous web interfaces such as ProofWeb [4], which is unsurprising because of the recent advances in web technology mentioned above. Comparing it with the two main other Isabelle interfaces (which are representative for other interactive theorem provers), Proof General [5] and Isabelle/jEdit [2], we find that Clide has a better rendering of mathematical notation. It equals them in terms of responsiveness, and is easier to set up and use. However, as this is still a research prototype the user and project management is rudimentary, and the data storage could be improved by integrating a source code management system or cloud storage on the server. Moreover, we see a great potential in *collaborative* proving with more than one user editing the same theory file at the same time.

In answer to the motivating question, however, we can offer the following: web technology is ready for theorem proving, but still needs to settle down (we had to use lots of different libraries, and expect none of them to be too stable); and Isabelle/Scala is a practically useful foundation to this end, but took some effort to get acquainted with. (We gratefully acknowledge the support of Makarius Wenzel here.) Readers are invited to validate this assessment on their own. A public test version of the system is online, so why not give it a spin?

References

1. Wenzel, M.: Isabelle as document-oriented proof assistant. In: Davenport, J.H., Farmer, W.M., Urban, J., Rabe, F. (eds.) Calculemus/MKM 2011. LNCS, vol. 6824, pp. 244–259. Springer, Heidelberg (2011)
2. Wenzel, M.: Isabelle/jEdit - a prover IDE within the PIDE framework. In: Jeuring, J., Campbell, J.A., Carette, J., Dos Reis, G., Sojka, P., Wenzel, M., Sorge, V. (eds.) CICM 2012. LNCS, vol. 7362, pp. 468–471. Springer, Heidelberg (2012)
3. Microsoft: Ux guidelines for windows store apps (November 2012), http://msdn.microsoft.com/en-us/library/windows/apps/hh465424.aspx.
4. Kaliszyk, C., Raamsdonk, F.V., Wiedijk, F., Hendriks, M., Vrijer, R.D.: Deduction using the ProofWeb system, http://prover.cs.ru.nl/
5. Aspinall, D.: Proof General: A generic tool for proof development. In: Graf, S. (ed.) TACAS 2000. LNCS, vol. 1785, pp. 38–42. Springer, Heidelberg (2000)

The ForMaRE Project –
Formal Mathematical Reasoning in Economics*

Christoph Lange[1], Colin Rowat[2], and Manfred Kerber[1]

[1] Computer Science, University of Birmingham, UK
[2] Economics, University of Birmingham, UK
http://cs.bham.ac.uk/research/projects/formare/

Abstract The ForMaRE project applies formal mathematical reasoning
to economics. We seek to increase confidence in economics' theoretical
results, to aid in discovering new results, and to foster interest in formal
methods, i.e. computer-aided reasoning, within economics. To formal
methods, we seek to contribute user experience feedback from new audiences, as well as new challenge problems. In the first project year, we continued earlier game theory studies but then focused on auctions, where
we are building a toolbox of formalisations, and have started to study
matching and financial risk. In parallel to conducting *research* that connects economics and formal methods, we organise events and provide
infrastructure to connect both *communities*, from fostering mutual awareness to targeted matchmaking. These efforts extend beyond economics,
towards generally enabling domain experts to use mechanised reasoning.

1   Motivation of the ForMaRE Project

The ForMaRE project applies formal mathematical reasoning to economics. Theoretical economics draws on a wide range of mathematics to explore and prove
properties of stylised economic environments. Mathematical formalisation and
computer-aided reasoning have been applied there before, most prominently to
social choice theory (cf., e.g., [3]) and game theory (cf., e.g., [15]). Immediately
preceding ForMaRE, we have ourselves formalised pillage games, a particular
form of cooperative games, and motivated this as follows at CICM 2011 [6]:
1. Economics, and particularly cooperative game theory, is a relatively new area
for mechanised reasoning (still in 2013) and therefore presents a new set of canonical examples and challenge problems. 2. Economics typically involves new
mathematics in that axioms particular to economics are postulated. One of the
intriguing aspects of cooperative game theory is that, while the mathematical
concepts involved are often intelligible to even undergraduate mathematicians,
general theories are elusive. This has made pillage games more amenable to
formalisation than research level mathematics. 3. In economics, as in any other
mathematical discipline, establishing new results is an error-prone process, even
for Nobel laureates (cf. [6] for concrete examples). As one easily assumes false

* This work has been supported by EPSRC grant EP/J007498/1.

J. Carette et al. (Eds.): CICM 2013, LNAI 7961, pp. 330–334, 2013.
© Springer-Verlag Berlin Heidelberg 2013

theorems or overlooks cases in proofs, formalisation and automated validation may increase confidence in results. Knowledge management facilities provided by mechanised reasoning systems may additionally help to reuse proof efforts and to explore theories to discover new results. Despite these potential benefits, economics has so far been formalised almost exclusively by computer scientists, not by economists.

2   The ForMaRE Strategy

The ForMaRE project, kicked off by the authors in May 2012 and further advised by more than a dozen of external computer scientist and economist collaborators, seeks to foster interest in formal methods *within* economics. Our strategy consists in using this technology to establish new results, building trust in formalisation technology and enabling economists to use it themselves.

2.1   Establishing New Results

In preparing one of our first activities, an overview of mechanised reasoning for economists (cf. sec. 3.1), we realised that exciting work was being done in areas with broader audiences than cooperative games. We therefore chose to study auctions, matching markets and financial risk. We have not yet established new results but have defined first research goals with experts in these fields, some of whose works we cite in the following: **auctions** are widely used for allocating goods and services. Novel auctions are constantly being designed – e.g. for allocating new top-level Internet domains [1] – but their complexity makes it difficult to establish basic properties, including their efficiency i.e. give a domain to the registrar who values it highest and is therefore expected to utilise it best. **Matching** problems occur, e.g., in health care (matching kidney donors to patients) and in education (children to schools) [14]. Impossibility results are of particular interest here; they rely on finding rich counter examples. Finally, modern **finance** relies on models to price assets or to compute risk, but banks and regulators still validate and check such models manually. One research challenge is to develop minimal test portfolios that ensure that capital models incorporate relevant risk factors [16].

2.2   Building Trust in Formalisation Technology

Economic theorists typically have a solid mathematics background. There is a field 'computational economics'; however, it is mainly concerned with *numerical* computation of solutions or simulations [4]. Contemporary economists still prove their theorems using pen and paper. While we aim at establishing new results to showcase the potential of formal methods (see above), we also seek to establish confidence in formal methods within the economics community. Thus, as a first step, we have demonstrated the reliability of formal methods by *re-establishing known results*. Computer scientists have previously done so by formalising some of the many known proofs of Arrow's impossibility theorem, a central result of social choice theory [13, 18]. We have started to formalise the review of an influential auction theory textbook [12] in four theorem proving systems in parallel, collaborating with their developers or expert users [10]. This formalisation, currently covering Vickrey's theorem on second price auctions, constitutes the core of an Auction Theory Toolbox (ATT [11]). The review covers 12 more canonical results for single good auctions. We plan to extend the ATT, including new auction designs as well, and welcome contributions from the community.

2.3   Enabling Economists to Use Mechanised Reasoning

Ultimately we aim at enabling economists to formalise their own designs and validate them themselves, or at least to train specialists beyond the core mechanised reasoning community, who will assist economists with formalisation – just like lawyers assist with legal issues. For users without a strong mechanised reasoning background the complexity and abundance of formalised languages and proof assistants poses an adoption barrier. In selected fields, we will provide toolboxes of ready-to-use formalisations of basic concepts, including definitions and essential properties, and guides to extending and applying these toolboxes. Concretely, this means: 1. identifying languages that are (a) sufficiently expressive while still exhibiting efficient reasoning tasks, that are (b) learnable for people used to informal textbook notation, and that (c) have rich libraries of mathematical foundations, and 2. identifying proof assistants that (a) assist with formalisation in a cost-effective way, (b) facilitate reuse from the toolbox, (c) whose output is sufficiently comprehensible to help non-experts understand, e.g., why a proof attempt failed, and (d) whose community is supportive towards non-experts. In building the ATT, we are comparing four different systems, whose philosophies cover a large subset of the spectrum: Isabelle (interactive theorem prover, HOL, accessible via a document-oriented IDE), CASL/Hets (uniform GUI frontend to a wide range of automated FOL provers), Theorema (automated but configurable theorem prover, HOL with custom FOL and set theory inference rules, Mathematica notebook interface with a textbook-like notation), and Mizar (automated proof checker, FOL plus set theory). For details on these systems and how well they satisfy the requirements, see [10].

3   Building, Connecting, and Serving Communities

In parallel to our research on connecting economics and formal methods, we are conducting community building efforts.

3.1   Connecting Computer Science and Economics

With this CICM paper, with an invited lecture at the British Automated Reasoning Workshop [5], and an upcoming tutorial at the German annual computer science meeting themed 'computer science adapted to humans, organisation and the environment' [8], we aim at making developers and users of mechanised reasoning systems, aware of 1. new, challenging problems in the application domain of economics, of 2. new target audiences not having the same background knowledge about formal languages, logics, etc., and thus of 3. the necessity of enhancing the usability and documentation of the systems for a wider audience. Conversely, our message to economists, e.g. in a mechanised reasoning invited lecture at the 2012 summer school of the Initiative for Computational Economics (ICE [4]), is that there is a wide range of tools to assist with reliably solving relevant problems in economics.

3.2 Infrastructure for the Community

With the `ForMaRE-discuss@cs.bham.ac.uk` mailing list and a project community site (both linked from our homepage), we furthermore provide infrastructure to the communities we intend to connect. The main purpose of the community site is to collect pointers to existing formalisations of theorems, models and theories in economics [2], inspired by Wiedijk's list of formalisations of 100 well-known theorems [17], and to give a home to economics formalisations not published online otherwise. The site is powered by Planetary [7], a mathematics-aware web content management system with LaTeX input, a format familiar to economists.

3.3 Reaching Out to Application Domains beyond Economics

Finally, we are reaching out to further application domains beyond economics. At our symposium on *enabling domain experts to use formalised reasoning* (Do-Form [9]), economics and its formalisation was a strong showcase, with our expert collaborators working on auctions, matching and finance giving hands-on tutorials (cf. sec. 2.1), but we also attracted submissions on domains as diverse as environmental models and autonomous systems and on tools from controlled natural language to formal specification. Do-Form has aimed at connecting domain experts having problems ('nails') and computer scientists developing systems ('hammers') from the start of its novel submission and review process, which involved match-making. We initially invited short hammer and nail descriptions. We published the accepted submissions with editorial summaries and indications of possible matches[1] online and then called for the second round of submissions: revisions of the initial submissions (now elaborating on possible matches), regular research papers or system descriptions, particularly encouraging new authors to match the initial submissions. This finally resulted in 12 papers.

We believe that such community-building efforts, which originated from ForMaRE's goal to apply formal mathematical reasoning in economics, will also help to achieve closer collaboration *within* the CICM community[2]: In future, CICM attendees and reviewers reading this paper might point us to the best tools for formalising auctions, matching markets, and financial risk.

---

[1] E.g., we pointed out to the authors of a hammer description that their system might be applicable to the problem mentioned in some nail description.

[2] This was one of the topics discussed in the 2012 MKM trustee election.

References

1. Applicant auction for top-level domains, http://applicantauction.com
2. Formalising '100' theorems/models/theories in economics,
   http://cs.bham.ac.uk/research/projects/formare/planetary/content/100-theorems
3. Geist, C., Endriss, U.: Automated search for impossibility theorems in social choice theory: ranking sets of objects. Artificial Intelligence Research 40 (2011)
4. Initiative for Computational Economics, http://ice.uchicago.edu
5. Kerber, M.: Automated Reasoning for Economics. Invited lecture at the Automated Reasoning Workshop (ARW) (2013)
6. Kerber, M., Rowat, C., Windsteiger, W.: Using *Theorema* in the Formalization of Theoretical Economics. In: Davenport, J.H., Farmer, W.M., Urban, J., Rabe, F. (eds.) Calculemus/MKM 2011. LNCS (LNAI), vol. 6824, pp. 58–73. Springer, Heidelberg (2011)
7. Kohlhase, M.: The PLANETARY Project: Towards eMath3.0. In: Jeuring, J., Campbell, J.A., Carette, J., Dos Reis, G., Sojka, P., Wenzel, M., Sorge, V. (eds.) CICM 2012. LNCS (LNAI), vol. 7362, pp. 448–452. Springer, Heidelberg (2012)
8. Lange, C., Kerber, M., Rowat, C.: Applying Mechanised Reasoning in Economics – Making Reasoners Applicable for Domain Experts. Tutorial at the annual meeting of the German Informatics Society (2013),
   http://cs.bham.ac.uk/research/projects/formare/events/informatik2013
9. Lange, C., Rowat, C., Kerber, M. (eds.): Enabling Domain Experts to use Formalised Reasoning. Symposium at the AISB Annual Convention (2013),
   http://cs.bham.ac.uk/research/projects/formare/events/aisb2013/
10. Lange, C., Caminati, M.B., Kerber, M., Mossakowski, T., Rowat, C., Wenzel, M., Windsteiger, W.: A Qualitative Comparison of the Suitability of Four Theorem Provers for Basic Auction Theory. In: Carette, J., Aspinall, D., Lange, C., Sojka, P., Windsteiger, W. (eds.) CICM 2013. LNCS (LNAI), vol. 7961, pp. 200–215. Springer, Heidelberg (2013); arXiv:1303.4193 [cs.LO]
11. Lange, C., et al.: Auction Theory Toolbox (2013),
    http://cs.bham.ac.uk/research/projects/formare/code/auction-theory/
12. Maskin, E.: The unity of auction theory: Milgrom's master class. Economic Literature 42(4) (2004)
13. Nipkow, T.: Social choice theory in HOL: Arrow and Gibbard-Satterthwaite. Automated Reasoning 43(3) (2009)
14. Sönmez, T., Unver, M.U.: Matching, Allocation, and Exchange of Discrete Resources. In: Handbook of Social Economics, vol. 1A. North-Holland (2011)
15. Tang, P., Lin, F.: Discovering theorems in game theory: two-person games with unique pure Nash equilibrium payoffs. Artificial Intelligence 175(14-15) (2011)
16. Vosloo, N.: Model Validation and Test Portfolios in Financial Regulation. In: Enabling Domain Experts to use Formalised Reasoning. AISB (2013)
17. Wiedijk, F.: Formalizing 100 Theorems, http://cs.ru.nl/~freek/100/
18. Wiedijk, F.: Formalizing Arrow's theorem. Sādhanā 34(1) (2009)

LaTeXml 2012 – A Year of LaTeXml*

Deyan Ginev[1] and Bruce R. Miller[2]

[1] Computer Science, Jacobs University Bremen, Germany
[2] National Institute of Standards and Technology, Gaithersburg, MD, USA

Abstract. LaTeXML, a TeX to XML converter, is being used in a wide range of MKM applications. In this paper, we present a progress report for the 2012 calendar year. Noteworthy enhancements include: increased coverage such as Wikipedia syntax; enhanced capabilities such as embeddable JavaScript and CSS resources and RDFa support; a web service for remote processing via web-sockets; along with general accuracy and reliability improvements. The outlook for an 0.8.0 release in mid-2013 is also discussed.

1 Introduction

LaTeXML [Mil] is a TeX to XML converter, bringing the well-known authoring syntax of TeX and LaTeX to the world of XML. Not a new face in the MKM crowd, LaTeXML has been adopted in a wide range of MKM applications. Originally designed to support the development of NIST's Digital Library of Mathematical Functions (DLMF), it is now employed in publishing frameworks, authoring suites and for the preparation of a number of large-scale TeX corpora.

In this paper, we present a progress report for the 2012 calendar year of LaTeXML's master and development branches. In 2012, the LaTeXML Subversion repository saw 30% of the total project commits since 2006.

Currently, the two authors maintain a developer and master branch of LaTeXML, respectively. The main branch contains all mature features of LaTeXML.

2 Main Development Trunk

LaTeXML's processing model can be broken down into two phases: the basic conversion transforms the TeX/LaTeX markup into a LaTeX-like XML schema; a post-processing phase converts that XML into the target format, usually some format in the HTML family. The following sections highlight the progress made in support for these areas.

2.1 Document Conversion

There has been a great deal of general progress in LaTeXML's processing: the fidelity of TeX and LaTeX simulation is much improved; the set of control sequences covered is more complete. The I/O code has been reorganized to more

---

* The rights of this work are transferred to the extent transferable according to title 17 U.S.C. 105.

J. Carette et al. (Eds.): CICM 2013, LNAI 7961, pp. 335–338, 2013.

closely track TEX's behavior and to use a more consistent path searching logic. It also provides opportunities for more security hardening, while allowing flexibility regarding the data sources, needed by the planned web-services. Together these changes allow the direct processing of many more 'raw' style files directly from the TEX installation (i.e., not requiring a specific LATEXML binding). This mechanism is, in fact, now used for loading input encoding definitions and multilanguage support (`babel`). Additionally, it provides a better infrastructure for sTeX.

The support for colors and graphics has been enhanced, with a more complete color model that captures the capabilities of the `xcolor` package and a move towards generation of native SVG [FFJ03]. A summer student, Silviu Oprea, now at Oxford, developed a remarkable draft implementation supporting the conversion of `pgf` and `tikz` graphics markup into SVG; this code will be integrated into the 0.8 release.

Native support for RDFa has been added to the schema, along with an optional package, `lxRDFa`, allowing the embedding of the semantic annotations within the TEX document. Various other LATEX packages have also been implemented: `cancel`, `epigraph`. Additionally, the `texvc` package provides for the emulation of the `texvc` program used by Wikipedia for processing math markup; this allows LATEXML to be used to generate MathML from the existing wiki markup.

2.2   Document Post-processing

The conversion of the internal math representation to common external formats such as MathML and OpenMath has been improved. In particular, the framework fully supports parallel math markup with cross-referencing between the alternative formats. Thus presentation and content MathML can be enclosed within a `m:semantics` element, with the corresponding `m:mi` and `m:ci` tokens connected to each other via `id` and `xref` attributes.

The evolution of MathML version 3 has also been tracked, as well as the current trends in implementations. Thus, we have shifted towards generating SMP (Supplemental Multilingual Plane, or Plane 1) Unicode and avoiding the `m:mfenced` element. Content MathML generation has been improved, particularly to cover the common (with LATEXML) situation where the true semantics are imperfectly recognized.

Finally, a comprehensive overhaul of the XSLT processing was carried out which avoids the divergence between generation of the various HTML family of markup. The stylesheets are highly parameterized so that they are both more general, and yet allow generation of HTML5 specific markup; they should allow extension to further HTML-like applications like ePub. Command-line options make these parameters available to the user.

While the stylesheets are much more consistent and modular, allowing easy extension and customization, other changes lessen the need to customize. The set of CSS class names have been made much more consistent and predictable, if somewhat verbose, so that it should be easier for users to style the generated HTML as they wish. Additionally, a `resource` element has been defined which allows binding developers to request certain CSS or JavaScript files or fragments to be associated with the document. A converted AMS article, now finally looks (somewhat) like an AMS article!

2.3 Unification

Although the separation of the conversion and post-processing phases is a natural one from the developer's document processing point of view, it is sometimes artificial to users. Moreover, keeping the phases too far separated inhibits interesting applications, such as envisioned by the Daemon (see section 3) and automated document processing systems such as the one used for arXMLiv. Thus, we have undertaken to bring all processing back under a single, consistent, umbrella, whether running in command-line mode, or in client/server mode. The goal is to simplify the common use-case of converting a single document to HTML, while still enabling the injection of intermediate processing.

Some steps in that direction include more consistent error reporting at all phases of processing, with embedded 'locator' information so that the original source of an error can (usually) be located in the source. Additionally, logs include the current SVN revision number to better enable tracking and fixing bugs.

3 Daemon Experimental Branch

The Daemon branch [Gina] hosts experimental developments, primarily the development of client/server modules that support web services, optimize processing and improve the integration with external applications. Since the last report in CICM's S&P track [GSK11], the focus has fallen on increasing usability, security and robustness.

The daemonized processing matured into a pair of robust HTTP servers, one optimized for local batch conversion jobs, the other for a real-time web-service, and a turnkey client executable that incorporates all shapes and sizes of LATEXML processing. Showing a commitment to maintaining prominent conversion scenarios, shorthand user-defined **profiles** were introduced in order to simplify complex LATEXML configurations, e.g. those of sTeX and PlanetMath[Pla]. An internal redesign of the configuration setup and option handling of LATEXML contributed to facilitating these changes and promises a consistent internal API for supporting both the core and post-processing conversion phases.

The RESTful [Fie00] web service offered via the Mojolicious [Rie] web framework now also supports multi-file LATEX manuscripts via a ZIP archive workflow, also facilitated by an upload interface. Furthermore, the built-in web editor and showcase [Ginb] is available through a websocket route and enjoys an expanded list of examples, such as a LATEX Turing machine and a PSTricks graphic.

A significant new experimental feature is the addition of an ambiguous grammar for mathematical formulas. Based on Marpa [Keg], an efficient Earley-style parser, the grammar embraces the common cases of ambiguity in mathematical expressions, e.g. that induced by invisible operators and overloaded operator symbols, in an attempt to set the stage for disambiguation to a correct operator tree. The current grammar in the main development trunk is heuristically geared to unambiguously recognize the mathematical formulas commonly used in DLMF and parts of arXiv. The long-term goal is for the ambiguous grammar to meet parity in coverage and implement advanced semantic techniques in order to establish the correct operator trees in a large variety of scientific domains.

It is anticipated that the bulk of these developments will be merged back into the main trunk for the 0.8 release. The new ambiguous grammar and Mojolicious web service are two notable exceptions, which will not make master prior to the 0.9 release.

4   Outlook

Although development was never stagnated, an official release is long overdue; a LaTeXML 0.8 release is planned for mid-2013. It will incorporate the enhancements presented here: support for several LaTeX graphics packages, such as Tikz and Xypic; an overhauled XSLT and CSS styling framework; and a merge of daemonized processing to the master branch.

References

[FFJ03] Ferraiolo, J., Fujisawa, J., Jackson, D.: Scalable Vector Graphics (SVG) 1.1 Specification. W3C Recommendation. World Wide Web Consortium (W3C) (January 14, 2003),
http://www.w3.org/TR/2008/REC-SVG11-20030114/

[Fie00] Fielding, R.T.: Architectural Styles and the Design of Network-based Software Architectures. PhD thesis. University of California, Irvine (2000),
http://www.ics.uci.edu/~fielding/pubs/dissertation/top.htm

[Gina] Ginev, D.: LaTeXML: A LaTeX to XML Converter, arXMLiv branch,
https://svn.mathweb.org/repos/LaTeXML/branches/arXMLiv (visited on March 12, 2013)

[Ginb] Ginev, D.: The LaTeXML Web Showcase,
http://latexml.mathweb.org/editor (visited on March 12, 2013)

[GSK11] Ginev, D., Stamerjohanns, H., Miller, B.R., Kohlhase, M.: The LaTeXML Daemon: Editable Math on the Collaborative Web. In: Davenport, J.H., Farmer, W.M., Urban, J., Rabe, F. (eds.) Calculemus/MKM 2011. LNCS (LNAI), vol. 6824, pp. 292–294. Springer, Heidelberg (2011),
https://svn.kwarc.info/repos/arXMLiv/doc/cicm-systems11/paper.pdf

[Keg] Kegler, J.: Marpa, A Practical General Parser. System homepage at
http://jeffreykegler.github.com/Marpa-web-site/

[Mil] Miller, B.: LaTeXML: A LaTeX to XML Converter,
http://dlmf.nist.gov/LaTeXML/ (visited on March 12, 2013)

[Pla] PlanetMath.org – Math for the people, by the people (March 2013),
http://www.planetmath.org

[Rie] Riedel, S.: Mojolicious - Perl real-time web framework. System homepage at
http://mojolicio.us/

The MMT API: A Generic MKM System

Florian Rabe

Computer Science, Jacobs University Bremen, Germany
http://trac.kwarc.info/MMT

Abstract. The MMT language has been developed as a scalable representation and interchange language for formal mathematical knowledge. It permits natural representations of the syntax and semantics of virtually all declarative languages while making MMT-based MKM services easy to implement. It is foundationally unconstrained and can be instantiated with specific formal languages.

The MMT API implements the MMT language along with multiple backends for persistent storage and frontends for machine and user access. Moreover, it implements a wide variety of MMT-based knowledge management services. The API and all services are generic and can be applied to any language represented in MMT. A plugin interface permits injecting syntactic and semantic idiosyncrasies of individual formal languages.

The MMT *Language.* Content-oriented representation languages for mathematical knowledge are usually designed to focus on either of two goals: (i) the automation potential offered by mechanically verifiable representations, as pursued in semi-automated proof assistants like Isabelle and (ii) the universal applicability offered by a generic meta-language, as pursued in XML-based content markup languages like OMDOC. The MMT language [11] (Module system for Mathematical Theories) was designed to realize both goals in one coherent system. It uses a minimal number of primitives with a precise semantics chosen to permit natural and adequate representations of many individual languages.

A key feature is *foundation-independence*: MMT systematically avoids a commitment to a particular type theory or logic. Instead, it represents every formal system as an MMT theory: domain theories (like the theory `Group`), logics (like first-order logic FOL), and logical frameworks (like LF [4]) are represented uniformly as MMT *theories*. These theories are related by the *meta-theory* relation, e.g., LF is the meta-theory of FOL, which in turn is the meta-theory of `Group`. MMT uses this relation to obtain the semantics of a theory from that of its meta-theory; thus, an external semantics (called the *foundation*), e.g., a research article or an implementation, only has to be supplied for the topmost meta-theories. For example, a foundation for LF can be given in the form of a type system.

Theories contain typed *symbol declarations*, which permit the uniform representation of constants, functions, and predicates as well as – via the Curry-Howard correspondence – judgments, inference rules, axioms, and theorems. Theories are related via *theory morphisms*, which subsume translations, functors, and models.

J. Carette et al. (Eds.): CICM 2013, LNAI 7961, pp. 339–343, 2013.

Finally, MMT provides a module system for building large theories and morphisms via reuse and inheritance.

Mathematical objects such as terms, types, formulas, and proofs are represented uniformly as OPENMATH objects [1], which are formed from the symbols available to the theory under consideration: For example, LF declares the symbols type and λ; FOL declares $\forall$ and $\Rightarrow$, and Group declares $\circ$ and e. MMT is agnostic in the typing relation between these objects and instead delegates the resolution of typing judgments to the foundation. Then all MMT results are obtained for arbitrary foundations. For example, MMT guarantees that theory morphisms translate objects in a typing- and truth-preserving way, which is the crucial invariant permitting the reuse of results in large networks of theories.

The MMT API. Exploiting the small number of primitives in MMT, the MMT API provides a comprehensive, scalable implementation of MMT itself and of MMT-based knowledge management (KM) services. The development is intentionally *application-independent*: It focuses on the data model of MMT and its KM services in a way that makes the integration into specific applications as easy as possible. But by itself, it provides only a basic user interface.

All algorithms are implemented generically and relegate all foundation-specific aspects to plugins. Concrete applications usually provide a few small plugins to customize the behavior to one specific foundation and a high-level component that connects the desired MMT services to a user interface.

The API is written in the functional and object-oriented language Scala [9], which is fully compatible with Java so that API plugins and applications can be written in either language. Excluding plugins and libraries, it comprises over $20,000$ lines of Scala code compiling into about 3000 Java class files totaling about 5 MB of platform-independent bytecode. Sources, binaries, API documentation, and user manual are available at http://trac.kwarc.info/MMT.

Knowledge Management Services. The MMT API provides a suite of coherently integrated KM services, which we only summarize here because they have been presented individually. A *notation language* based on [7] is used to serialize MMT in arbitrary output formats. Notations are grouped into *styles*, and a rendering engine presents any MMT concept according to the chosen style.

MMT content can be organized in *archives* [5], a light-weight project abstraction to integrate source files, content markup, narrative structure, notation indices, and RDF-style relational indices. Archives can be built, indexed, and browsed, and simplify distribution and reuse. A *query language* [10] integrates hierarchic, relational, and unification-based query paradigms. A *change management* infrastructure [6] permits detecting and propagating and changes at the level of individual OPENMATH objects.

User and System Interfaces. If run as a standalone application, the API responds with a shell that interacts via standard input/output. The shell is scriptable, which permits users and application developers to initialize and configure it conveniently. For example, to check the theory *Group*, the initialization script would first register the MMT theory defining the syntax of LF and then a plugin providing a foundation for LF, then register the theory FOL, and finally check the file containing the theory *Group*.

A second frontend is given by an HTTP server. For machine interaction, it exposes all API functionality and KM services via a RESTful interface, which permits developing MMT-based applications outside the Java/Scala world. For human interaction, the HTTP server offers an interactive *web browser* based on HTML+presentation MathML. The latter is computed on demand according to the style interactively selected by the user. Based on the JOBAD JavaScript library [3], user interaction is handled via JavaScript and Ajax. In particular, MMT includes a JOBAD module that provides interactive functionality such as definition lookup and type inference.

To facilitate distributing MMT content, all MMT declarations are referenced by canonical *logical identifiers* (the MMT URIs), and their physical locations (their URLs) remain transparent. This is implemented as a catalog that translates MMT URIs into URLs according to the registered knowledge repositories. MMT declarations are retrieved and loaded into memory transparently when needed so that storage and memory management are hidden from high-level services, applications, and users. Supported knowledge repositories are file systems, SVN working copies and repositories, and TNTBASE databases [12]. The latter also supports MMT-specific indexing and querying functions [8] permitting, e.g., the efficient retrieval of the dependency closure of an MMT knowledge item.

A Specific Application for a Specific Foundation. The LATIN project [2] built an atlas of logics and related formal systems. The atlas is realized as an MMT project, and MMT is used for building and interactively browsing the atlas.

All theories in the atlas use LF as their meta-theory, which defines the abstract syntax of LF and thus of the logics in the atlas.

For concrete syntax, the Twelf implementation of LF is used. To integrate Twelf with MMT, LATIN developed an MMT plugin that calls Twelf to read individual source files and convert them to OMDOC, which MMT reads natively.

Based on this import, MMT's foundation-independent algorithms can index and catalog the LATIN atlas and make it accessible to KM services and applications. Here the use of Twelf remains fully transparent: An application sends only an MMT URI (e.g., the one LATIN defines for the theory FOL) to MMT and receives the corresponding Scala object.

From the perspective of MMT, Twelf is an external tool for parsing and type reconstruction that is applicable only to theories whose meta-theory is LF. From the perspective of Twelf on the other hand, the MMT theory LF does not exist. Instead, the symbols *type*, λ, etc. are implemented directly in Twelf's underlying programming language.

This is a typical situation: Generally, MMT uses the meta-theory to determine which plugin is applicable, and these plugins hard-code the semantics of the respective meta-theory. Similar concrete syntax plugins can be written for most languages and exist for, e.g., the ATP interface language TPTP, the ontology language OWL, and the Mizar language for formalized mathematics.

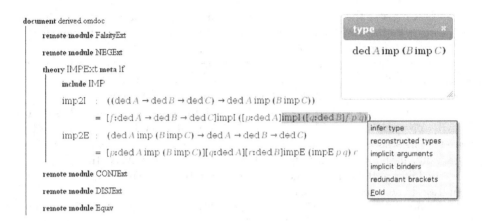

LATIN also customizes the MMT web server by providing a style that provides notations for objects from theories with meta-theory LF. The above screen shot shows the generic web server displaying a theory IMPExt: It imports the theory IMP of the implication connective imp and extends it with derived rules for the introduction and elimination of double implications, i.e., formulas of the form $A \operatorname{imp}(B \operatorname{imp} C)$. The symbol impI represents the derivation of the rule $\frac{A, B \vdash C}{\vdash A \operatorname{imp}(B \operatorname{imp} C)}$. Via the context menu, the user has called type inference on the selected subobject, which opened a dialog showing the *dynamically inferred type*.

The interactive type inference is implemented using the HTTP interface to the MMT API. First of all, the LATIN style is such that the rendered HTML includes parallel markup in the form of special attributes on the presentation MathML elements. JavaScript uses them to build an MMT query that is posted to the server as an Ajax request and whose response is shown in the dialog. This query bundles multiple MMT API calls into a single HTTP request-response cycle: First the parallel markup is used to retrieve the OPENMATH object corresponding to the selected expression (and its context), then type inference is called, and finally the rendering engine is called to render the type as presentation MathML.

For the type inference, LATIN provides one further plugin: a foundation plugin that supplies the typing relation for theories with meta-theory LF. MMT uses it to perform type inference directly in memory without having to call external tools like Twelf. Such foundation plugins are easy to write because they can focus on the logical core of the type system and, e.g., parsing and module system remain transparent to the plugin. For example, the plugin for LF comprises only 200 lines of code

Except for the concrete syntax plugin, the presentation style, and the foundation plugin, all steps of the above example are foundation-independent and are immediately available for MMT content written in any other meta-theory. Moreover, being a logical framework, these plugins LF are immediately inherited by all logics defined in LF: We obtain, e.g., type inference for FOL and *Group* (in fact: for all logics defined in LATIN) without writing additional plugins.

Furthermore, all implementations are application-independent and can be immediately integrated into any application, e.g., a Wiki containing LF objects. This customization of MMT to specific foundations and specific applications occurs at minimal cost, a principle we call *rapid prototyping for formal ystems*.

Acknowledgements. Over the last 6 years, contributions to the API or to individual plugins have been made by Maria Alecu, Alin Iacob, Catalin David, Stefania Dumbrava, Dimitar Misev, Fulya Horozal, Füsun Horozal, Mihnea Iancu, Felix Mance, and Vladimir Zamdzhiev. Some of them were partially supported by DFG grant KO-2428/9-1.

References

1. Buswell, S., Caprotti, O., Carlisle, D., Dewar, M., Gaetano, M., Kohlhase, M.: The Open Math Standard, Version 2.0. Technical report, The Open Math Society (2004), http://www.openmath.org/standard/om20
2. Codescu, M., Horozal, F., Kohlhase, M., Mossakowski, T., Rabe, F.: Project Abstract: Logic Atlas and Integrator (LATIN). In: Davenport, J.H., Farmer, W.M., Urban, J., Rabe, F. (eds.) Calculemus/MKM 2011. LNCS, vol. 6824, pp. 289–291. Springer, Heidelberg (2011)
3. Giceva, J., Lange, C., Rabe, F.: Integrating Web Services into Active Mathematical Documents. In: Carette, J., Dixon, L., Coen, C.S., Watt, S.M. (eds.) Calculemus/MKM 2009. LNCS, vol. 5625, pp. 279–293. Springer, Heidelberg (2009)
4. Harper, R., Honsell, F., Plotkin, G.: A framework for defining logics. Journal of the Association for Computing Machinery 40(1), 143–184 (1993)
5. Horozal, F., Iacob, A., Jucovschi, C., Kohlhase, M., Rabe, F.: Combining Source, Content, Presentation, Narration, and Relational Representation. In: Davenport, J.H., Farmer, W.M., Urban, J., Rabe, F. (eds.) Calculemus/MKM 2011. LNCS, vol. 6824, pp. 212–227. Springer, Heidelberg (2011)
6. Iancu, M., Rabe, F.: Management of Change in Declarative Languages. In: Jeuring, J., Campbell, J.A., Carette, J., Dos Reis, G., Sojka, P., Wenzel, M., Sorge, V. (eds.) CICM 2012. LNCS, vol. 7362, pp. 326–341. Springer, Heidelberg (2012)
7. Kohlhase, M., Müller, C., Rabe, F.: Notations for Living Mathematical Documents. In: Autexier, S., Campbell, J., Rubio, J., Sorge, V., Suzuki, M., Wiedijk, F. (eds.) AISC/Calculemus/MKM 2008. LNCS (LNAI), vol. 5144, pp. 504–519. Springer, Heidelberg (2008)
8. Kohlhase, M., Rabe, F., Zholudev, V.: Towards MKM in the Large: Modular Representation and Scalable Software Architecture. In: Autexier, S., Calmet, J., Delahaye, D., Ion, P.D.F., Rideau, L., Rioboo, R., Sexton, A.P. (eds.) AISC 2010. LNCS, vol. 6167, pp. 370–384. Springer, Heidelberg (2010)
9. Odersky, M., Spoon, L., Venners, B.: Programming in Scala. artima (2007)
10. Rabe, F.: A Query Language for Formal Mathematical Libraries. In: Jeuring, J., Campbell, J.A., Carette, J., Dos Reis, G., Sojka, P., Wenzel, M., Sorge, V. (eds.) CICM 2012. LNCS, vol. 7362, pp. 143–158. Springer, Heidelberg (2012)
11. Rabe, F., Kohlhase, M.: A Scalable Module System. Information and Computation (2013), conditionally accepted http://arxiv.org/abs/1105.0548
12. Zholudev, V., Kohlhase, M.: TNTBase: a Versioned Storage for XML. In: Proceedings of Balisage: The Markup Conference 2009. Balisage Series on Markup Technologies, vol. 3, Mulberry Technologies, Inc. (2009)

Math-Net.Ru as a Digital Archive of the Russian Mathematical Knowledge from the XIX Century to Today

Dmitry E. Chebukov, Alexander D. Izaak, Olga G. Misyurina,
Yuri A. Pupyrev, and Alexey B. Zhizhchenko

Steklov Mathematical Institute of the Russian Academy of Sciences

Abstract. The main goal of the project Math-Net.Ru is to collect scientific publications in Russian and Soviet mathematics journals since their foundation to today and the authors of these publications into a single database and to provide access to full-text articles for broad international mathematical community. Leading Russian mathematics journals have been comprehensively digitized dating back to the first volumes.

1 Introduction

Math-Net.Ru (http://www.mathnet.ru) is an information system developed at the Steklov Mathematical Institute of the Russian Academy of Sciences and designed to provide online access to Russian mathematical publications for the international scientific community. It is a non-profit project supported by the Russian Academy of Sciences and working in the first place with journals founded by the RAS, but covering also other high-quality math journals. The project was started in 2006. Its main idea is to digitize the full archives of leading Russian and Soviet mathematics journals going back to the first volumes. Old Russian and Soviet mathematics journals especially published before the 1930th years could only be found in several libraries, i.e. in fact were hardly available to the public.

The Journals section is a key component of the system. Other sections include Persons, Organizations, Conferences and Video Library [1]. The mobile version of the database http://m.mathnet.ru reproduces the most important functionality of the system but adopted for viewing on smart phones and other mobile devices.

The system has two-server architecture which includes a MSSQL database server powered by Windows 2008 server and an Apache web server powered by Linux. The servers are connected by a 1Gb direct network line and are located in the same server rack. All webserver scripts are written on PHP, also MSSQL stored procedures are used in SQL logic. The database including statistics has size about 80Gb, the total size of the full-text PDF files is about 110 Gb. The total size of videofiles is 2600 Gb.

J. Carette et al. (Eds.): CICM 2013, LNAI 7961, pp. 344–348, 2013.

2 Journals

The section contains a collection of 120 000 articles published in 86 mathematical and physical journals. The number of journals and papers is constantly growing. Most articles were published in Russia, but there are also journals from the former USSR: Ukraine, Belorussia, Kazakhstan, Republic of Moldova.

The page of a journal provides information about its founder, publisher and the editorial board. The archive of the journal represents both the current issues and its historic archives, including full texts articles. Access to full-text PDF files is specified in an agreement signed with each journal, normally access is free except for the recent (2–3 years) issues.

We have comprehensively digitized historic archives of the leading Russian and Soviet mathematics journals back to the fist volumes, the list includes: *Algebra i Analiz* (since 1989); *Zhurnal Vychislitel'noĭ Matematiki i Matematicheskoĭ Fiziki* (since 1961); *Diskretnaya Matematika* (since 1989); *Funktsional'nyi Analiz i ego Prilozheniya* (since 1967); *Bulletin de l'Académie des Sciences* (1894–1937); *Izvestiya Akademii Nauk. Seriya Matematicheskaya* (since 1937); *Matematicheskoe Modelirovanie* (since 1989); *Matematicheskii Sbornik* (since 1866); *Matematicheskie Zametki* (since 1967); *Trudy Matematicheskogo Instituta im. V. A. Steklova* (since 1931); *Uspekhi Matematicheskikh Nauk* (since 1936); *Teoreticheskaya i Matematicheskaya Fizika* (since 1969).

The original title, abstract and keywords, English translation title, abstract and keywords, a link to the English version, a list of references and a list of forward links are supplied for every paper. Titles, abstracts, keywords, references and forward links are stored in the database in the LATEX format. We use MathJax technology (`http://www.mathjax.org`) to output mathematics on the website. For every paper we provide external links to all possible representations of the publication in Internet including links to Crossref, MathSciNet, Zentralblatt MATH, ADS NASA, ISI Web of Knowledge, Google Scholar links to the references cited and related papers. Enhanced search facilities include search for publications by keywords in the title, abstract or full-text paper, by the authors' and/or institutions names.

For most journals we provide information about their citation statistics and impact factors [2]. Impact factors are calculated on the basis of forward links (back references) stored in the database. English version journals are supplied with the classical Impact Factors calculated by the Institute for Scientific Information (ISI) of the Thomson Reuters Corporation (ISI Web of Knowledge). It is important to note that the classical (ISI) impact factors do not include citations of the Russian versions. We take into account citations of both versions and calculate the integral impact factor of the journal. This includes citations in classical scientific journals, but also citations in conference proceedings, electronic publications. Table 1 provides examples of the number of references to the volumes of years 2009–2010 and the values of Impact Factors 2011 of some journals provided by the Math-Net.Ru and ISI Web of Knowledge. A significant difference between the citation numbers and impact factors of Math-Net.Ru and ISI Web of Knowledge is explained by the fact that the latter does not take

Table 1. Citation number for volumes of years 2009–2010 and Impact Factors 2011, provided by Math-Net.Ru and ISI

Journal	Math-Net.Ru values		ISI values	
	Citations number	Impact Factor	Citations number	Impact factor
Matematicheskii Sbornik	**130**	0.813	**85**	0.567
Trudy Matem. Instituta im. V. A. Steklova	**75**	0.455	**42**	0.171
Avtomatika i Telemekhanika	**227**	0.698	**96**	0.246
Diskretnaya Matematika	**43**	0.483	–	–
Siberian Electronic Mathematical Reports	**37**	0.378	–	–
Russian Journal of Nonlinear Dynamics	**35**	0.407	–	–

into account references to the Russian versions of papers. Proofs of the data stated in the Table 1 can be found in the section "Impact factor" of the page of the corresponding journal on Math-Net.Ru and in Journal Citations Reports provided by ISI Web of Knowledge system. Our system calculates one-year and 2-year impact factors (similar to classical) and also 5-year ones.

It is noteworthy that for many Russian journals ISI does not provide impact factors so Math-Net.Ru data is a single way to evaluate the citation indexes of the journal. Table 1 provides some examples of Russian mathematical journals having no classical (ISI) impact factor but a significant number of citations in Russian and international sources.

3   Persons and Institutions

Portal Math-Net.Ru also includes comprehensive information about Russian and foreign mathematicians and institutions where authors of publications work or study. Up to now the database includes 52 000 individual persons and 4 000 institutions. Visitors of the website are free to register online and to contribute to the database in case when they have at least one published article in a scientific mathematics or physics journal. Personal web page provides the list of personal publications and presentations, keywords, the list of scientific interests and biography, web-links to additional personal resources. Special tools are available to arrange a full list of personal publications, including papers not available within the Math-Net.Ru system. The web-page of an institution contains general information, a link to its original web-page and a list of authors whose papers are presented in Math-Net.Ru.

4   Citation and Forward Link Database

The citation database accumulates the reference lists and forward links of all the publications available at Math-Net.Ru as well as personal lists of publications of the authors. All references are collected into a single database and stored in the format AMSBIB [2] developed in the Division of Mathematics of the Russian Academy of Sciences. The bibliography is stored using special LaTeX commands dividing a reference into several parts: journal name, authors, publication year, volume, issue, pages information, additional information about the volume/issue and other possible publication details. By means of these commands it is possible to avoid manual markup of the list of references; it also enables an automatic creation of the bibliography in PDF/HTML/XML formats, arranging hyperlinks to various publication databases including MathSciNet, Crossref and ZentalBlatt Math. Since all the references are collected into a single database an advanced cited-reference search by various terms is arranged: publication title, year, author, pages. The reference database is much wider than the database of Math-Net.Ru publications and a search through the reference database results in additional information about articles.

5 Video-Library, Conferences, Seminars

Our project thoroughly collects information about mathematical events occuring in Russia and the states of the former Soviet Union. This concerns scientific conferences and seminars, public lectures. Most information about such events is provided by the organizers. The system software allows arranging an event home page, which includes general information, the list of organizers, the event schedule and a list of presentations with links to own web pages. A presentation webpage contains the title, abstract, date and place of the event and includes additional materials such as a list of references, PowerPoint files and a video-record when available. We encourage conference/seminar organizers to record videos of the presentations and we take on post-processing of the video files. Our system accepts the most popular video formats and enables viewing them online in all operation systems, including mobile devices. The mobile version of the system provides full access to the video-library. The system offers viewing videos in normal and full High Definition quality. An online access to all video records is free, all video files can be downloaded for home viewing.

6 Manuscript Submission and Tracking System

The website is managed by a contents management system, which provides necessary functionality to add/update/remove any information available. The content management system is used to manage current and archive publications of the journals, to communicate with authors and to create reports for editorial needs. The content management system resolves the problem of the creation of a document processing system in the editorial office of a Russian scientific journal. Most western publishers provide such systems for their journals but they cannot be used in Russian journals due to lack of Russian language fields. The content management system includes all kinds of editorial activities from the submission of a manuscript to the publication of the peer-reviewed paper in print and online.

The main features of the system include: submission of a manuscript by the author in electronic form at the journal website; registration of the authors in the database of persons; registration of the manuscript in the paper database and arranging a paper flow process, which includes *classification, peer review, authors' revision, scientific editing, translation into English, editing of the English version publication, publication in print and online*; communication facilities between authors, referees, translators, typesetters, editorial board members and other people involved into publication process; personal access of the authors, referees, editors to editorial information necessary for publication process; arranging a list of forthcoming papers; sending email notifications from the database; creation of automatic reports for editorial needs.

Manuscript submission is available at the journal home page for registered authors only. New authors should first fill in a registration form. Manuscript submission process consists of filling in several online forms providing information about the manuscript title, abstract, authors, keywords; then the author is asked to supply a full-text manuscript in LaTeX and PDF formats. The editor is notified about a new submission by email, examines it on the subject of compliance with the journal rules and starts the peer-review process. Every editorial paper flow record can be supplied with a number of documents (files) specifying details of the process. The system generates comprehensive reports about all kinds of editorial activities.

Access to the manuscript submission system depends on the user's role in the publishing process: author, referee, editor, journal administrator. Authors can only see paper flow details with hidden referees names. Authors are able to submit a revised version of the manuscript and download the final PDF of the published paper. Referees can download full texts of papers and upload reviews. They have access only to those papers they are working with. Editors normally register new papers, add paper flow records, communicate with the authors, referees, typesetters. Journal administrators can amend anything within their journals.

References

1. Zhizhchenko, A.B., Izaak, A.D.: The information system Math-Net.Ru. Application of contemporary technologies in the scientific work of mathematicians. Russian Math. Surveys 62(5), 943–966 (2007)
2. Zhizhchenko, A.B., Izaak, A.D.: The information system Math-Net.Ru. Current state and prospects. The impact factors of Russian mathematics journals. Russian Math. Surveys 64(4), 775–784 (2009)

A Dynamic Symbolic Geometry Environment Based on the GröbnerCover Algorithm for the Computation of Geometric Loci and Envelopes

Miguel A. Abánades[1] and Francisco Botana[2,*]

[1] CES Felipe II, Universidad Complutense de Madrid
C/ Capitán 39, 28300 Aranjuez, Spain
abanades@ajz.ucm.es
[2] Depto. de Matemática Aplicada I, Universidad de Vigo
Campus A Xunqueira, 36005 Pontevedra, Spain
fbotana@uvigo.es

Abstract. An enhancement of the dynamic geometry system GeoGebra for the automatic symbolic computation of algebraic loci and envelopes is presented. Given a GeoGebra construction, the prototype, after rewriting the construction as a polynomial system in terms of variables and parameters, uses an implementation of the recent GröbnerCover algorithm to obtain the algebraic description of the sought locus/envelope as a locally closed set. The prototype shows the applicability of these techniques in general purpose dynamic geometry systems.

Keywords: Dynamic Geometry, Locus, Envelope, GröbnerCover Algorithm, GeoGebra, Sage.

1 Introduction

Most dynamic geometry systems (DGS) implement loci generation just from a graphic point of view, returning a locus as a set of points in the screen with no algebraic information. A simple algorithm based on elimination theory to obtain the equation of an algebraic plane curve from its description as a locus set was described in [1]. This new information expands the algebraic knowledge of the system, allowing further transformations of the construction elements, such as constructing a point on a locus, intersecting the locus with other elements, etc. The same consideration can be made with respect to the envelope of a family of curves. This algebraic approach is a significant improvement over the numeric-graphic method mentioned above. An implementation of the algorithm in a system embedding GeoGebra in the Sage notebook was described at CICM 2011 [2]. In fact, the algorithm is already behind the *LocusEquation* command in the beta version of the next version of the DGS GeoGebra [3] (see

---

* Both authors partially supported by the project MTM2011-25816-C02-(01,02) funded by the Spanish *Ministerio de Economía y Competitividad* and the European Regional Development Fund (ERDF).

J. Carette et al. (Eds.): CICM 2013, LNAI 7961, pp. 349–353, 2013.

http://wiki.geogebra.org/en/LocusEquation_Command). It has also recently been implemented by the DGS JSXGraph [4] to determine the equation of a locus set using remote computations on a server [5], an idea previously developed by the authors in [6].

Unfortunately, this algorithm does not discriminate between regular and special components of a locus (following the definitions in [7])[1]. More concretely, the obtained algebraic set may contain extra components sometimes due to the fact that the method returns only Zariski closed sets (i.e. zero sets of polynomials) and sometimes due to degenerate positions in the construction (e.g. two vertices being coincidental for a triangle construction).

There is little that can be done to solve these problems with the simple elimination approach. Concerning degeneration, there is no alternative except explicitly requesting information from the user about the positions producing special components. However, the recent GröbnerCover algorithm [8] has opened new possibilities for the automated processing of these problems. From the canonical decomposition of a polynomial system with parameters returned by the algorithm, and following a remark by Tomás Recio concerning the dimensions of the spaces of variables and parameters, a protocol has been established to distinguish between regular and special components of a locus set. For example, a circle, a variety of dimension 1, is declared to be a special component of a locus by the protocol if it corresponds to a point, a variety of dimension 0. This heuristic in the protocol improves the automatic determination of loci but does not fully resolve it. It is not difficult to find examples where this general rule does not suit the user's interests. This is a delicate issue because, in some situations, these special components are the relevant parts of the sought set (the study of bisector curves is a good source for such examples).

As an illustration, let us consider the following problem included in [9] together with a remark about its difficult synthetic treatment: *Given a triangle ABC. Take a point M on BC. Consider the orthogonal projections N of M onto AC, and P onto AB respectively. The lines AM and PN meet at X. What is the locus set of points X when M moves along the line BC?*

When the vertices of the triangle ABC are the points $(2, 3)$, $(1, 0)$ and $(0, 1)$, the locus set is a conic from which a point has to be removed. That is, the locus set is not an algebraic variety but a locally closed set. Figure 1 shows the plotting of the conic in GeoGebra together with its precise algebraic description as provided by the prototype.

If we consider this same construction with $A(0, 0)$, $B(1, 0)$ and $C(0, 1)$, a standard DGS will plot a straight line as locus, while ordinary elimination will give the true locus $2x + 2y = 1$ plus two other lines, namely, the coordinate axes $x = 0$ and $y = 0$. These extra lines correspond to two degenerate positions for the mover: $M = B$ and $M = C$. Applying the criterion sketched above, the system identifies these two lines as special components and hence removes them

---

[1] A special component of a locus is basically a one-dimensional subset of the locus corresponding to a single position of the moving point.

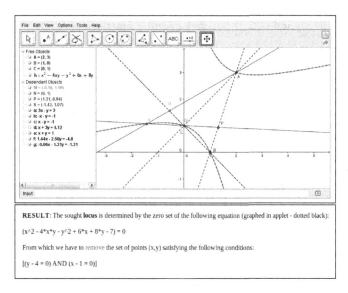

Fig. 1. Locus as a constructible set

from the final description. In tables 1 and 2 we find the parametric systems and outputs from the GröbnerCover algorithm for the two considered instances respectively.

2   Prototype Description

The system (accessible at [10]) consists of a web page with a GeoGebra applet where the user constructs a locus or a family of linear objects depending on a point. For any of these constructions (specified using a predetermined set of GeoGebra commands) the prototype provides the algebraic description of the locus/envelope by just pressing one button. Note that in its current state, the system does not provide the equation of the envelope, but the one of the discriminant line. A note stating this fact should be given if using the system for teaching purposes.

The process is roughly as follows. First, the XML description of the GeoGebra construction is sent to a Server where an installation of a Sage Cell Server ([11]) is maintained by the authors. There, the construction follows an algebraization process, as specified by a Sage library [12]. The communication Sage-GeoGebra is made possible by the JavaScript GeoGebra functions that allow the data transmission to and from the applet. In particular, the XML description of any GeoGebra diagram can be obtained. The processing of the XML description of the diagram is made by some ad-hoc code by the authors that use Sage through the Sage cell server, a general service by Sage. More concretely, Singular, included in Sage and with an implementation of the GröbnerCover algorithm, is used. The Gröbner cover of the obtained parametric polynomial system is analyzed, and the accepted components of the locus/envelope are incorporated into the applet.

Table 1. Parametric system, GröbnerCover output and returned locus for instance with $A(2,3)$, $B(1,0)$ and $C(0,1)$.[*] Includes regular functions (see [8]).

Parametric System	$-x_1 - x_2 + 1, 2x_1 + 2x_2 - 2x_3 - 2x_4, -2x_3 + 2x_4 - 2, x_1 + 3x_2 - x_5 - 3x_6, -3x_5 + x_6 + 3, -(x_4 - y)(x_3 - x_5) + (x_4 - x_6)(x_3 - x), -(y - 3)(x_1 - 2) + (x - 2)(x_2 - 3)$
Basis segment 1	$\{1\}$
Segment 1	$\mathbb{V}(0) \setminus \mathbb{V}(x^2 - 4xy + 6x - y^2 + 8y - 7)$
Basis segment 2 *	$\{\{(5y - 20)x_6 + (-3x + 6y + 3), (x - 2y + 7)x_6 + (-3y)\}, \{(5y-20)x_5+(-x-3y+21), (x-7y+27)x_5+(4y-28)\}, x_4-1, x_3, \{(y-4)x_2+(-x+2y+1), (x-2y+7)x_2+(-5y)\}, \{(y-4)x_1+(x-3y+3), (x-y+3)x_1+(4y-4)\}\}$
Segment 2	$\mathbb{V}(x^2 - 4xy + 6x - y^2 + 8y - 7) \setminus \mathbb{V}(y - 4, x - 1)$
Basis segment 3	$\{1\}$
Segment 3	$\mathbb{V}(y - 4, x - 1) \setminus \mathbb{V}(1)$
Locus (after heuristic step)	$\mathbb{V}(x^2 - 4xy - y^2 + 6x + 8y - 7) \setminus \mathbb{V}(y - 4, x - 1)$

Table 2. Parametric system, GröbnerCover output and returned locus for instance with $A(0,0)$, $B(1,0)$ and $C(0,1)$

Parametric System	$-x_1 - x_2 + 1, x_3, -x_2 + x_4, -x_1 + x_5, -x_6, -x_1y + x_2x, -(x_4 - y)(x_3 - x_5) + (x_4 - x_6)(x_3 - x)$
Basis segment 1	$\{1\}$
Segment 1	$\mathbb{V}(0) \setminus (\mathbb{V}(2x + 2y - 1) \cup \mathbb{V}(x) \cup \mathbb{V}(y))$
Basis segment 2	$\{x_6, (x + y)x_5 - x, (x + y)x_4 - y, x_3, (x + y)x_2 - y, (x + y)x_1 - x\}$
Segment 2	$(\mathbb{V}(2x+2y-1) \setminus \mathbb{V}(1)) \cup (\mathbb{V}(x) \setminus \mathbb{V}(x,y)) \cup (\mathbb{V}(y) \setminus \mathbb{V}(x,y))$
Basis segment 3	$\{x_6, x_4 + x_5 - 1, x_3, x_2 + x_5 - 1, x_1 - x_5, x_5^2 - x_5\}$
Segment 3	$\mathbb{V}(x,y) \setminus \mathbb{V}(1)$
Locus (after heuristic step)	$\mathbb{V}(2x + 2y - 1)$

Note that the goal is not to provide a final tool but a proof-of-concept prototype showing the feasibility of using sophisticated algorithms like GröbnerCover to supplement the symbolic capabilities of existing dynamic geometry systems, as well as to show the advantage of connecting different systems by using web services.

References

1. Botana, F., Valcarce, J.L.: A software tool for the investigation of plane loci. Mathematics and Computers in Simulation 61(2), 139–152 (2003)
2. Botana, F.: A symbolic companion for interactive geometric systems. In: Davenport, J.H., Farmer, W.M., Urban, J., Rabe, F. (eds.) Calculemus/MKM 2011. LNCS, vol. 6824, pp. 285–286. Springer, Heidelberg (2011)
3. GeoGebra: `http://www.geogebra.org` (last accessed January 2013)
4. JSXGraph: `http://jsxgraph.org/` (last accessed January 2013)
5. Gerhäuser, M., Wassermann, A.: Automatic calculation of plane loci using Gröbner bases and integration into a dynamic geometry system. In: Schreck, P., Narboux, J., Richter-Gebert, J. (eds.) ADG 2010. LNCS, vol. 6877, pp. 68–77. Springer, Heidelberg (2011)
6. Escribano, J., Botana, F., Abánades, M.A.: Adding remote computational capabilities to dynamic geometry systems. Mathematics and Computers in Simulation 80, 1177–1184 (2010)
7. Sendra, J.R., Sendra, J.: Algebraic analysis of offsets to hypersurfaces. Matematische Zeitschrift 234, 697–719 (2000)
8. Montes, A., Wibmer, M.: Gröbner bases for polynomial systems with parameters. Journal of Symbolic Computation 45, 1391–1425 (2010)
9. Guzmán, M.: La experiencia de descubrir en Geometría. Nivola (2002)
10. Locus/Envelope Prototype: (2012),
    `http://193.146.36.205:8080/GgbSageDirect/LocusEnvelope/`
11. Simple sagecell server: `https://github.com/sagemath/sagecell` (last accessed January 2013)
12. Botana, F.: On the parametric representation of dynamic geometry constructions. In: Murgante, B., Gervasi, O., Iglesias, A., Taniar, D., Apduhan, B.O. (eds.) ICCSA 2011, Part IV. LNCS, vol. 6785, pp. 342–352. Springer, Heidelberg (2011)

ML4PG in Computer Algebra Verification*

Jónathan Heras and Ekaterina Komendantskaya

School of Computing, University of Dundee, UK
{jonathanheras,katya}@computing.dundee.ac.uk

Abstract. ML4PG is a machine-learning extension that provides statistical proof hints during the process of Coq/SSReflect proof development. In this paper, we use ML4PG to find proof patterns in the CoqEAL library – a library that was devised to verify the correctness of Computer Algebra algorithms. In particular, we use ML4PG to help us in the formalisation of an efficient algorithm to compute the inverse of triangular matrices.

Keywords: ML4PG, Interactive Theorem Proving, Coq, SSReflect, Machine Learning, Clustering, CoqEAL.

1 Introduction

There is a trend in interactive theorem provers to develop general purpose methodologies to aid in the formalisation of a family of related proofs. However, although the application of a methodology is straightforward for its developers, it is usually difficult for an external user to decipher the key results to import such a methodology into a new development. Therefore, tools which can capture methods and suggest appropriate lemmas based on proof patterns would be valuable. ML4PG [5] – a machine-learning extension to Proof General that interactively finds proof patterns in Coq/SSReflect – can be useful in this context.

In this paper, we use ML4PG to guide us in the formalisation of a fast algorithm to compute the inverse of triangular matrices using the CoqEAL methodology [4] – a method designed to verify the correctness of efficient Computer Algebra algorithms.

Availability. ML4PG is accessible from [5], where the reader can find related papers, examples, the links to download ML4PG and all libraries and proofs we mention here.

2 Combining the CoqEAL Methodology with ML4PG

Most algorithms in modern Computer Algebra systems are designed to be efficient, and this usually means that their verification is not an easy task. In order to overcome this problem, a methodology based on the idea of *refinements* was

---

* The work was supported by EPSRC grant EP/J014222/1.

J. Carette et al. (Eds.): CICM 2013, LNAI 7961, pp. 354–358, 2013.

presented in [4], and was implemented as a new library, built on top of the SSReflect libraries, called *CoqEAL*. The approach [4] to formalise efficient algorithms can be split into three steps:

S1. define the algorithm relying on rich dependent types, as this will make the proof of its correctness easier;

S2. refine this definition to an efficient algorithm described on high-level data structures; and,

S3. implement it on data structures which are closer to machine representations.

The CoqEAL methodology is clear and the authors have shown that it can be extrapolated to different problems. Nevertheless, this library contains approximately 400 definitions and 700 lemmas; and the search of proof strategies inside this library is not a simple task if undertaken manually. Intelligent proof-pattern recognition methods could help with such a task.

In order to show this, let us consider the formalisation of a fast algorithm to compute the inverse of triangular matrices over a field with 1s in the diagonal using the CoqEAL methodology. SSReflect already implements the matrix inverse relying on rich dependent types using the `invmx` function; then, we only need to focus on the second and third steps of the CoqEAL methodology. We start defining a function called `fast_invmx` using high-level data structures.

Algorithm 1. *Let M be a square triangular matrix of size n with 1s in the diagonal; then* $fast\_invmx(M)$ *is recursively defined as follows.*

- *If $n = 0$, then* $fast\_invmx(M)=1\%M$ *(where $1\%M$ is the notation for the identity matrix in SSReflect).*
- *Otherwise, decompose M in a matrix with four components: the top-left element, which is 1; the top-right line vector, which is null; the bottom-left column vector C; and the bottom-right $(n - 1) \times (n - 1)$ matrix N; that is,* $M = \left(\begin{array}{c|c} 1 & 0 \\ \hline C & N \end{array}\right).$ *Then define* $fast\_invmx(M)$ *as:*

$$fast\_invmx(M) = \left(\begin{array}{c|c} 1 & 0 \\ \hline -\texttt{fast\_invmx(N)} \texttt{ *m } C & \texttt{fast\_invmx(N)} \end{array}\right)$$

where $*m$ is the notation for matrix multiplication in SSReflect.

Subsequently, we should prove the equivalence between the functions `invmx` and `fast_invmx` – Step S2 of the CoqEAL methodology. Once this result is proven, we can focus on the third step of the CoqEAL methodology. It is worth mentioning that neither `invmx` nor `fast_invmx` can be used to actually compute the inverse of matrices. These functions cannot be executed since the definition of matrices is locked in SSReflect to avoid the trigger of heavy computations during deduction steps. Using Step S3 of the CoqEAL methodology, we can overcome this pitfall. In our case, we implement the function `cfast_invmx` using lists of lists as the low level data type for representing matrices and to finish the formalisation we should prove the following lemma.

Lemma 1. *Let M be a square triangular matrix of size n with 1s in the diagonal; then given M as input, fast_invmx and cfast_invmx obtain the same result but with different representations. The statement of this lemma in SSReflect is:*

> Lemma cfast_invmxP : forall (n : nat) (M : 'M_n),
>   seqmx_of_mx (fast_invmx M) = cfast_invmx (seqmx_of_mx M).

where the function seqmx_of_mx transforms matrices represented as functions to matrices represented as lists of lists.

The proof of Lemma 1 for a non-expert user of CoqEAL is not direct, and, after applying induction on the size of the matrix, the developer can get easily stuck when proving such a result.

Problem 1. *Find a method to proceed with the inductive case of Lemma 1.*

In this context, the user can invoke ML4PG to find some common proof-pattern in the CoqEAL library. ML4PG generated solutions is presented in Figure 1.

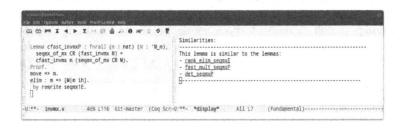

Fig. 1. Suggestions for Lemma cfast_invmxP. The Proof General window has been split into two windows positioned side by side: the left one keeps the current proof script, and the right one shows the suggestions provided by ML4PG.

ML4PG suggests three lemmas which are the equivalent counterparts of Lemma 1 for the algorithms computing the rank, the determinant and the fast multiplication of matrices. Inspecting the proof of these three lemmas, the user can find Proof Strategy 1 which is followed by those three lemmas and which can also be applied in Lemma 1.

Proof Strategy 1. *Apply the morphism lemma to change the representation from abstract matrices to executable ones. Subsequently, apply the translation lemmas of the operations involved in the algorithm – translation lemmas are results which state the equivalence between the executable and the abstract counterparts of several operations related to matrices.*

It is worth remarking that the user is left with the task of finding a proof strategy from the suggestions provided by ML4PG. In the future, we could apply symbolic machine-learning techniques such as Rippling [1] and Theory Exploration [3] to automatically conceptualise the proof strategies from the suggestions provided by ML4PG.

3 Applying ML4PG to the CoqEAL Library

In the section, we show how ML4PG discovers the lemmas which follow Proof Strategy 1. This process can be split into 4 steps: extraction of significant features from library-lemmas, selection of the machine-learning algorithm, configuration of parameters, and presentation of the output.

Step 1. Feature Extraction. During the proof development, ML4PG works on the background of Proof General, and extracts (using the algorithm described in [5]) some simple, low-level features from interactive proofs in Coq/SSReflect. In addition, ML4PG extends Coq's compilation procedure to extract lemma-features from already-developed libraries.

In the example presented in the previous section, we have extracted the features from the 18 files included in the CoqEAL library (these files involve 720 lemmas). Any number of additional Coq libraries can be be selected using the ML4PG menu. Unlike e.g. [6], scaling is done at the feature extraction stage, rather than on the machine-learning stage of the process.

Step 2. Clustering Algorithm. On user's request, ML4PG sends the gathered statistics to a chosen machine-learning interface and triggers execution of a clustering algorithm of the user's choice – clustering algorithms [2] are a family of unsupervised learning methods which divide data into n groups of similar objects (called clusters), where the value of n is provided by the user.

We have integrated ML4PG with several clustering algorithms available in MATLAB (K-means and Gaussian) and Weka (K-means, FarthestFirst and Expectation Maximisation). In the CoqEAL example, ML4PG uses the MATLAB K-means algorithm to compute clusters – this is the algorithm used by default.

Step 3. Configuration of Granularity. The input of the clustering algorithms is a file that contains the information associated with the lemmas to be analysed, and a natural number n, which indicates the number of clusters. The file with the features of the library-lemmas is automatically extracted (see [5]).

To determine the value of n, ML4PG has its own algorithm that calculates the optimal number of clusters interactively, based on the library size. As a result, the user does not provide the value of n directly, but just decides on granularity in the ML4PG menu. The granularity parameter ranges from 1 to 5, where 1 stands for a low granularity (producing a few large clusters with a low correlation among their elements) and 5 stands for a high granularity (producing many smaller clusters with a high correlation among their elements). By default, ML4PG works with the granularity value of 3 and this is the value presented in the previous section.

Step 4. Presentation of the Results. Clustering algorithms output contains not only clusters but also a measure which indicates the proximity of the elements of the clusters. In addition, results of one run of a clustering algorithm may differ from another; then ML4PG runs the clustering algorithm 200 times, obtaining the frequency of each cluster as a result. These two measures (proximity and frequencies) are used as thresholds to decide on the single "most reliable" cluster to be shown to the user, cf. Figure 1.

These 4 steps are the workflow followed by ML4PG to obtain clusters of similar proofs. Let us present now the results that ML4PG will obtain if the user varies the different parameters – these results are summarised in Table 1.

Table 1. A series of clustering experiments discovering Proof Strategy 1. The table shows the sized of clusters containing: a) Lemma `cfast_invmxP`, b) Lemma about rank (`rank_elim_seqmxE`), c) Lemma about fast multiplication (`fast_mult_seqmxP`), and d) Lemma about determinant (`det_seqmxP`).

Algorithm:	$g = 1$ $(n = 72)$	$g = 2$ $(n = 80)$	$g = 3$ $(n = 90)$	$g = 4$ $(n = 102)$	$g = 5$ $(n = 120)$
Gaussian	$24^{a,b,c,d}$	$12^{a,b,c,d}$	$10^{a,b,c,d}$	$10^{a,b,c,d}$	$10^{a,b,c,d}$
K-means (Matlab)	$20^{a,b,c,d}$	$14^{a,b,c,d}$	$4^{a,b,c,d}$	0	0
K-means (Weka)	$16^{a,b,c,d}$	$11^{a,b,c,d}$	$4^{a,b,c,d}$	0	0
Expectation Maximisation	$52^{a,b,c,d}$	$45^{a,b,c,d}$	$43^{a,b,c,d}$	$39^{a,b,c,d}$	$14^{a,b,c,d}$
FarthestFirst	$30^{a,b,c,d}$	$27^{a,b,c,d}$	$27^{a,b,c,d}$	$26^{a,b,c,d}$	$20^{a,b,c,d}$

As can be seen in Table 1, the clusters obtained by almost all variations of the learning algorithms and parameters include the lemmas which led us to formulate Proof Strategy 1. However, there are some remarkable differences among the results. First of all, the results obtained with the Expectation Maximisation and FarthestFirst algorithms include several additional lemmas that make difficult the discovery of a common pattern. The same happens with the other algorithms for granularity values 1 and 2; however the clusters can be refined when increasing the granularity value. The results are clusters of a sensible size which contain lemmas with a high correlation; allowing us to spot Proof Strategy 1.

References

1. Basin, D., Bundy, A., Hutter, D., Ireland, A.: Rippling: Meta-level Guidance for Mathematical Reasoning. Cambridge University Press (2005)
2. Bishop, C.: Pattern Recognition and Machine Learning. Springer (2006)
3. Claessen, K., Johansson, M., Rosén, D., Smallbone, N.: Automating inductive proofs using theory exploration. In: Bonacina, M.P. (ed.) CADE 2013. LNCS, vol. 7898, pp. 392–406. Springer, Heidelberg (2013)
4. Dénès, M., Mörtberg, A., Siles, V.: A Refinement Based Approach to Computational Algebra in Coq. In: Beringer, L., Felty, A. (eds.) ITP 2012. LNCS, vol. 7406, pp. 83–98. Springer, Heidelberg (2012)
5. Heras, J., Komendantskaya, E.: ML4PG: downloadable programs, manual, examples (2012-2013), www.computing.dundee.ac.uk/staff/katya/ML4PG/
6. Kühlwein, D., Blanchette, J.C., Kaliszyk, C., Urban, J.: MaSh: Machine Learning for Sledgehammer. In: Proceedings of ITP 2013. LNCS (2013)

Pervasive Parallelism in Highly-Trustable Interactive Theorem Proving Systems

Bruno Barras[3], Lourdes del Carmen González Huesca[2], Hugo Herbelin[2],
Yann Régis-Gianas[2], Enrico Tassi[3], Makarius Wenzel[1], and Burkhart Wolff[1]

[1] Univ. Paris-Sud, Laboratoire LRI, UMR8623, Orsay, F-91405, France
CNRS, Orsay, F-91405, France
[2] INRIA, Univ. Paris Diderot, Paris, France
[3] INRIA, Laboratoire d'Informatique de l'Ecole Polytechnique

1 Background

Interactive theorem proving is a technology of fundamental importance for mathematics and computer-science. It is based on expressive logical foundations and implemented in a highly trustable way. Applications include huge mathematical proofs and semi-automated verifications of complex software systems. Interactive development of larger and larger proofs increases the demand for computing power, which means explicit parallelism on current multicore hardware [6].

The architecture of contemporary interactive provers such as Coq [13, §4], Isabelle [13, §6] or the HOL family [13, §1] goes back to the influential LCF system [4] from 1979, which has pioneered key principles like correctness by construction for primitive inferences and definitions, free programmability in userspace via ML, and toplevel command interaction. Both Coq and Isabelle have elaborated the prover architecture over the years, driven by the demands of sophisticated proof procedures, derived specification principles, large libraries of formalized mathematics etc. Despite this success, the operational model of interactive proof checking was limited by sequential ML evaluation and the sequential read-eval-print loop, as inherited from LCF.

2 Project Aims

The project intends to overcome the sequential model both for Coq and Isabelle, to make the resources of multi-core hardware available for even larger proof developments. Beyond traditional processing of proof scripts as sequence of proof commands, and batch-loading of theory modules, there is a vast space of possibilities and challenges for pervasive parallelism. Reforming the traditional LCF architecture affects many layers of each prover system, see figure 1.

Parallelization of the different layers is required on the level of the execution environments (SML, OCaml), which need to include some form of multithreading or multi-processing supported by multi-core architectures. Isabelle can build on parallel Poly/ML by David Matthews [5] and earlier efforts to support

J. Carette et al. (Eds.): CICM 2013, LNAI 7961, pp. 359–363, 2013.
© Springer-Verlag Berlin Heidelberg 2013

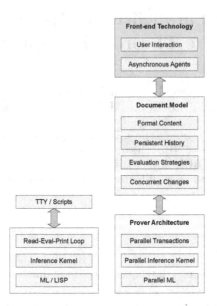

Fig. 1. Reformed LCF-architecture for parallel proof-document processing

parallel proof checking [8]. For Coq, some alternatives with separate OCaml processes need to be investigated, because early support for parallel threads in Caml [3] was later discontinued.

Further reforms carry over to the **inference kernel**, which has to be extended by means to decompose proof checking tasks into independent parts that can be evaluated in parallel. The tactic code of proof procedures or derived specification packages needs to be reconsidered for *explicit* parallelism, while the inherent structure of the proof command language can be exploited for *implicit* parallelism. The latter is particularly appealing: the prover acts like system software and schedules proofs in parallel without user (or programmer) intervention. Some of these aspects need to be addressed for Coq and Isabelle in slightly different ways, to accommodate different approaches in either system tradition.

Our approach is *document-centric*: the user edits a document containing text, code, definitions, and proofs to be checked incrementally. This means that checking is split into parallel subtasks reporting their results asynchronously. The **document model** and its protocols need to support this natively, as part of the primary access to the prover process. Finally, a system **front-end** is required to make all these features accessible to users, both novices and experts. Instead of a conventional proof-script editor, the project aims to provide a full-scale Prover-IDE following the paradigm of "continuous build — continuous check".

These substantial extensions of the operational aspects of interactive theorem proving shall retain the trustability of LCF-style proving at the very core. The latter has to be demonstrated by **formal analysis** of some key aspects of the prover architecture.

The theoretic foundation of the document model is directed by a *fine-grained analysis of the impact of changes* made by the user on the formal text. This analysis not only helps the parallelization of the verification of the document but also the reuse of already checked parts of the document that are *almost unimpacted by the user edits*. To give a formal account on this notion of proof reuse and to implement this mechanism without compromising the system trustability, we must assign a precise static semantics to the changes. One foundational part of the project will consist of studying what kind of logical framework is adapted to the specification and verification of proof reuses. By the end of the project, we expect to get a language of semantically-aware and mechanically-verifiable annotations for the document model.

3   Current Research and First Results

Project results are not just paper publications, but actual implementations that are expected to be integrated into Coq and Isabelle, respectively. Thus users of these proof assistants will benefit directly from the project results.

3.1   A State Transaction Machine for Coq

Parallelizing a sequential piece of purely functional code is a relatively easy task. On the contrary parallelizing an already existing piece of imperative code is known to be extremely hard. Unfortunately Coq stores much of its data in global imperative data structures that can be accessed and modified by almost any component of the system

For example some tactics, while building the proof, may generate support lemmas on the fly and add them to the global environment. The kernel, that will verify the entire proof once completed, needs to find these lemmas in order to validate the proof. Hence distributing the work of building and checking the proof among different partners is far from being trivial, given that the lack of proper multi-threading in OCaml forces these partners to live in different address spaces.

In the prototype under implementation [7] all side effects have been eliminated or tracked and made explicit in a state-transaction data structure. This graph models a collection of states and the transactions needed to perform in order to obtain a particular state given another one. Looking at this graph one can deduce the minimum set of transactions needed to reach the state the user is interested in, and postpone unnecessary tasks. While this is already sufficient to increase the reactivity of the system, the execution of the tasks is still sequential.

Running postponed tasks in concurrent processes is under implementation, but we are confident that the complete tracking of side effects done so far will make this work possible.

3.2   Logical Framework for Semantic-Aware Annotations

During the POPLmark challenge, Coq has been recognized as a metalanguage of choice to formalize the metatheory of formal languages. Hence, it can semantically represent the very specific relations between the entities of a proof development. Using Coq as a logical framework (for itself and for other theorem provers) is ambituous and requires: (i) to represent partial (meta)programs; (ii) to design a programming artefact to automatically track dependencies between computations; (iii) to reflect the metatheory of several logics; (iv) to implement a generic incremental proof-checker. The subgoal (i) has been achieved thanks to a new technique of *a posteriori simulation of effectful computations* based on an extension of monads to *simulable* monads [2]. The goal (ii) is investigated through a generalization of adaptative functional programming [1].

3.3   Parallel Isabelle and Prover IDE

The first stage of multithreaded Isabelle, based on parallel Poly/ML by David Matthews, already happened during 2006–2009 and was reported in [8,9]. In the project so far, the main focus has been improved scalalibity and more uniformity of parallel batch-mode wrt. asynchronous interaction. Cumulative refinements have lead to saturation of 8 CPU cores (and a bit more): see [12] for an overview of the many aspects of the prover architecture that need to be reconsidered here.

The Isabelle2011-1 release at the start of the project included the first officially "stable" release of the Isabelle/jEdit Prover IDE [9], whose degree of parallelism was significantly improved in the two subsequent releases Isabelle2012 (May 2012) and Isabelle2013 (February 2013). The general impact of parallelism on interaction is further discussed in [11].

Ongoing work investigates further sub-structural parallelism of proof elements, and improved real-time reactivity of the implementation. Here the prover architecture and the IDE front-end are refined hand-in-hand, as the key components that work with the common document model. The combination of parallel evaluation by the prover with asynchronous and erratic interactions by the user is particularly challenging. We also need to re-integrate tools like Isabelle/Sledgehammer into the document model as *asynchronous agents* that do not block editing and propose results from automated reasoning systems spontaneously.

3.4   Prover IDE for Coq

Once that the Coq prover architecture has become sufficiently powerful during the course of the project, we shall investigate how the Isabelle/PIDE front-end and Coq as an alternative back-end can be integrated to make a practically usable system. Some experiments to bridge OCaml and Scala in the same spirit as for Isabelle have been conducted successfully [10]. An alternative (parallel) path of development is to re-use emerging Prover IDE support in Coq to improve its existing CoqIde front-end, to become more stateless and timeless and overcome the inherently sequential TTY loop at last.

4   Project Partners

The project involves three sites in the greater Paris area:

- The *LRI ForTesSE* team at UPSud (coordinator: **B. Wolff**), including members from the *Cedric* team (CNAM),

– the *INRIA Pi.r2* team at PPS / UParis-Diderot (site leader: **H. Herbelin**), including members from the *INRIA Gallium* team, and
– the *INRIA Marelle-TypiCal* team at LIX / Ecole Polytechnique (site leader: **B. Barras**)

Research is supported by under grant *Paral-ITP (ANR-11-INSE-001)* with formal start in November 2011 and duration of 40 months total. Further information is available from the project website http://paral-itp.lri.fr/.

References

1. Acar, U.A., Blelloch, G.E., Harper, R.: Adaptive functional programming. ACM Trans. Program. Lang. Syst. 28(6) (November 2006)
2. Claret, G., Gonzalez Huesca, L.D.C., Regis-Gianas, Y., Ziliani, B.: Lightweight proof by reflection by a posteriori simulation of effectful computations. In: Blazy, S., Paulin-Mohring, C., Pichardie, D. (eds.) Interactive Theorem Proving (ITP 2013). LNCS. Springer (2013)
3. Doligez, D., Leroy, X.: A concurrent, generational garbage collector for a multi-threaded implementation of ML. In: 20th ACM Symposium on Principles of Programming Languages (POPL). ACM Press (1993)
4. Gordon, M.J., Milner, A.J., Wadsworth, C.P.: Edinburgh LCF. LNCS, vol. 78. Springer, Heidelberg (1979)
5. Matthews, D., Wenzel, M.: Efficient parallel programming in Poly/ML and Isabelle/ML. In: ACM SIGPLAN Workshop on Declarative Aspects of Multicore Programming (DAMP 2010) (2010)
6. Sutter, H.: The free lunch is over — a fundamental turn toward concurrency in software. Dr. Dobb's Journal 30(3) (2005)
7. Tassi, E., Barras, B.: Designing a state transaction machine for Coq. In: The Coq Workshop 2012 (co-located with ITP 2012) (2012)
8. Wenzel, M.: Parallel proof checking in Isabelle/Isar. In: Dos Reis, G., Théry, L. (eds.) ACM SIGSAM Workshop on Programming Languages for Mechanized Mathematics Systems (PLMMS 2009). ACM Digital Library (2009)
9. Wenzel, M.: Isabelle/jEdit – A prover IDE within the PIDE framework. In: Jeuring, J., Campbell, J.A., Carette, J., Dos Reis, G., Sojka, P., Wenzel, M., Sorge, V. (eds.) CICM 2012. LNCS, vol. 7362, pp. 468–471. Springer, Heidelberg (2012)
10. Wenzel, M.: PIDE as front-end technology for Coq. ArXiv (April 2013), http://arxiv.org/abs/1304.6626
11. Wenzel, M.: READ-EVAL-PRINT in parallel and asynchronous proof-checking. In: User Interfaces for Theorem Provers (UITP 2012). EPTCS (2013)
12. Wenzel, M.: Shared-memory multiprocessing for interactive theorem proving. In: Blazy, S., Paulin-Mohring, C., Pichardie, D. (eds.) Interactive Theorem Proving (ITP 2013). LNCS. Springer (2013)
13. Wiedijk, F. (ed.): The Seventeen Provers of the World. LNCS (LNAI), vol. 3600. Springer, Heidelberg (2006)

The Web Geometry Laboratory Project

Pedro Quaresma[1], Vanda Santos[2], and Seifeddine Bouallegue[3,*]

[1] CISUC/Department of Mathematics, University of Coimbra
3001-454 Coimbra, Portugal
pedro@mat.uc.pt
[2] CISUC, 3001-454 Coimbra, Portugal
vsantos7@gmail.com
[3] Innov'Com / University of Carthage, Tunisia
saief.bouallegue@gmail.com

Abstract. The *Web Geometry Laboratory* (*WGL*) project's goal is to build an adaptive and collaborative blended-learning Web-environment for geometry.

In its current version (1.0) the WGL is already a collaborative blended-learning Web-environment integrating a dynamic geometry system (DGS) and having some adaptive features. All the base features needed to implement the adaptive module and to allow the integration of a geometry automated theorem prover (GATP) are also already implemented.

The actual testing of the WGL platform by high-school teachers is underway and a field-test with high-school students is being prepared.

The adaptive module and the GATP integration will be the next steps of this project.

Keywords: adaptive, collaborative, blended-learning, geometry.

1 Introduction

The use of intelligent computational tools in a learning environment can greatly enhance its dynamic, adaptive and collaborative features. It could also extend the learning environment from the classroom to outside of the fixed walls of the school.

To build an adaptive and collaborative blended-learning environment for geometry, we claim that we should integrate dynamic geometry systems (DGSs), geometry automated theorem provers (GATPs) and repositories of geometric problems (RGPs) in a Web system capable of individualised access and asynchronous and synchronous interactions. A system with that level of integration will allow building an environment where each student can have a broad experimental learning platform, but with a strong formal support. In the next paragraphs we will briefly explain what do we mean by each of these features and how the Web Geometry Laboratory (WGL) system cope, or will cope, with that.

A blended-learning environment is a mixing of different learning environments, combining traditional face-to-face classroom (synchronous) methods with more modern computer-mediated (asynchronous) activities. A Web-environment is appropriate for both situations (see Figure 1).

---

* IAESTE traineeship PT/2012/71.

J. Carette et al. (Eds.): CICM 2013, LNAI 7961, pp. 364–368, 2013.
© Springer-Verlag Berlin Heidelberg 2013

An adaptive environment is an environment that is able to adapt its behaviour to individual users based on information acquired about its user(s) and its environment and also, an important feature in a learning environment, to adapt the learning path to the different users needs. In the *WGL* project this will be realised through the registration of the geometric information of the different actions made by the users and through the analysis of those interactions [4].

A collaborative environment is an environment that allows the knowledge to emerge and appear through the interaction between its users. In WGL this is allowed by the integration of a DGS and by the users/groups/constructions relationships.

Using a DGS, the constructions are made from free objects, objects not defined by construction steps, and constructed objects using a finite set of property preserving manipulations. These property preserving manipulations allow the development of "visual proofs", these are not formal proofs. The integration in the WGL of a GATP will give its users the possibility to reason about a given DGS construction, this is an actual formal proof, eventually in a readable format. They can be also used to test the soundness of the constructions made by a DGS [1,2].

As said above to have an adaptive and collaborative blended-learning environment for geometry we should integrate intelligent geometric tools in a Web system capable of asynchronous and synchronous interactions. This integration is still to be done, there are already many excellent DGSs [7], some of them have some sort of integration with GATPs, others with RGP [1,5]. Some attempts to integrate these tools in a learning management system (LMS) have already been done, but, as far as we know, all these integrations are only partial integrations. A learning environment where all these tools are integrated and can be used in a fruitful fashion does not exist yet [6].

2 The Web Geometry Laboratory Framework

A class session using *WGL* is understood as a Web laboratory where all the students (eventually in small groups) and the professor will have a computer running *WGL* clients. Also needed is a *WGL* server, e.g. in a school Web-server (see Figure 1).

The *WGL* server is the place where all the information is kept: the login information; the group definition; the geometric constructions of each user; the users activity registry; etc. In the *WGL* server is also kept the DGS applet and the GATP will also execute

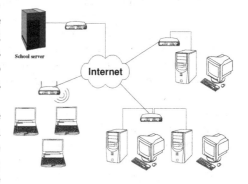

Fig. 1. School Server

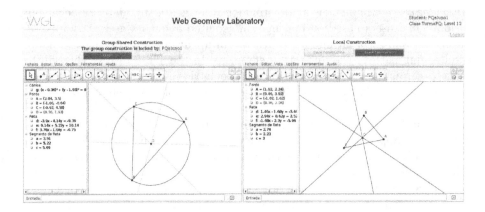

Fig. 2. Students' Interface

there. Each client will have an instance of the DGS applet, using the server to all the needed information exchange.

After installing a *WGL* server the administrator of the system should define all the teachers that will be using the system. The teachers will be privileged users in the sense that they will be capable of define other users, their students. In the beginning of each school year the teachers will define all his/her students as regular users of the *WGL*. The teacher may also define groups of users (students), these groups can be define at any given time, e.g. for a specific class, and it will be within this groups that the collaboration between its members will be possible. The definition of the groups and the membership relation between groups and its members will be the responsibility of the teachers that could create groups, delete groups and/or modify the membership relation at any given time.

Each user will have a "space" in the server where he/she can keep all the geometric construction that he/she produces. Each user will have full control over this personal scrapbook, having the possibility of saving, modifying and deleting each and every construction he/she produces using the DGS applet.

To allow the collaborative work a permissions system was implemented. This system is similar to the "traditional Unix permissions" system. The users will own the geometric construction defining the reading, **writing** and **visibility** permissions (rwv) per geometric construction. The users to groups and the constructions to groups relationships can be established in such a way that the collaborative working, group-wise, is possible.

By default, the teacher will own all the groups he/she had created granting him/her, in this way, access to all the constructions made by the students. The default setting will be rwvr-v---, meaning that the creator (owner) will have all the permissions, other users belonging to his/her groups will have "read" access and all the others users will have none. At any given moment he/she can download (read) the construction into the DGS, modify it and, eventually, upload the modified version into the database.

The collaborative module of *WGL* distinguishes students having the lock over the group construction from those without the lock. The students with the lock will have a full-fledged DGS applet, and they will be working with the group construction (see Figure 2). The students without the lock will have also the two DGS applets, but the construction in the "group shared construction" one is a synchronised version of the one being developed by the student with the lock, and a full-fledged version that can be used to develop his/her own efforts. A text-chat will be available to exchange information between group members. The teacher could always participate in this efforts having for that purpose an interface where he/she can follow the students and groups activities.

The *WGL* collaborative features are thought mostly for a blended-learning setting, that is, a classroom/laboratory where the computer-mediated activities are combined with a face-to-face classroom interaction. Nevertheless given the fact that the *WGL* is a Web application the collaborative work can extend itself to the outside of the classroom and be used to develop collaborative work at home, e.g. solving a given homework. In this setting the only drawback it will be a slow connection to the *WGL* server. We estimate that a normal bandwidth ($\geq$ 20Mb) will be enough.

The *WGL* as a Web client/server application; the database (to keep: constructions; users information; constructions; permissions; user's logs); the DGS applet; the GATP and the synchronous and asynchronous interaction are all implemented using free cross-platform software, namely PHP, Javascript, Java, AJAX, JQuery and MySQL, and also Web-standards like XHTML, CSS stylesheets and XML. The WGL is a internationalised tool (i18n/l10n) with already translations for Portuguese and Serbian, apart from default support for English. All this will allow to build a collaborative learning environment where the capabilities of tools such as the DGS and the GATP can be used in a more rich setting that in an isolated environment where (eventually) every students could have a computer with a DGS but where the communication between them would be non-existent. The exchange of text, oral and geometric information between members of a group will enrich the learning environment.

Learning environments supported by computer are seen as an important means for distance education. The DGS are also important in classroom environments, as a much enhanced substitute for the ruler and compass physical instruments, allowing the development of experiments, stimulating learning by experience. There are several DGS available, such as: *GeoGebra*, *Cinderella*, *Geometric Supposer*, *GeometerSketchpad*, *CaR*, *Cabri*, *GCLC* but none of then defines a Web learning environment with adaptive and collaborative features [6]. The program *Tabulæ* is a DGS with Web access and with collaborative features. This system is close to *WGL*, the permissions system and the fact that the DGS is not "hardwired" to the system but it is an external tool incorporated into the system, are features that distinguish positively *WGL* from *Tabulæ*. The adaptive features, the connection to the GATP and the internationalisation/localisation are also features missing in *Tabulæ* [6].

3   Conclusions and Future Work

When we consider a computer system for an educational setting in geometry, we feel that a collaborative, adaptive blended-learning environment with DGS and GATP integration is the most interesting solution. That leads to a Web system capable of being used in the classroom but also outside the classroom, with collaborative and adaptive features and with a DGS and GATPs integrated.

The *WGL* system is a work-on-progress system. It is a client/server modular system incorporating a DGS, some adaptive features, i.e., the individualised scrapbook where all the users can keep their own constructions and with a collaborative module. Given the fact that it is a client/server system the incorporation of a GATP (on the server) it will not be difficult. One of the authors has already experience on that type of integration [2,3,5].

A first case study, involving two high-schools (in the North and Center of Portugal) three classes, two teachers and 44 students and focusing in the use of *WGL* in a classroom, is already being prepared and it will be implemented in the spring term of 2013.

The next task will be the adaptive module, the logging of all the steps made by students and teacher and the construction of student's profiles on top of that. The last task will be the integration of the GATP in the *WGL*. We hope that at the end the *WGL* can became an excellent learning environment for geometry.

A prototype of the *WGL* system is available at `http://hilbert.mat.uc.pt/WebGeometryLab/`. You can enter as "anonymous/anonymous", a student-level user, or as "cicm2013/cicm", a teacher-level user.

References

1. Janičić, P., Narboux, J., Quaresma, P.: The Area Method: a recapitulation. Journal of Automated Reasoning 48(4), 489–532 (2012)
2. Janičić, P., Quaresma, P.: Automatic verification of regular constructions in dynamic geometry systems. In: Botana, F., Recio, T. (eds.) ADG 2006. LNCS (LNAI), vol. 4869, pp. 39–51. Springer, Heidelberg (2007)
3. Quaresma, P.: Thousands of Geometric problems for geometric Theorem Provers (TGTP). In: Schreck, P., Narboux, J., Richter-Gebert, J. (eds.) ADG 2010. LNCS (LNAI), vol. 6877, pp. 169–181. Springer, Heidelberg (2011)
4. Quaresma, P., Haralambous, Y.: Geometry Constructions Recognition by the Use of Semantic Graphs. In: Atas da XVIII Conferência Portuguesa de Reconhecimento de Padrões, RecPad 2012, Tipografia Damasceno, Coimbra (2012)
5. Quaresma, P., Janičić, P.: Integrating dynamic geometry software, deduction systems, and theorem repositories. In: Borwein, J.M., Farmer, W.M. (eds.) MKM 2006. LNCS (LNAI), vol. 4108, pp. 280–294. Springer, Heidelberg (2006)
6. Santos, V., Quaresma, P.: Collaborative aspects of the WGL project. Electronic Journal of Mathematics & Technology (to appear, 2013)
7. Wikipedia. List of interactive geometry software (April 2013), `http://en.wikipedia.org/wiki/List_of_interactive_geometry_software`

swMATH – A New Information Service for Mathematical Software

Sebastian Bönisch[1], Michael Brickenstein[2], Hagen Chrapary[1],
Gert-Martin Greuel[3], and Wolfram Sperber[1]

[1] FIZ Karlsruhe/Zentralblatt MATH, Franklinstr. 11, 10587 Berlin, Germany
[2] Mathematisches Institut Oberwolfach, Schwarzwaldstr. 9-11, 77709
Oberwolfach-Walke, Germany
[3] Technische Universität Kaiserslautern, Fachbereich Mathematik, Postfach 3049,
67653 Kaiserslautern, Germany

Abstract. An information service for mathematical software is presented.
Publications and software are two closely connected facets of mathematical knowledge. This relation can be used to identify mathematical software and find relevant information about it. The approach and the state of the art of the information service are described here.

1 Introduction

In 1868, the first autonomous reviewing journal for publications in mathematics – the "Jahrbuch über die Fortschritte der Mathematik" – was started, a reaction of the mathematical community to the increasing number of mathematical publications. The new information service should inform the mathematicians about recent developments in mathematics in a compact form. Today, we encounter a similar situation with mathematical software. Until now, a comprehensive information service for mathematical software is still missing. We describe an approach towards a novel kind of information service for mathematical software. A core feature of our approach is the idea of systematically connecting mathematical software and relevant publications.

2 The State of the Art

There have already been some activities towards the development of mathematical software information services. A far-reaching concept for a semantic web service for mathematical software was developed within the MONET project [1] which tries to analyze the specific needs of a user, search for the best software solution and organize the solution by providing a web service. However, the realization of such an ambitious concept requires a lot of resources. Also, a number of specialized online portals and libraries for mathematical software were developed. One of the most important portals for mathematical software is the Netlib [2] provided by NIST. Netlib provides not only metadata for a software but also hosts the software. Netlib has developed an own classification scheme, the GAMS [3] system, which allows for browsing in the Netlib. Other important manually maintained portals, e.g., ORMS [4], Plato [5] or the mathematical part of Freecode [6], provide only metadata about software.

J. Carette et al. (Eds.): CICM 2013, LNAI 7961, pp. 369–373, 2013.

3   The Publication-Based Approach

Mathematical software and publications are closely interconnected. Often, ideas and algorithms are first presented in publications and later implemented in software packages. On the other hand, the use of software can also inspire new research and lead to new mathematical results. Moreover, a lot of publications in applied mathematics use software to solve problems numerically. The use of the publications which reference a certain software is a central building block of our approach.

Identification of Software References in the zbMATH Database. There are essentially two different types of publications which refer to a software, publications describing a certain software in detail, and publications in which a certain software is used to obtain or to illustrate a new mathematical result. In a first step, the titles of publications were analyzed to identify the names of mathematical software. Heuristic methods were developed to search for characteristic patterns in the article titles, e.g., 'software', 'package', 'solver' in combination with artificial or capitalized words. It was possible to detect more than 5,000 different mathematical software packages which were then evaluated manually.

Software References in Publications – Indirect Information of Software. The automatically extracted list of software names (see above) can be used as a starting point for searching software references in the abstracts: More than 40,000 publications referring to previously identified software packages were found in the zbMATH database. Of course, the number of articles referring to a given software is very different, ranging from thousands of publications for the 'big players' (e.g. Mathematica, Matlab, Maple) to single citations for small, specialized software packages. An evaluation of the metadata of the publications, especially their keywords and MSC classifications has shown that most of the information is also relevant for the cited software and can therefore be used to describe the latter. For instance, we collect the keywords of all articles referring to a certain software and present them in swMATH as a keyword cloud, common publications and the MSC are used to detect similar software.

More Information about Software. Web sites of a software – if existing – are an important source for direct information about a software. As mentioned above, there are also special online portals which sometimes provide further information about certain mathematical software packages.

Metadata Scheme for Software. The formal description of software can be very complex. There are some standard metadata fields which are also used for publications, like authors, summary, key phrases, or classification. For software however, further metadata are relevant, especially the URL of the homepage of a software package, version, license terms, technical parameters, e.g. , programming languages, operating systems, required machine capacities, etc., dependencies to other software packages (some software is an extension of another software), or granularity. Unfortunately, often a lot of this metadata information is not available or can only be found with big manual effort. The focus of the metadata in swMATH is therefore on a short description of the software package, key phrases, and classification. For classification, we use the MSC2010 [7] scheme even though the Mathematics Subjects Classification is not optimal for software.

Quality Filter for Software. swMATH aims at listing high-quality mathematical software. Up to now, no peer-reviewing control system for software is established. However, the references to software in the database zbMATH can be used as an indirect criterion for the quality of a software: The fact that a software package is referred in a peer-reviewed article also implies a certain quality of the software.

4 Further Software

There are several reasons which suggest an extension of the publication-based approach. A major drawback of the publication-based approach is the time-delay between the release of software and the publication of an article describing the software. This delay can be up to several years for peer-reviewed journals. A second reason, not every software is referenced in peer-reviewed publications. Often, software is described in technical reports or conference proceedings.

Also, not all publications describing or using mathematical software are contained in the zbMATH database, e.g., if a software was developed for a special application and articles on it were published in a journal outside the scope of Zentralblatt MATH.

In order to build a comprehensive information service about mathematical software, we therefore still use other sources of information as online portals for mathematical software, contacts to renowned mathematical institutions, research in Google and other search engines with heuristic methods. One problem here is the quality control of this software. Being listed on a renowned portal for mathematical software should be a clear indicator for the quality of a software, whereas a mere Google hit does not mean much with respect to quality.

5 Sustainability

swMATH is a free open-access information service for the community. The development and maintenance of it, however, are not for free. For sustainability, the resources needed for the maintenance of the service must be minimized. Automatic methods and tools are under development to search for mathematical software in the zbMATH database, and to maintain and update the information on software (e.g. an automatic homepage verification tool).

In order to ease the maintenance of the service, the developments of the user interface and the retrieval functionalities are carried out in close coordination with the corresponding developments in zbMATH. The swMATH service enhances the existing information services provided by FIZ Karlsruhe/Zentralbatt MATH. The integration of the database swMATH in the information services of Zentralblatt MATH contributes to its sustainability. At the moment, links from software-relevant articles to zbMATH are provided. In the near future, back links from zbMATH to swMATH will be added too.

6 The swMATH Prototype

The first prototype of the swMATH service was published in autumn 2012. Currently, the service contains information about nearly 5,000 mathematical software packages.

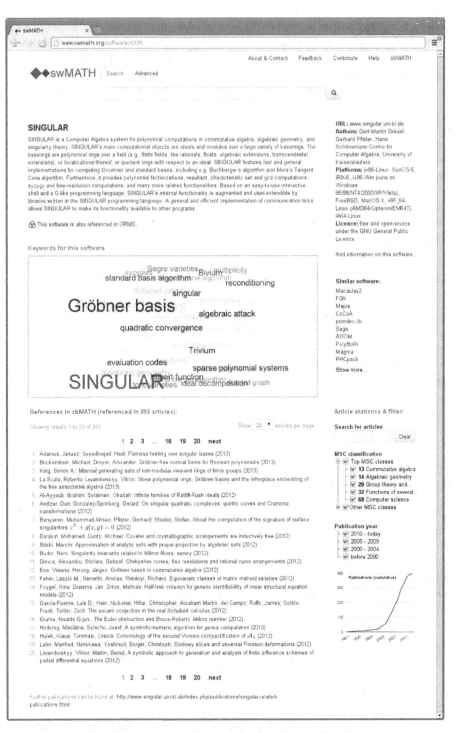

Fig. 1. The detailed information for the software "Singular"

It can be found at `http://www.swmath.org`. The user interface of swMATH concentrates on the essentials, containing simple search and an advanced search mask. Then a list of the relevant software is presented.

The detailed information about this software is shown if the name is clicked. It contains a description of the software, a cloud representation of key phrases (auto-generated from the key phrases of the publications), the publications referring to the software, the most important MSC sections, similar software and a plot showing the number of references over time. The latter is an indicator for usefulness, popularity and acceptance of a package within the mathematical community.

7 swMATH – An Information Service under Development

swMATH is a novel information service on mathematical software basing on the analysis of mathematical publications. Automatic tools periodically check the availability of URLs. Further heuristic methods to automatically extract relevant information from software websites are currently developed. Another possibility to keep the software metadata up-to-date is direct contact with (selected) software authors and providers.

So far, the software identifiers in swMATH are not persistent. However, for the productive release of swMATH persistent identifiers are planned.

The user interface is under permanent development; we recently added a browsing feature and will further enhance the usability of the swMATH web application. In order to meet the demands of the mathematical software community, we created an online questionnaire which has recently been distributed to several thousand participants, `https://de.surveymonkey.com/s/swMATH-survey`.

We hope that the swMATH service will be a useful and broadly accepted information service for the mathematical community.

References

1. MONET project, `http://www.ist-world.org/ProjectDetails.aspx?ProjectId=bcfbe93045764208a1c5173cc4614852&SourceDatabaseId=9cd97ac2e51045e39c2ad6b86dce1ac2`
2. Netlib, `http://www.netlib.org`
3. Guide to Available Mathematical Software (GAMS), `http://gams.nist.gov/`
4. Oberwolfach References to Mathematical Software (ORMS), `http://orms.mfo.de`
5. Decision Tree for Optimization Software (Plato), `http://plato.asu.edu/guide.html`
6. Math part of Freecode,
   `http://freecode.com/tags/mathematics`
7. MSC2010, `http://www.msc2010.org`

Software for Evaluating Relevance
of Steps in Algebraic Transformations

Rein Prank

University of Tartu, Estonia

Abstract. Students of our department solve algebraic exercises in mathematical logic in a computerized environment. They construct transformations step by step and the program checks the syntax, equivalence of expressions and completion of the task. With our current project, we add a program component for checking relevance of the steps.

1 Introduction

Computerized exercise environments for algebraic transformations try to preserve equivalence of expressions but they usually do not evaluate whether solution steps are relevant (for the actual task type) or not. Some versions of Algebra Tutors of Carnegie Mellon University in the nineties required a prescribed solution path to be followed. For example, Equation Solving Tutor [6] counted division before subtraction in $2x = 11 - 3$ as an error. But the review article "Cognitive Tutors: Lessons Learned" [1] summarizes: "Our earlier tutors required students to always stay on path. More recent tutors allow the student to go off path but still focus instruction on getting student back on path ...". There is one commonly known algebra environment, Aplusix [3], where the program displays the ratios of what part of the syntactic goals *factored, expanded, reduced, sorted* is already reached and what part remains. However the ratios in itself are not of much help for a student. For example, if the student does not reduce the fraction ba/bc but converts it to ab/bc then the ratios simply indicate some improvement with regard to the goal *sorted*.

Students of our department have solved technical exercises in Mathematical Logic on computers since 1991. One of our programs is an environment for algebraic transformations [5,4]. For many years it seemed that checking of syntax, order of logical operations and equivalence of expressions is sufficient for training and assessment. Some years ago the introductory part of propositional logic containing also tasks on expressing of given formulas using $\{\&, \neg\}$, $\{\vee, \neg\}$ or $\{\supset, \neg\}$ only and on disjunctive normal form (DNF) was moved into the first-term course Elements of Discrete Mathematics. We saw that, besides students who solved our exercises very quickly, there were others who were in real trouble. Most problematic were DNF exercises where many solutions had a length of 50–70 steps or more. The instructors were not able to analyze long solutions (note that the main program does not record the marking and conversion rule but only displays rows with formulas). We decided to write an additional program that checks the relevance of solution steps and annotates the solutions.

J. Carette et al. (Eds.): CICM 2013, LNAI 7961, pp. 374–378, 2013.

Our main program, analysis tool and some other necessary files are available at http://vvv.cs.ut.ee/~prank/rel-tool.zip. The paper describes the basic environment (Section 2) and our supplementary tool for normal form exercises (Section 3). Section 4 provides some discussion of further opportunities.

2   Correctness Checking in the Main Program

Working in our formula transformation environment, the student creates the solution step by step. Each conversion step consists of two substeps. At the first substep the student marks a subformula to be changed. For the second substep the program has two different modes. In the INPUT mode the program opens an input box and the student enters a subformula that replaces the marked part. In the RULE mode the student selects a rule from the menu and the program applies it. Figure 1 demonstrates a DNF exercise in the RULE mode.

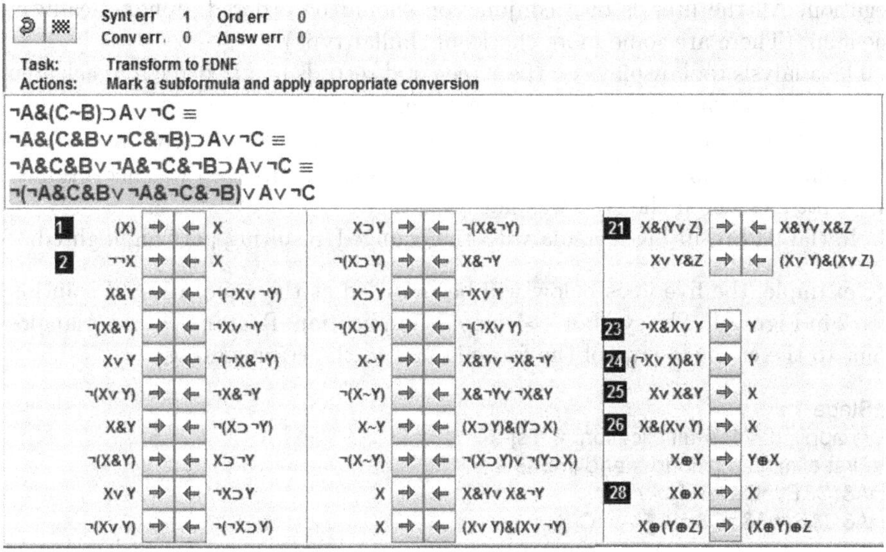

Fig. 1. Solution window of the main program. The student has performed three steps and marked a subformula for moving the negation inside.

At the first substep the program checks whether the marked part is a proper subformula. At the second substep in the INPUT mode the program checks syntactical correctness of the entered subformula and equivalence. In the RULE mode the program checks whether the selected rule is applicable to the marked part. In case of an error the program requires correction. However, our main program does not evaluate the relevance of conversions.

In our course the exercises on expression of formulas using given connectives are solved in the INPUT mode and exercises on DNF in the RULE mode.

3   A Tool for Solution Analysis

Our lectures contain the following six-stage version of the algorithm for conversion of formulas to full disjunctive normal form:

1. Eliminate implications and biconditionals from the formula.
2. Move negations inside.
3. Use distributive law to expand the conjunctions of disjunctions.
4. Exclude contradictory conjunctions and redundant copies of literals.
5. Add missing variables to conjunctions.
6. Order the variables alphabetically, exclude double conjunctions.

We now describe how the analysis tool treats relevance of solution steps. The program accepts the choice of the rule if it corresponds to the algorithm stage or is one of the simplification rules (rules in positions 1–2, 23–26 and 28 in Figure 1). For some conversions the tool checks additionally that the rule is applied reasonably. Elimination of biconditional should not duplicate implications and biconditionals. Negations should be moved inside starting from the outermost negation. All the literals of a conjunction should be ordered alphabetically in one step. (There are some more checks of similar type).

The analysis tool displays on the screen and records in a text file for each step an annotation that contains the following information:

1. Number of the stage in the FDNF algorithm [+ a clue about the conversion].
2. Number and meaning of the applied rule + OK if the step was acceptable.
3. Error message if the step was not acceptable.
4. Initial and resulting formula with the changed/resulting part highlighted.

For example, the five lines below will be recorded as the annotation of solution step 2 in Figure 1. The symbol '≫' denotes implication. Rectangles and triangles point to the changed part of the formula and to the error message.

```
2: Stage 1:»
Rule applied: 21: multiplication of disj-s
■◀First eliminate bicond-s and impl-s!◀
▮¬A&(C&Bv¬C&¬B) ▮»Av¬C
▮¬A&C&Bv¬A&¬C&¬B ▮»Av¬C
```

The tool also compiles statistics of error messages in the whole solution file of the student and statistics of the group of students. This statistics is recorded in the form of tables where the rows correspond to separate solution attempts of the tasks and the columns are for particular error types and for some other characteristics (number of steps, number of steps taken back, total number of errors, stage reached in the solution algorithm). This output can be copied into a spreadsheet environment for further statistical treatment.

The analysis tool gives an error message when the formula contains independent parts that are in different stages of the algorithm and the applied conversion does not correspond to the stage of the whole formula. However, in such cases it is quite easy to understand whether the step makes the solution longer or not.

Does the tool find all reasons for long solutions? Our initial count of the possibilities for rule misapplication gave us 15 error types for full DNF tasks. A comparison of solutions and received annotations disclosed several additional unwise approaches to performing the *right* conversions: incomplete reordering of variables, addition of one variable instead of two etc. After including them we ended up with 19 error types. The most frequent errors are presented in Table 1.

From the scanned solution files we learned about a further, 'more delicate' solution economy problem. The algorithm prompts the user to apply the distributivity law at stage 3 and to eliminate redundant members at stage 4. Such ordering enables a very straightforward proof of the feasibility of the algorithm. However, it is often useful to perform some conversions of stage 4 before stage 3. Our analyzer does not require nor prohibit this. Conversions of stage 4 use only simplification rules and they do not evoke error messages.

Table 1. Results and numbers of diagnosed errors in final tests in 2011 and 2012

Quantity/error	Test 2011	Test 2012
Number of solutions (completed/total)	131/162	150/169
Steps	5766/7270	4096/4764
Steps taken back	500/933	112/186
Relevance errors diagnosed	1097/1481	321/502
4. Negation moved into brackets at stage 1	39/56	32/33
5. Negation moved out of brackets	55/157	12/36
6. Inner negation processed first	142/219	68/94
7. Distributive law applied too early	70/84	35/44
9. Members reordered too early	274/327	21/52
10. Members of FALSE conjunction reordered	67/81	13/39
11. Reordering together with redundant members	59/68	12/13
12.Members of disjunction reordered (as for CNF)	192/197	32/33
13. Variables added too early	102/147	45/74
16. Only a part of conjunction reordered	45/45	15/16
Average number of steps in completed solutions	44.0	27.3

Table 1 presents data about solutions of a full DNF task in the final tests of 2011 and 2012. Randomly generated initial formulas contained four different binary connectives and 2–3 negations (like Fig. 1). The results of 2011 looked rather disappointing. With 185 students taking the test, 162 of them submitted the solution file of formula transformation tasks, and the full DNF task was completed in 131 files. The average number of steps in completed solutions was 44 when the optimal number was 15–25. Very often several steps had been taken back (using Undo).

In the autumn term of 2012 we made the analyzer available to the students, although it does not have a developed user interface. We added a small task file with only two full DNF tasks and required that they submit a solution file where each of the two solutions can only contain one diagnosed relevance error. The students could also use the annotation tool when preparing for the final test.

The results of 2012 in Table 1 demonstrate that the annotation tool is useful for the students as well.

4   Extending the Approach to Other Situations

It seems that our current program is able to produce satisfactory explicit diagnosis of the relevance of steps in solutions of DNF and CNF tasks in RULE mode. There is an obvious extension to the algorithmically less interesting tasks on expression of formulas using negation and one binary connective.

Is it possible to apply relevance checking to the conversions in INPUT mode? Our students solve some exercises in INPUT mode. The relevance tool is designed to determine what rule is used for the step and so we had the opportunity to scan the input-based solutions. We discovered that virtually all steps were performed using the same rules 1–29, sometimes removing double negations from the result of the step. Nevertheless it is clear that for understanding free conversions we should replace the indirect identification of a single rule by direct modelling of one or more sequentially applied rules. It probably also means replacing our string representations with structured representations of mathematical objects and using the tools that work in these representations.

There exists a very powerful rule-based conversion environment, Mathpert, for algebra and calculus exercises [2] (later versions are called MathXpert). It could be a quite interesting task to complement MathXpert with relevance checking.

Acknowledgments. Current research is supported by Targeted Financing grant SF0180008s12 of the Estonian Ministry of Education.

References

1. Anderson, J.R., Corbett, A.T., Koedinger, K.R., Pelletier, R.: Cognitive tutors: Lessons learned. The Journal of the Learning Sciences 4, 167–207 (1995)
2. Beeson, M.: Design Principles of Mathpert: Software to support education in algebra and calculus. In: Kajler, N. (ed.) Computer-Human Interaction in Symbolic Computation, pp. 89–115. Springer, Heidelberg (1998)
3. Nicaud, J., Bouhineau, D., Chaachoua, H.: Mixing Microworld and CAS Features in Building Computer Systems that Help Students Learn Algebra. International Journal of Computers for Mathematical Learning 5, 169–211 (2004)
4. Prank, R., Vaiksaar, V.: Expression manipulation environment for exercises and assessment. In: 6th International Conference on Technology in Mathematics Teaching, pp. 342–348. New Technologies Publications, Athens (2003)
5. Prank, R., Viira, H.: Algebraic Manipulation Assistant for Propositional Logic. Computerised Logic Teaching Bulletin 4, 13–18 (1991)
6. Ritter, S., Anderson, J.R.: Calculation and Strategy in the Equation Solving Tutor. In: Proceedings of the Seventeenth Annual Conference of the Cognitive Science Society, pp. 413–418. Lawrence Erlbaum Associates, Hillsdale (1995)

The DeLiVerMATH Project
Text Analysis in Mathematics

Ulf Schöneberg and Wolfram Sperber

FIZ Karlsruhe/Zentralblatt MATH, Franklinstr. 11, 10587 Berlin, Germany

Abstract. A high-quality content analysis is essential for retrieval functionalities but the manual extraction of key phrases and classification is expensive. Natural language processing provides a framework to automatize the process. Here, a machine-based approach for the content analysis of mathematical texts is described. A prototype for key phrase extraction and classification of mathematical texts is presented.

1 Introduction

The database zbMATH [1] provided by FIZ Karlsruhe/Zentralblatt MATH is the most comprehensive bibliographic reviewing service in mathematics. Both key phrases and classification of the mathematical publications are central features of content analysis in zbMATH. Up to now, these data are created by expert which means time and labor-expensive work.

In the last years, computational linguistics has developed concepts for natural language processing by combining linguistic analysis and statistics. These concepts and tools were used as a platform for our activities to develop machine-based methods for key phrase extraction and classification according to the Mathematical Subject Classification (MSC2010) [2].

The DeLiVerMATH project funded by the Deutsche Forschungsgemeinschaft is a common activity of the library TIB Hannover, the research center L3S Hannover and FIZ Karlsruhe. It started in March 2012.

2 The Prototype

We are starting with a presentation of a prototype extracting key phrases and classifying a mathematical text. Snapshot on Figure 1 demonstrates its functionality.

The original text, here a review from zbMATH, is located in the left box on the top. The input can be – in principle – an arbitrary mathematical text.

The extracted candidates for key phrases and their frequencies are presented in the list on the right side and are also highlighted in the original text.

The proposed MSC classes calculated with Naive Bayes (nb) and Support Vector Machines (sv) are shown below the input text. Currently, the classification is restricted to the top level of the MSC.

Moreover, a list of unknown tokens (tokens outside of dictionary) together with a proposed word class is given.

J. Carette et al. (Eds.): CICM 2013, LNAI 7961, pp. 379–382, 2013.

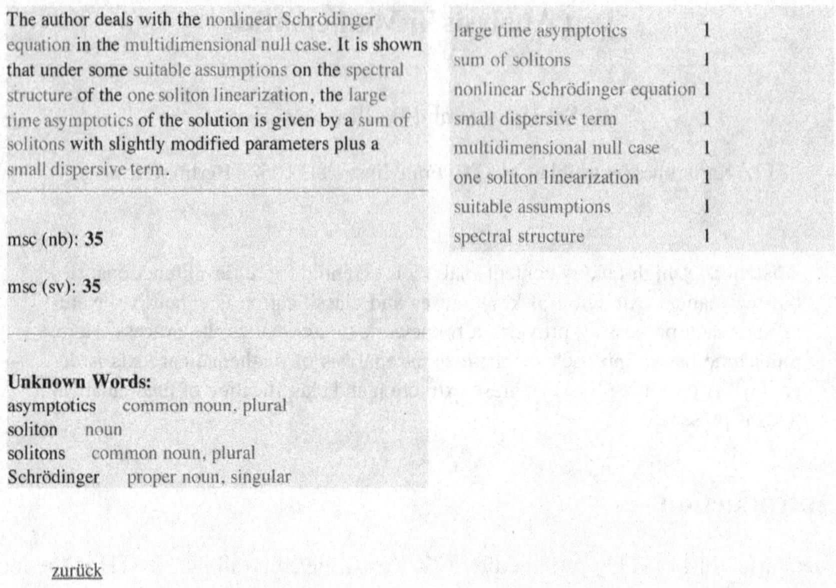

The author deals with the nonlinear Schrödinger equation in the multidimensional null case. It is shown that under some suitable assumptions on the spectral structure of the one soliton linearization, the large time asymptotics of the solution is given by a sum of solitons with slightly modified parameters plus a small dispersive term.

large time asymptotics	1
sum of solitons	1
nonlinear Schrödinger equation	1
small dispersive term	1
multidimensional null case	1
one soliton linearization	1
suitable assumptions	1
spectral structure	1

msc (nb): **35**

msc (sv): **35**

Unknown Words:
asymptotics   common noun, plural
soliton   noun
solitons   common noun, plural
Schrödinger   proper noun, singular

zurück

Fig. 1. The user interface of the prototype

3   Natural Language Processing (NLP) in Mathematical Publications

Existing Open Source tools and dictionaries from NLP were adapted to the special needs of mathematical texts. NLP provides a broad spectrum of methods for text analysis, especially

- Segmentation to identify text units
- Tokenization, the process of breaking a text stream into words, symbols and formulae, or other meaningful elements called tokens
- Morphological Parsing for a linguistic analysis of tokens
- Part-Of-Speech (PoS) tagging, the classification of a token within a text, e.g., 'convergence' as a noun.
- Parse Tree, the identification of associated text fragments, e.g., of a noun phrase
- Named Entity Recognition, the detection of phrases typically used by a community

Especially, the PoS tagging is of fundamental importance for the text analysis. PoS tagging requires a classification scheme for the tokens. Here we use the Penn Treebank PoS scheme [3] consisting of 45 tags for words and punctuation symbols. This scheme has relevant drawback for mathematical texts: no special tag for mathematical formulae. In our approach, mathematical formulae will be handled with an auxiliary construct: formulae (which are available as TeX code) are transformed to special nouns. This allows us to extract phrases which contain formulae beside the English text. A more detailed analysis of the mathematical formulae will be done in the MathSearch project of FIZ Karlsruhe and Jacobs University of Bremen. The PoS tagging use large dictionaries to assign a tag to a token. Here, the Brown corpus [4] is used as a dictionary covering more than 1,000,000 English words which are classified according to the Penn Treebank scheme. Some NLP tools are provided as Open Source software. Within the DeLiVerMATH project, the Stanford PoS Tagger [5] is used.

There are two problems: the ambiguity of PoS tags (many tokens of the corpus can belong to more than one word class) and unknown words (mathematics has a lot of tokens outside the Brown corpus). For both problems, a suitable PoS tag of a token in a phrase can be determined using the Viterbi algorithm, a dynamic programming technique. Moreover, mathematics relevant dictionaries are under development

Acronyms. Often special spellings are used for acronyms (capitals) which can be easily identified in texts by heuristic methods. A database of 3,000 acronyms with their possible resolutions was build up and implemented.

Mathematicians. The author database of zbMATH covers the names of more than 840,000 mathematicians which can be used to identify names of mathematicians in the text phrases.

Named Mathematical Entities. Named mathematical entities are phrases which are used for a special mathematical object by the mathematical community. It is planned to integrate existing lists of named mathematical entities and other vocabularies, especially the Encyclopedia of Mathematics [6], PlanetMath [7], and the vocabularies of the mathematical part of Wikipedia and the MSC.

The analysis of single tokens is only the first step for the linguistic analysis. Also in mathematics, phrases are often much more important for the content analysis than single tokens. Key phrase extraction requires the identification of chunks. Noun phrases are the most important candidates for key phrases. The first version of our tool searches for distinguished PoS tag patterns of noun phrases in the zbMATH items (abstracts or reviews). Some problems may arise, e.g., longer key phrases are extracted only partially. A more general approach is to use, instead of that, syntactic parsers for the identification of key phrases being based on grammars for the English language. A syntactic parser will be implemented into the prototype as the next step. Not all extracted noun phrases are meaningful for the content analysis as the example shows. Here we plan to build up filters (dictionaries of often used irrelevant noun phrases). For the classification, different approaches and classifiers already are implemented on different corpora: abstract and reviews, key phrases, and also full texts (in planning). Up to now, the classification is restricted to the top level of the MSC, a finer classification is under way.

4   Evaluation

The prototype will be evaluated by the editorial staff of Zentralblatt MATH to assign key phrases and classification to mathematical publications. The additional expense to evaluate the key phrases and classification is low because the tool can be integrated and used in the daily workflow. The editors can reject a proposed key phrase and classification in an easy way. We state that the number of key phrases seems to be significantly higher compared with the number of manually created key phrases which could be used for a better retrieval (e.g., to find similar publications) and the ranking. Some techniques to reject irrelevant phrases and detect similar phrases must be developed. For the classification, different methods are under development, the prototype shows the classification calculated with the Naive Bayes and the Support Vector Machine approach. It seems that the quality for the top level of the MSC is sufficient. But, we need a finer MSC classification in the future. We will also analyze the influence on the corpora to the classification and seek an answer to the question: Is it better to use key phrases instead of reviews and abstracts or fulltexts for the classifiers?

5 Outlook

The methods and tools are developed and checked for database zbMATH but can also be used for the content analysis of full texts. Up to now, the method is restricted to English texts. An extension to other languages is possible by adding dictionaries and grammars of other languages. A production version of the tool could be provided on the Web site of the FIZ Karlsruhe/Zentralblatt Math as an additional service for the mathematical community. One objective of the project is to build up a controlled vocabulary of mathematics and to match this with the MSC. Therefore, the most frequent key phrases for the MSC classes will be determined. These phrases could be the base of a controlled vocabulary for mathematics, a helpful tool for the standardization of mathematical language and communication and a starting point for developing a thesaurus for mathematics. The prototype is a first step towards a machine-based content analysis of mathematical publications. We are optimistic that the approach described above can contribute to a less expensive content analysis and allows an improved retrieval.

References

1. The database zbMATH, http://www.zentralblatt-math.org/zbmath/
2. Mathematics Subject Classification (MSC), http://www.msc2010.org
3. Santorini, B.: Part-of-Speech-Tagghing guidelines for the Penn Treebank Project (3rd Revision, 2nd printing) (June 1990),
   ftp://ftp.cis.upenn.edu/pub/treebank/doc/tagguide.ps.gz
4. Greene, B.B., Rubin, G.M.: Automatic grammatical tagging of English. Brown University, Providence (1981)
5. Samuelsson, C., Voutilainen, A.: Comparing a linguistic and a stochastic tagger. In: Proceedings of the 35th Annual Meeting of the Association for Computational Linguistics, pp. 246–253 (1997)
6. Encyclopedia of Mathematics,
   http://www.encyclopediaofmath.org/index.php/Main_Page
7. PlanetMath, http://planetmath.org/

Author Index